Springer Series in Game Theory

Official Series of the Game Theory Society

More information about this series at http://www.springer.com/series/8107

Pierre von Mouche • Federico Quartieri
Editors

Equilibrium Theory for Cournot Oligopolies and Related Games

Essays in Honour of Koji Okuguchi

Springer

Editors
Pierre von Mouche
Wageningen University
Wageningen, The Netherlands

Federico Quartieri
University Federico II of Naples
Naples, Italy

ISSN 1868-517X ISSN 1868-5188 (electronic)
Springer Series in Game Theory
ISBN 978-3-319-29253-3 ISBN 978-3-319-29254-0 (eBook)
DOI 10.1007/978-3-319-29254-0

Library of Congress Control Number: 2016937751

Printed on acid-free paper

This Springer imprint is published by Springer Nature
The registered company is Springer International Publishing AG Switzerland

On vous a sans doute souvent demandé à quoi servent les mathématiques et si ces délicates constructions que nous tirons tout entières de notre esprit ne sont pas artificielles et enfantées par notre caprice. Parmi les personnes qui font cette question, je dois faire une distinction: les gens pratiques réclament seulement de nous le moyen de gagner de l'argent. Ceux-là ne méritent pas qu'on leur réponde.

Henri Poincaré,
La Valeur de la Science, Chapitre V.

Preface

This book is about homogeneous good Cournot oligopolies and games with a similar structure. The theory was proposed some time ago: in the first half of the nineteenth century. One might expect that the passing of time would have brought about an exhaustion of the aspects of this peculiar class needing to be fully understood. This, however, appears not to be true. Various scholars throughout the world—including Prof. Okuguchi, to whom this book is dedicated—are still trying to find answers to the unsolved and new problems connected to this class. The diversity of the articles in the present book provides a balanced selection of these problems.

We are grateful to all authors, in particular to Dr. Takeshi Yamazaki, for the helpful and fruitful cooperation. Lastly, we would like to thank the founding editor of the series Dr. Bernhard von Stengel and also Dr. Johannes Glaeser from Springer for their positive role and valuable assistance in the process of publishing this book.

Wageningen, The Netherlands — Pierre von Mouche
Naples, Italy — Federico Quartieri
October 2015

Contents

List of Contributors

Gian Italo Bischi Department of Economics, Society, Politics, University of Urbino, Urbino, Italy

Luis C. Corchón Universidad Carlos III, Madrid, Spain

Richard Cornes Australian National University, Canberra, Australia

Rein Haagsma Wageningen UR, Wageningen, The Netherlands

Kojun Hamada Faculty of Economics, Niigata University, Niigata, Japan

Takuya Iimura Tokyo Metropolitan University, Tokyo, Japan

Martin Kaae Jensen Department of Economics, University of Leicester, Leicester, UK

Polina Kartunova Lomonosov Moscow State University, Moscow, Russia

Nikolai S. Kukushkin Dorodnitsyn Computing Center, Moscow, Russia

Fabio Lamantia University of Calabria, Rende, Italy

Akio Matsumoto Chuo University, Hachioji, Tokyo, Japan

Noriaki Matsushima Institute of Social and Economic Research, Osaka University, Osaka, Japan

Lourdes Moreno Universidad Complutense, Madrid, Spain

Federico Quartieri Dipartimento di scienze economiche e statistiche, Università degli studi di Napoli Federico II, Naples, Italy

Takashi Sato Shimonoseki City University, Shimonoseki, Japan

Ryusuke Shinohara Hosei University, Tokyo, Japan

Ferenc Szidarovszky Department of Applied Mathematics, University of Pécs, Pécs, Hungary

Ngo Van Long Department of Economics, McGill University, Montreal, QC, Canada

Alexander Vasin Lomonosov Moscow State University, Moscow, Russia

Elena Viganò University of Urbino, Urbino, Italy

Pierre von Mouche Wageningen UR, Wageningen, The Netherlands

Takahiro Watanabe Department of Business Administration, Tokyo Metropolitan University, Tokyo, Japan

Takeshi Yamazaki Faculty of Economics, Niigata University, Nishi-ku, Niigata-shi, Japan

Introduction

Pierre von Mouche and Federico Quartieri

Abstract In this Introduction to the Festschrift we identify a game structure that is commonly analyzed in many—albeit not all—works of Prof. Okuguchi. We define the games characterized by this structure as Cournotian games and we show that the archetypical Cournot model of oligopolistic competition and other models of economic interest are special instances of these games. We then pass to a brief summary of the contributions collected in this volume that are related to Cournotian games.

It is difficult to underestimate Prof. Okuguchi's contribution to Economics. His role of economic scholar has encompassed various fields, ranging from Economic Growth to Oligopoly Theory; but the game-theoretic analysis of the latter is probably the subject that has most characterised his research. The long bibliography on this matter[1] testifies to his deep intellectual interest for oligopolistic competition, which generated many influential works published in several prestigious journals. This Festschrift, on the occasion of his eightieth birthday, pays tribute to his intellectual research.

Prof. Okuguchi's contributions to Oligopoly Theory are wide-ranging, addressing various problems and multiple analytical techniques. Nonetheless a number of his works consider games that can be characterised by a certain common structure of strategic interaction. In this Introduction we shall refer to such games as Cournotian games and define them as follows: by a *Cournotian game* we mean a triple $(N, (S_i)_{i \in N}, (u_i)_{i \in N})$ where N is a finite set of players, player i's strategy set

[1] See the next contribution in this Festschrift.

P. von Mouche (✉)
Wageningen UR, Postbus 6700 EW, Wageningen, The Netherlands
e-mail: pvmouche@deds.nl

P. von Mouche, F. Quartieri (eds.), *Equilibrium Theory for Cournot Oligopolies and Related Games*, Springer Series in Game Theory,
DOI 10.1007/978-3-319-29254-0_1

S_i is a non-empty subset of $\mathbb{R}$ and, putting $S = \prod_{l \in N} S_l$, player i's payoff function $u_i : S \to \mathbb{R}$ satisfies

$$u_i(s) = g_i(s)\, s_i - h_i(s_i) \tag{1}$$

for some real-valued function h_i on S_i and for some real-valued function g_i on S.[2]

The virtues and the properties of the "functional form" that characterises the payoff functions of Cournotian games manifest themselves when some additional conditions are imposed. One of the specifications, which is perhaps one of the most studied and probably the oldest, is the one (more or less implicitly) considered by Augustin Cournot. In the archetypical Cournot model of oligopolistic competition (and hence without product differentiation), the function g_i assigns a price to the sum of the components of the quantity vector s, h_i is firm i's cost function and $g_j = g_l$ for all j and l. Other specifications that are compatible with the previous functional form of payoff functions are, for instance, models of Cournot competition with differentiated products, models of Bertrand competition with differentiated products and "linear" costs,[3] models of contesting and models of patenting. In this regard, we like to recall that Prof. Okuguchi was the first to point out[4] the structural equivalence between some rent-seeking games and the archetypical Cournot model of oligopolistic competition.

1 Organisation of the Festschrift

Prof. Okuguchi's analysis of Cournotian games pertains to many of the most fundamental issues: equilibrium existence, equilibrium uniqueness, equilibrium stability and comparative statics. The Festschrift contains fifteen contributions. The first one is on the works of Prof. Okuguchi and the second one is about Cournot. The other contributions are about the four above-mentioned fundamental issues. Some of these four are jointly considered in the same contribution and a neat distinction of the contributions by way of these matters is in fact impossible. For the reader's convenience we shall provide a brief summary thereof.

1. **Yamazaki** presents a short scientific biography of Prof. Okuguchi and a list of his most important publications.
2. **Haagsma** presents a short biography of Cournot and, after a brief description of the Recherches, discusses the acceptance of this pioneering book and its

[2]Although this formula is correct for homogeneous Cournot oligopolies, for related games like rent-seeking games the formula may need a modification at the boundary of S.

[3]Indeed—denoting by s the price vector, by D_i the demand at s and by C_i the average cost—one can put $g_i(s) = D_i(s) - C_i$ and $h_i = 0$, thus obtaining a Bertrand model of competition with differentiated products.

[4]See Okuguchi (1995) in the list of references in the next contribution.

mathematical method by fellow-economists during Cournot's life and after his death. The article is a refreshing contribution to the considerable amount of existing literature on Cournot.

3. **Kukushkin** provides sufficient conditions for the acyclicity of Cournot tatonnement in strategic games with aggregation and monotone best responses, and importantly—as a by-product—sufficient conditions for the existence of a Nash equilibrium for such games. His contribution allows for many economically important cases of oligopolistic competition (e.g., those of discrete entrepreneurial choices) and generalises old results (allowing for more abstract strategy spaces and dispensing with usual topological conditions on best response correspondences). Kukushkin's work helps us to understand the deepest reasons for the existence of Nash equilibria in games with best responses that admit order-reversing (as well as order-preserving) single-valued selections.
4. **Cornes and Sato** consider the problem of the existence and uniqueness of a Cournot equilibrium when the demand has constant elasticity, providing an elementary treatment that—by means of the replacement[5] and share functions—exploits the aggregative structure of a Cournot oligopoly. Interestingly, they derive a relation between the elasticity of the demand function and the number of firms that ensures the existence of a unique equilibrium.
5. **von Mouche** considers the Cournot equilibrium set in the case of concave industry revenue and convex cost functions. First the principles behind a recent simple proof of sufficient conditions that guarantee the existence of at most one equilibrium are refined and conceptualised. Next, results for the geometric structure of the equilibrium set E in the case where at least two equilibria exist are provided. In particular, sufficient conditions are given for E to be a polytope on which the equilibrium aggregator is constant, and such conditions for E to be a 1-dimensional polytope on which the equilibrium aggregator is injective.
6. **Iimura and Watanabe** prove the existence of Nash equilibria (in pure strategies) for symmetric games with a finite set of strategies. Employing a generalisation of the notion of concavity to discrete sets, they provide sufficient conditions that guarantee the existence of a (possibly non-symmetric) Nash equilibrium. They show that their general results apply to the discrete variant of the oligopolies considered in the contribution of von Mouche (i.e., to oligopolies with concave industry revenue and convex cost functions).
7. **Szidarovszky** and **Matsumoto** propose an oligopoly model with linear demand and a piecewise linear cost function that includes the cost for waste treatment and transportation. This particular and innovative choice of cost functions yields discontinuities of payoff functions. The equilibrium analysis of this model is complex. Using the technique of backward reply correspondences, they prove (distinguishing between many cases) equilibrium existence.

[5]Instead of 'replacement function (or correspondence)' other terminology like 'backward reply function (correspondence)' is used in the literature.

8. **Corchón** and **Moreno** consider an oligopoly model where firms face exponential demand functions. They provide a simple proof of equilibrium uniqueness using the structural analogy of their model with a Cournot oligopoly model where inverse demand has constant elasticity. In addition, they prove the existence of a monotone relation between the markup applied by firms and their production in equilibrium. Empirical evidence supports the conclusions of such a model.
9. **Hamada** deals with the privatisation neutrality theorem that states that when the government gives the optimal subsidy to both public and private firms in a mixed oligopoly, social welfare is exactly the same before and after privatisation. It has been shown that this theorem fails when public and private firms have different costs but the subsidy must be the same for both types of firms. Hamada shows that if a discriminatory subsidy is adopted even when there is firm asymmetry between public and private firms, the privatisation neutrality theorem continues to hold.
10. **Vasin** and **Kartunova** deal with the interesting application of oligopolies to electricity markets. Besides the functional form, given by (1) above, now other objects have to be added to the model, which heavily complicates the formal analysis, even for the simplest cases. Their article first reconsiders the classical Cournot model and in particular an equilibrium uniqueness result for non-decreasing demand elasticity. Then previous results for two-node electricity markets are surveyed and generalised.
11. **Van Long** considers a situation of oligopolistic competition in which all firms fix quantities but some firms maximize profits according to the usual "Nashian" fashion and the others according to some "Kantian" fashion. Each Kantian firm does not maximize own profit on its own production set but with respect a positive scaling factor applied to the production vector of the whole industry. The grounds for such behavior are justified by one of the Kantian categorical imperatives. Van Long defines two equilibrium notions (the Inclusive Kant-Nash Equilibrium and the Exclusive Kant-Nash Equilibrium), compares the aggregate outputs in the two respective equilibria, and shows various results that relate the two novel concepts and establish an equivalence between Exclusive Kant-Nash Equilibria and Nash equilibria with mergers.
12. **Bischi**, **Lamantia** and **Viganò** provide an overview of various models of commercial fisheries and the recent trends of this literature. The understanding of the environmental problems they consider is of importance for regulatory purposes of harvesting policies and fishing limitations. They consider three types of models of harvesting which explain the possible effects on the most endangered species, the spatial exploitations of some marine regions, and different fishery behaviours and practices.

13. **Jensen** contributes in two ways. First he extends some results of the literature on contests by allowing for the possibility that the probability that no contender wins the prize is positive. Then he extends previous results on comparative statics about a class of games with upper semicontinuous payoffs to a class of games in which—like in many contest games—payoff functions are neither lower nor upper semicontinuous at the origin. Also his analysis is based on backward reply correspondences.
14. **Quartieri** and **Shinohara** consider socially bipartite games and formalise the notion of a group-proof Nash equilibrium (a refinement of Nash equilibrium for games with players partitioned into groups, which prescribes the same communication possibilities as those required by a coalition-proof Nash equilibrium only among the players of the same group and between the groups). Then, by means of fictitious players, they prove the existence of a unique group-proof Nash equilibrium in two-group contest games with logit contest success functions, weakest-link impact functions, and convex cost functions (which typically possess infinitely many Nash equilibria). Their work also provides evidence of the importance of a non-differentiable analysis of contests.
15. **Yamazaki**, like the previous authors, considers contests between groups. Using the technique of backward reply and share functions, he proves the existence of a unique type of Nash equilibrium under general assumptions on contest success functions and additive impact functions. More precisely, Yamazaki proves the existence of at least one Nash equilibrium and that in any Nash equilibrium—of the possibly many Nash equilibria of the games he considers—the sum of the contributions is unique. Some comparative static analysis follows.
16. **Matsushima** and **Yamazaki** set up a patent-race game with one efficient player and identical inefficient players who engage in R&D and then set quantities à la Cournot. This makes that the value of the prize in general depends on the number of players. They analyze how an increase in heterogeneity among players or the number of players changes the winning rate of the efficient player. One of the main results is that if the number of inefficient players is larger than some critical value, an increase in the number of players always increases the winning rate of the efficient player.

On the Works of Professor Koji Okuguchi

Takeshi Yamazaki

Abstract This chapter briefly explains life and works of Professor Koji Okuguchi. This chapter also contains a list of his selected publications.

Professor Koji Okuguchi was born in Hokkaido, Japan in 1935. He earned B.A., M.A. and Ph.D. degrees all from Hitotsubashi University in Tokyo. He has also had short graduate education at the University of Rochester and the University of Cambridge under the Fulbright and British Council grants, respectively. He started teaching at Yokohama City University, then taught at Tokyo Metropolitan University and Nanzan University. In 2002 he has retired from Gifu Shotoku Gakuen University at the age of 72. He has been professor emeritus, Tokyo Metropolitan University since 1995. Outside of Japan, Professor Okuguchi has been associated with University of New South Wales, the Australian National University, the National Taiwan University, Bonn University, Munich University, Budapest University of Economics, Bilkent University (Turkey) and Shiraz University (Iran). Professor Okuguchi's well-known paper on Cournot oligopoly[1] has appeared in *Review of Economic Studies* in 1964 when he was a Ph.D. student at Hitotsubashi University. Many of his papers are joint works with economists such as F. Szidarovszky, M. C. Kemp, N. V. Long, C. Chiarella, T. Yamazaki, M. Hadjimichalakis, T. Negishi, and K. Suzumura. Professor Okuguchi's main research topic has been Cournot, Bertrand and Stackelberg oligopolies. His further research topics include economic growth theory, international trade theory, fisheries, exhaustible and renewable resources in relation to the Hartwick savings rule for intergenerational equity, optimal pollution tax under imperfect competition, input-output system, the Averch-Johnson model of firm's rate of return regulation, and static and dynamic analysis

[1]For life and works of Cournot, see the article of Haagsma in this book.

T. Yamazaki
Faculty of Economics, Niigata University, 8050 Ikarashi 2-no-cho, Nishi-ku, Niigata-shi 950-2181, Japan
e-mail: tyamazak@econ.niigata-u.ac.jp

P. von Mouche, F. Quartieri (eds.), *Equilibrium Theory for Cournot Oligopolies and Related Games*, Springer Series in Game Theory,
DOI 10.1007/978-3-319-29254-0_2

of rent-seeking model à la Tullock which is mathematically identical to a specific Cournot oligopoly without product differentiation. Professor Okuguchi's current interest is privatization in oligopoly. The detail of his publications is listed below.

Selected Publications of Koji Okuguchi

Okuguchi K (1964) The stability of the Cournot oligopoly solution: a further generalization. Rev Econ Stud 31(2):143–146
Okuguchi K (1966) On adaptive expectations and Cobweb phenomena. Zeitschrift für Nationalökonomie/J Econ 26:470–472
Okuguchi K (1968) The labour participation ratio and the speed of adjustment. Economica 35:445–450
Okuguchi K (1968) Taxation and portfolio choice. Kikan Rironkeizaigaku (Econ Stud Q, Journal of former Japan Association of Theoretical Economics, now renamed Japanese Economic Association) 19(3):58–62
Okuguchi K (1969) On the stability of price adjusting oligopoly equilibrium under product differentiation. South Econ J 35(3):244–246
Okuguchi K (1969) The stability of the balanced growth path under factor-augmenting technical progress. Can J Econ/Revue canadienne d'Economique 2(1):135–137
Okuguchi K (1970) Adaptive expectations in an oligopoly model. Rev Econ Stud 37(2):233–237
Okuguchi K, Suzumura K (1971) Uniqueness of the Cournot oligopoly equilibrium. Econ Stud Q 22(1):81–83
Okuguchi K (1971) The stability of the Stackelberg duopoly solutions: extensions of Kamerschen-Smith's results. Econ Stud Q 22(3):50–53
Okuguchi K (1971) Theory of oligopoly. Sobunsha, Tokyo (In Japanese)
Negishi, T., Okuguchi, K., (1972) A model of duopoly with Stackelberg equilibrium. Zeitschrift für Nationalökonomie/J Econ 32(2):153–162. Reprinted in Collected essays of Takashi Negishi, vol 1 (1994) General Equilibrium Theory. Elgar, Aldershot, pp 116–25
Okuguchi K (1972) A note on 'Inferior' input in production. Swed J Econ 74(3):398–399
Okuguchi K (1972) A dynamic model of firm entry: comment. Rev Econ Stud 39(4):521–522
Okuguchi K (1973) Quasi-competitiveness and cournot oligopoly. Rev Econ Stud 40(1):145–148
Okuguchi K (1973) Magnification effect in a two-good model with intermediate inputs. Zeitschrift für Nationalökonomie/J Econ 33(3):413–418
Okuguchi K (1974) Global stability of a dynamic pigovian policy. Econ Stud Q 25(2):59–60
Okuguchi K (1974) Input choices under rate of return regulation. Keizai Kenkyu (Econ Rev, Journal of Institute of Economics, Hitotsubashi University) 25(3):272–275
Okuguchi K (1975) A model of oligopoly with unknown demand function and adaptive expectations. Econ Stud Q 26(1):59–63
Okuguchi K (1975) The implications of regulation for induced technical change: comment. Bell J Econ 6(2):703–705
Okuguchi K (1975) Power-positive matrices and global stability of competitive equilibrium. Keio Econ Stud 12(1):37–40
Okuguchi K (1976) Natural equilibrium in the Bush-Mayer model of anarchy: a dynamic analysis. Public Choice 26(1):129–134
Okuguchi K (1976) Further note on matrices with Quasi-Dominant diagonals. Econ Stud Q 27(2):151–154
Okuguchi K (1976) Product price change and intersectoral re-allocation of specific factors. Econ Rec 52(4):497–504
Okuguchi K (1976) Expectations and stability in Oligopoly models. Springer, Berlin/Heidelberg

Kemp MC, Kimura Y, Okuguchi K (1977) Monotonicity properties of a dynamical version of the Heckscher-Ohlin model of production. Econ Stud Q 28(3):249–253
Okuguchi K (1977) Mathematical foundations for economic analysis. Kogakusha-McGraw-Hill, Tokyo (In Japanese)
Okuguchi K (1977) Input price uncertainty and the theory of the firm. Econ Stud Q 28(1):25–30
Okuguchi K (1977) Trade, savings and a nontraded good. J Int Econ 7(4):379–383
Okuguchi K (1977) Factor specificity and the Rybczinski theorem. In: Sato R (ed) Resource allocation and division of space. Springer, Berlin/Heidelberg, pp 83–94
Okuguchi K (1978) The stability of price adjusting oligopoly with conjectural variations. Zeitschrift für Nationalökonomie/J Econ 38(1):55–60
Okuguchi K (1978) Matrices with dominant diagonal blocks and economic theory. J Math Econ 5(1):43–52
Okuguchi K (1978) On the global stability of the Morishima system. Keio Econ Stud 15(2):69–72
Okuguchi K (1978) A note on systems of simultaneous linear difference equations. Nav Res Logist Q 25(4):727–728
Hadjimichalakis MG, Okuguchi K (1979) The stability of a generalized Tobin model. Rev Econ Stud 46(1):175–178
Kemp MC, Okuguchi K (1979) Optimal policies for exhaustible resources in open economies. Zeitschrift für die gesamte Staatswissenschaft/J Inst Theor Econ 135(2):207–215
Okuguchi K (1979) Monopoly, intermediate goods and the pure theory of international trade. Zeitschrift für Nationalökonomie/J Econ 39(1–2):119–130
Okuguchi K (1979) Technical progress, population growth and intergenerational equity in a model with many exhaustible and renewable resources. Econ Lett 3(1):57–60
Okuguchi K (1979) On the stability of the Stackelberg oligopoly equilibrium. Econ Lett 3(4):321–325
Kemp MC, Okuguchi K (1980) Exhaustible resource and optimal consumption in an open economy. Econ Stud Q 31(1):79–83
Okuguchi K (1980) Stability of equilibrium in a factor market distortion. Econ Stud Q 31(2):156–158
Okuguchi K (1980) Exhaustible resources and savings rule for intergenerational equity. Zeitschrift für die gesamte Staatswissenschaft/J Inst Theor Econ 136(4):673–677
Kemp MC, Long NV, Okuguchi K (1981) On the possibility of deriving conclusions of Stolper-Samuelson type when commodity prices are random. Econ Stud Q 32(2):111–116
Okuguchi K (1981) Stability of the arms race models. Zeitschrift für Nationalökonomie/J Econ 41(3–4):353–360
Okuguchi K (1981) Population growth, costly innovation and modified Hartwick's rule. Int Econ Rev 22(3):657–661
Okuguchi K (1981) Joint production and specific factor: a dynamic analysis. Econ Stud Q 32(3):267–271
Okuguchi K (1981) Innovation and intergenerational equity in a model with many exhaustible and renewable resources. Econ Stud Q 32(3):272–275
Okuguchi K (1981) A Dynamic Cournot-Nash equilibrium in fishery: the effects of entry. Rivista di matematica per le scienze economiche e sociali 4(2):59–64
Chiarella C, Okuguchi K (1982) Comparison of optimal, intergenerational equity and hartwick savings rule dynamic paths in a growth model with exhaustible resources. Zeitschrift für die gesamte Staatswissenschaft/J Inst Theor Econ 138(4):711–721
Okuguchi K (1982) Global stability of a generalized morishima system. Keio Econ Stud 19(1):101–104
Okuguchi K (1983) Equilibria in an industry with a cartel and a competitive fringe. Econ Stud Q 34(1):38–43
Okuguchi K (1983) The Cournot oligopoly and competitive equilibria as solutions to non-linear complementarity problems. Econ Lett 12(2):127–133

Chiarella C, Kemp MC, Long NV, Okuguchi K (1984) On the economics of international fisheries. Int Econ Rev 25(1):145–148; Also in Kemp MC, Long NV (eds) Essays in the economics of exhaustible resources. North-Holland, Amsterdam, pp 189–198
Okuguchi K (1984) Utility function, group size, and the aggregate provision of a pure public good. Public Choice 42(3):247–255
Okuguchi K (1984) Commercial fishing with predator-prey interaction. Keio Econ Stud 21(2):37–44
Chiarella C, Okuguchi K (1985) The Cournot-Nash and cooperative solutions in the harvesting of a fish stock. Rivista di matematica per le scienze economiche e sociali 8(2):93–101
Okuguchi K (1985) Nash-Cournot equilibrium for an industry with oligopoly and a competitive fringe. Keio Econ Stud 22:51–56
Okuguchi K (1986) Labor-managed Bertrand and Cournot oligopolies. Zeitschrift für Nationalökonomie/J Econ 46(2):115–122
Szidarovszky F, Okuguchi K (1986) Linear and nonlinear oligopoly models. In: Prekopa A et al. (eds) System modelling and optimization. Springer, Berlin/Heidelberg, pp 949–956
Fluck Z, Okuguchi K, Szidarovszky F (1987) Contribution to oligopoly theory: the case of uncertain collusions. Keio Econ Stud 24(1):13–23
Okuguchi K (1987) Equilibrium prices in the bertrand and Cournot oligopolies. J Econ Theory 42(1):128–139. (Appeared earlier as, Discussion Paper, Sonderforschungsbereich 21, Universität Bonn, 1985)
Okuguchi K, Szidarovszky F (1987) Stability of the linear Cournot oligopoly with multi-product firms. Econ Stud Q 38(2):184–187
Okuguchi K, Szidarovszky F (1987) On a nonlinear input-output system. Math Soc Sci 13(3):277–281
Szidarovszky F, Okuguchi K (1987) Notes on the stability of quadratic games. Keio Econ Stud 24(2):33–45
Okuguchi K, Szidarovszky F (1988) Dynamics of the Cournot oligopoly with multi-product firms. Math Soc Sci 16(2):159–169
Szidarovszky F, Okuguchi K (1988) A note on global asymptotic stability of non-linear difference equations. Econ Lett 26(4):349–352
Szidarovszky F, Okuguchi K (1988) A linear oligopoly model with adaptive expectations: stability reconsidered. Zeitschrift für Nationalökonomie/J Econ 48(1):79–82
Okuguchi K (1989) The Bertrand and Cournot equilibria for international duopoly with multimarkets. Int Econ J 3(4):61–68
Okuguchi K, Szidarovszky F (1989) An adaptive model of oligopoly with multi-product firms. Econ Stud Q 40(1):48–52
Szidarovsky F, Okuguchi K (1989) A non-differentiable input-output model. Math Soc Sci 18(2):187–190
Okuguchi K (1990) Cournot duopoly with several markets. Reg Sci Urban Econ 20(3):305–311
Okuguchi K (1990) Transport cost in Cournot duopoly with several markets. Stud Reg Sci (the Japan Section of the Regional Science Association International) 20(2):95–105
Okuguchi K, Irie K (1990) The Schur and Samuelson conditions for a cubic equation. Manch Sch 58(4):414–418. Reprinted in Wood JC, McLure M (eds) (2005) Paul A. Samuelson: critical assessments of contemporary economists, Second series. Routledge, London, pp 169–173
Okuguchi K, Szidarovszky F (1990) The theory of oligopoly with multi-product firms. Springer, Berlin/Heidelberg (Translated into Persian)
Okuguchi K, Szidarovszky F (1990) Global asymptotic stability of nonlinear difference equations with non-differentiable functions. Econ Lett 33(4):333–338
Szidarovszky F, Okuguchi K (1990) Dynamic oligopoly: models with incomplete information. Appl Math Comput 38(2):161–177
Okuguchi K (1991) Labor-managed and capitalistic firms in international duopoly: the effects of export subsidy. J Comp Econ 15(3):476–484
Okuguchi K, Szidarovszky F (1991) Entry in Bertrand and Cournot oligopolies with product differentiation. Keio Econ Stud 28(1):43–49

Okuguchi K (1992) Labor-managed Cournot oligopoly with product differentiation. Zeitschrift für Nationalökonomie/J Econ 56(2):197–208
Okuguchi K (1993) Cournot oligopoly with profit-maximizing and labor-managed firms. Keio Econ Stud 30(1):27–38
Okuguchi K (1993) Follower's advantage in labor-managed duopoly. Keio Econ Stud 30(2):1–5
Okuguchi K (1993) Trade policy for international duopoly with product differentiation. In: Herberg H, Long NV (eds) Trade, welfare, and economic policies: essays in honor of Murray C. Kemp. University of Michigan Press, Ann Arbor
Okuguchi K (1993) Unified approach to Cournot models: oligopoly, taxation and aggregate provision of a pure public good. Eur J Pol Econ 9(2):233–245
Okuguchi K (1993) Comparative statics for international duopoly with profit-maximizing and labor-managed firms. J Int Comp Econ 2:65–75
Okuguchi K (1993) Labor-managed firms under imperfect competition (and related problems): introduction. Manag Decis Econ 14(5):379–381. Guest edited by Okuguchi K
Okuguchi K (1993) Comparative statics for profit-maximizing and labor-managed Cournot oligopolies. Manag Decis Econ 14(5):433–444
Okuguchi K (1994) Indirect taxation in profit-maximizing and labor-managed Cournot oligopolies. J Int Comp Econ 3:101–113
Okuguchi K, Yamazaki T (1994) Bertrand and hierarchical Stackelberg oligopolies with product differentiation. Keio Econ Stud 31(1):75–80
Okuguchi K, Yamazaki T (1994) Ad valorem and specific taxes, and optimal Pigouvian tax within Cournot Oligopoly. Keio Econ Stud 31(2):25–32
Okuguchi K, Yamazaki T (1994) General analysis of horizontal merger. Seoul J Econ 7(3):279–290
Okuguchi K (1995) Effects of tariff on international mixed duopoly with several markets. CES working paper number 93, Universität München
Okuguchi K (1995) Decreasing returns and existence of nash equilibrium in rent-seeking games. Working paper, Department of Economics, Nanzan University, Nagoya, Japan
Okuguchi K (1996) Dynamic analysis of Cournot oligopoly with workers' enterprises. In: Sertel M (ed) Workers' enterprises – alternative in privatization. North-Holland, Amsterdam, pp 57–70
Okuguchi K, Serizawa N (1996) Effects of trade policy for international duopoly. Keio Econ Stud 33(2):13–22
Okuguchi K, Yamazaki T (1996) Two-stage Cournot oligopolies with industry-wide externalities. Seoul J Econ 9(1):1–16
Chiarella C, Okuguchi K (1997) A dynamic analysis of Cournot duopoly in imperfectly competitive product and factor markets. Keio Econ Stud 34(1):21–33
Szidarovszky F, Okuguchi K (1997) On the existence and uniqueness of pure Nash equilibrium in rent-seeking games. Games Econ Behav 18(1):135–140. Reprinted in Walker DA (ed) (2000) Some recent types of equilibrium models. Elgar, Cheltenham/Northampton, pp 512–17. Reprinted also in Congleton RD, Hillman AL, Konrad KA (eds) 40 years of research on rent seeking, vol 1, 2008. Springer, Berlin/Heidelberg, pp 271–276
Okuguchi K (1998) Long-run fish stock and imperfectly competitive international commercial fishing. Keio Econ Stud 35(1):9–17
Okuguchi K (1998) Existence of equililbrium for Cournot oligopoly-oligopsony. Keio Econ Stud 35(2):45–53
Okuguchi K, Serizawa N (1998) Cournot vs Stackelberg: the case of labor-managed duopoly. Keio Econ Stud 35(2):37–43
Szidarovszky F, Okuguchi K (1998) An oligopoly model of commercial fishing. Seoul J Econ 11(3):321–330
Okuguchi K (1999) Cournot and Stackelberg duopolies revisited. Jpn Econ Rev (former Economic Studies Quarterly) 50(3):363–367
Okuguchi K, Szidarovszky F (1999) The theory of oligopoly with multi-product firms, 2nd edn. Springer, Heidelberg/New York
Okuguchi K (2000) Comparative statics for oligopoly, oligopsony and oligopsonistic oligopoly. J Econ Res 5(2):173–187

Szidarovszky F, Okuguchi K (2000) A dynamic model of international fishing. Seoul J Econ 13(4):471–476

Okuguchi K (2001) International competition in one country: comparative static analysis. Pac Econ Rev 6(1):25–35

Okuguchi K (2002) Comparative static analysis for Cournot oligopoly. In: Woodland AD (ed) Economic theory and international trade: essays in honour of Murray C. Kemp. Elgar, Cheltenham/Northampton, pp 208–217

Szidarovszky F, Ilieva V, Okuguchi K (2002) Entry and merger in commercial fishing with multiple markets. Zeitschrift für Nationalökonomie/J Econ 76(3):247–259

Okuguchi K (2003) Optimal pollution tax in Cournot oligopoly. Keio Econ Stud 40(1):85–89

Okuguchi K (2003) Dynamic and comparative static analysis of imperfectly competitive international fishery. Zeitschrift für Nationalökonomie/J Econ 80(3):249–265

Okuguchi K (2003) Imperfect competition and optimal pollution tax. In: Bars R, Gyurkovics E (eds) Control applications of optimisation 2003, a proceedings volume from the 12th IFAC workshop, Visegrád. Elsevier Science Limited, Oxford, pp 123–126

Okuguchi K, Szidarovszky F (2003) Oligopoly with intertemporal demand interaction. J Econ Res 8(1):51–62

Szidarovszky F, Okuguchi K (2003) Competition and cooperation in international commercial fishing. In: Neck R (ed) Modeling and control of economic systems 2001, a proceedings volume from the 10th IFAC symposium, Klagenfurt, 2001. Elsevier Science Limited, Oxford, pp 65–68

Okuguchi K, Yamazaki T (2004) Stability of equilibrium in Bertrand and Cournot Duopolies. Int Game Theory Rev 6(3):381–390

Okuguchi K (2005) Revisiting perverse effects on exchange rate pass-through. Int J Bus Econ 4(1):67–76

Okuguchi K (2005) Entry in Bertrand oligopoly. PU M A (Pure Math Appl) 16(4):467–474

Okuguchi K (2005) Existence of Nash equilibria in endogenous rent-seeking games. In: Nowak AS, Szajowski K (eds) Advances in dynamic games. Birkhäuser, Boston, pp 445–453

Okuguchi K, Szidarovszky F (2005) Changes in demand function in Cournot oligopoly. Pac Econ Rev 10(3):371–378

Szidarovszky F, Okuguchi K, Kopel M (2005) International fishery with several countries. PU M A 16(4):493–514

Erjaee HG, Okuguchi K (2006) Bifurcation and stability in imperfectly competitive international commercial fishing. Keio Econ Stud 43(2):61–71

Okuguchi K (2006) Aggregate game and international fishery with several countries. In: Haasis H-D, Kopfer H, Schönberger J (eds) Operations research proceedings 2005. Springer, Berlin/Heidelberg, pp 461–466

Okuguchi K (2007) Quasi-competitive Cournot oligopoly with product differentiation and symmetric firms. Rev Dev Econ 11(2):404–411

Okuguchi K, Szidarovszky F (2007) Existence and uniqueness of equilibrium in labor-managed Cournot oligopoly. Riv Pol Econ 97(3):9–16. Reprinted in Lambertini L (ed) Firms' objective and internal organization in a global economy: positive and normative analysis, 2009. St. Martin's Press/Palgrave Macmillan, New York, pp 9–16

Okuguchi K, Szidarovszky F (2007) Environmental R and D in Cournot oligopoly with emission or performance standards. PU M A 18(1–2):111–118

Okuguchi K, Yamazaki T (2008) Global stability of unique Nash equilibrium in Cournot oligopoly and rent-seeking game. J Econ Dyn Control 32(4):1204–1211

Okuguchi K (2010) Inferior factor in Cournot oligopoly. Zeitschrift für Nationalökonomie/J Econ 101(2):125–131

Okuguchi K (2011) Optimal pollution tax under international competition. Keio Econ Stud 47(1):153–159

Okuguchi K (2012) Commercial fishing with predator-prey interaction revisited. Keio Econ Stud 48(1):139–144

Okuguchi K (2012) General analysis of Cournot mixed oligopoly with partial privatization. Eurasian Econ Rev 2(1):48–62

Erjaee GH, Astadzad MH, Okuguchi K, Rahemi A (2013) Fractional differential equations system for comercial fishing under predator-prey interaction. J Appl Nonlinear Dyn 2(4)
Okuguchi K (2013) Existence of unique equilibrium and quasi-competitiveness in Cournot mixed oligopoly. Presented at PET13, Lisbon
Okuguchi K, Yamazaki T (2014) Global stability of Nash equilibrium in aggregative games. Int Game Theory Rev 16(4)
Okuguchi K (2015) Analysis of privatization in Stackelberg mixed oligopoly. Presented at DIEM (Dubrovnik International Economic Meeting) 2015, Dubrovnik

Cournot, a Non-strategic Economist

Rein Haagsma

Abstract After a brief biography of Augustin Cournot, this note gives a selective outline of his *Recherches sur les principes mathématiques de la théorie des richesses* and then discusses the acceptance of the book and its mathematical method by fellow-economists during Cournot's life and after his death.

1 Introduction

He could not have wished for more than to have the support of the two founding fathers of microeconomics, Alfred Marshall and Léon Walras. Marshall had read Cournot's pioneering work as early as in 1868, some 30 years after its publication (Shubik, 1998). In the preface of his *Principles* (1890), he acknowledged that Cournot had been a major influence. On the views expressed in his book, Marshall wrote that "their form has been most affected by mathematical conceptions of continuity, as represented in Cournot's *Principes mathématiques de la théorie des richesses*" (Marshall, 1890, p. x). Walras had read Cournot's book as a teenager. In the preface of his *Élements* (1874), he credited the source of his ideas: "I am indebted to my father Auguste Walras for the fundamental principles of my doctrine; and to Augustin Cournot for the idea of using the calculus of functions in the elaboration of this doctrine" (Walras, 1874, p. 37). The praise of Marshall and Walras rescued Cournot's economics from oblivion, but a full revival did not materialise until the 1950s. In terms of scope and method, Cournot clearly was generations ahead of his time.

R. Haagsma
Wageningen UR, Postbus 6700 EW, Wageningen, The Netherlands
e-mail: rein.haagsma@wur.nl

P. von Mouche, F. Quartieri (eds.), *Equilibrium Theory for Cournot Oligopolies and Related Games*, Springer Series in Game Theory,
DOI 10.1007/978-3-319-29254-0_3

2 Life and Work

Antoine Augustin Cournot was born in Gray (Haute Saône, France) on 28 August 1801 and died in Paris on 30 March 1877, where he was buried in the Montparnasse cemetery.[1] In daily life, Cournot was known as a high-ranking administrative official of the French public school system. By those with whom he mingled in intellectual debate, he was mainly seen as a mathematician and philosopher, rather than as an economist. Also today, the plate next to the main entrance of the house where he lived and passed away (2 Rue de Tournon, Paris) states "Augustin Cournot, Mathématicien et Philosophe". Cournot's interest in mathematics and philosophy arose during his teenage years, while working "en amateur" in a lawyer's office. Reading Fontenelle and Laplace convinced him of the use of mathematics for scientific progress and Leibniz influenced his philosophical thinking (Moore, 1905). At the age of 19, Cournot entered the Collège Royal de Besançon to read mathematics. This allowed him to be admitted to the École Normale Supérieure in Paris and, a year later, when the school was abruptly closed for political reasons, to the Sorbonne for a degree in sciences.

His student days in Paris were rich in education, friendships, and experiences. The air was filled with excitement about the recent outstanding achievements in mathematics and physics (de Villé and Ménard, 1989). Cournot was introduced to Laplace and Lagrange, and attended lectures by Poisson, Legendre, Fourier, and Cauchy. He became lifelong friends with fellow student Gustav Lejeune Dirichlet, the German mathematician, and attended classes with Auguste Walras, to whom he would later send a copy of his *Recherches*, which would end up in the hands of his son, Léon. In 1823, as his small allowance was running out, Cournot accepted a job from one of Napoleon's retired generals, Marshall Gouvion Saint-Cyr, as tutor to his young son (just as his friend Dirichlet did, who was hired as a tutor by another general). It was a lucrative position. He was able to stay in Paris and would have enough spare time to complete his doctoral studies and pursue his mathematical work. The engagement lasted ten years. Cournot quickly became the private secretary and confidant of Saint-Cyr, helping him to edit and publish his military memoirs, taking time to finish this work after his employer died in 1830. Through Saint-Cyr, he had the opportunity to meet people from the highest echelons of French society and to ruminate upon matters of history and politics. He also intensified his contacts with leading intellectuals of the day, mainly physical scientists and engineers. In 1827 he obtained a degree in law and in 1829 a doctorate in sciences with a main thesis in mechanics and a supplementary one in astronomy.

From then on, Cournot made rapid advancement as official of the highly centralized university system. His career started when a series of mathematical papers written as a student caught Poisson's eye. Following the death of Laplace,

[1]A classic biography is Moore (1905), which is based on Cournot's memoirs called *Souvenirs* (which ignores the last 18 years of his life), and another is Nichol (1938). See also Shubik (1998), de Villé and Ménard (1989), Friedman (1999), Magnan de Bornier (2007), and Baloglou (2012).

Poisson had become the authority among French mathematicians and he was soon in charge of the instruction of mathematics throughout France. Impressed by Cournot's achievements, Poisson arranged his appointment to the chair of Mechanics and Analysis in Lyon in 1834. The professorship mostly consisted of teaching differential calculus to a small group of students and lasted only one university year, when Poisson persuaded Cournot to accept the position of rector of the Academy of Grenoble. Though short, the Lyon period was very productive. Besides sketching the main outlines of his mathematical theory of chance and probability (*Exposition de la théorie des chances et des probabilités* appeared in 1843), Cournot wrote the major part of the book that was to become the *Recherches*. Just after his arrival in Grenoble, he was invited to combine the post of rector temporarily with the prestigious duties of Inspector General of Education, succeeding Ampère, who had died suddenly. The position became permanent and full time in 1838, the year the *Recherches* was published, and would keep him occupied for the next 16 years. Also in that year, he married Colombe-Antoinette Petitguyot in his birthplace Gray with whom he had a son, Pierre. With headquarters in Paris, he was sad to leave the Rhône region.

In 1844 Cournot was forced to take a year's leave of absence due to his diminished eyesight. His weak eyes had increasingly troubled him since his student days. When he returned from a long stay in Italy, he seemed sufficiently recovered to resume his regular duties, but the character of his work would change (Nichol, 1938). It became virtually impossible for him to engage in long, concentrated periods of writing. Nearing the end of his life, he wrote to Léon Walras: "Thirty years ago I had to renounce all mathematics" (Walras et al., 1935, p. 120, as cited in Nichol, 1938).

During his years as Inspector General, Cournot held other respectable offices in the education system, including membership of the Commission for Superior Studies and the presidency of the jury overseeing the competitive examinations for mathematics teachers. He was made Knight and later Officer of the *Légion d'honneur*. In 1854, weary of his function and disappointed with the changes in educational policy under the Second Empire, Cournot gladly accepted the post of Rector of the Academy in Dijon in his native country. He stayed in Dijon until his retirement in 1862. A few years earlier he had completed his personal memoirs, *Souvenirs*. Though in many respects illuminating—they are the chief source of what is known about his life—his memoirs shy away from disclosing any intimate personal details. This is not far off from Moore's picture of Cournot as an unassuming, modest man with a marked dislike of self-advertisement (Moore, 1905). Cournot returned to Paris, where he lived for 15 years in retirement without further public functions. Much energy was devoted to philosophical work until he died suddenly at the age of 75. Toward the end of his life he was almost blind.

Cournot's administrative duties left him ample time for private study. He started with mathematics and mechanics, was then drawn to economics, and later in his life turned to philosophy. His main contributions to mathematics are the above-mentioned *Théorie des chances et des probabilités*, which Keynes found novel but disappointing,[2] and a study of the relationship between algebra and geometry (*De l'origine et les limites de la correspondance entre l'algèbre et la géométrie* appeared in 1847). More than for his mathematical originality, Cournot is known for his contributions to the philosophy of science. His major works in this field are an essay on the foundations of scientific knowledge (*Essai sur les fondements de nos connaissances et sur les caractères de la critique philosophique* appeared in 1851) and a study on empiricism and rationalism (*Matérialisme, Vitalisme, Rationalisme: études sur l'emploi des données de la science en philosophie* appeared in 1875). In the latter, Cournot tried to reconcile rationalism and empiricism.[3] If consistently applied, he said, rationalism would lead to loss of contact with reality and empiricism would lead to scepticism. Yet reality is characterised by both continuity and discontinuity, by both law and brute empirical fact, which is unresponsive to law. Rationalism is preoccupied with the former, empiricism with the latter. Continuity is real because the world is not a chaos, but discontinuity is also real, because the order we see in the various events is not a single, predictable order. Cournot's philosophy is closely interwoven with his thinking on probability and, as will be seen, on the use of mathematics for the study of economic and social reality.

In the field of economics, Cournot wrote three books: *Recherches sur les principes mathématiques de la théorie des richesses* appeared in 1838, *Principes de la théorie des richesses* in 1863, and *Revue sommaire des doctrines économiques* in 1877. The *Principes* is largely a literary version of the *Recherches* without the mathematics. The *Revue is* also a non-mathematical treatise. None of the three books was given a warm reception during Cournot's lifetime. The *Revue* appeared in print just after his death. The publication of the *Recherches* was scarcely noticed. It did get some attention when the *Principes* appeared, though not with much acclaim. A reviewer of the *Principes* at the time wondered about the change in Cournot's methodology: "Why have you amused yourself to talk to us in scary hieroglyphics, since you could present all this to us, and even better than this, in simple French prose and without algebra!".[4] The recognition of Cournot's contributions did not materialise until the early 1950s, when the mathematical *Recherches* turned out to be an important precursor of some new developments in economic theory. It is safe to say that without this book, Cournot's name in the history of economic thought would have been limited to a footnote.

The following discusses the acceptance of the *Recherches* and its mathematical method by fellow-economists during Cournot's life and after his death. Before doing so, a brief characterisation of the *Recherches* is given to get a feel for the book.

---

[2] "Cournot never touched any subject without contributing something to it, but, on the whole, his work on Probability is, in my opinion, disappointing" (Keynes, 1921, p. 324). Keynes discussed Cournot's *Théorie* in his PhD thesis on probability.

[3] The next lines are based on Mounce (1998).

[4] de Fontenay (1864, p. 232), as cited in Baloglou (2012, p. 459). The reviewer was Roger de Fontenay (see below).

3 A Look at the *Recherches*

The *Recherches* can be split up into three parts: Chapters 1–3 contain a general discussion of exchange value and wealth, Chapters 4–10 study the determination of prices under different degrees of competition (Chapter 10 is on market integration), and Chapters 11 and 12 are more macroeconomic in nature and discuss what determines social income.[5] Chapters 4–9 form the original core of the book, about 80 pages. Chapter 4, entitled "Of the law of demand", contains the first accepted mathematical function in economics:[6]

> Let us admit therefore that the sales or the annual demand D is, for each article, a particular function $F(p)$ of the price p of such article. (Cournot, 1897, p. 47)

Cournot was interested in the demand that is followed by an actual sale, i.e., an observable and measurable quantity. It thus differs from Marshallian demand, where quantity demanded depends on various hypothetical prices. Though Cournot's demand function essentially is an empirical concept rather than an a priori one, he thought that even in the case where sufficient reliable observations cannot be obtained to construct the demand curve, working with its general mathematical expression is a wise approach. Indeed, "one of the most important functions of analysis consists precisely in assigning determinate relations between quantities to which numerical values and even algebraic forms are absolutely unassignable" (Cournot, 1897, p. 48). Chapter 4 goes on to discuss the effect of price variations on total revenue, $pF(p)$. The first and second-order conditions for a maximum are formulated, and the analysis implicitly introduces the notion of price elasticity of demand.

Chapters 5–8 develop Cournot's theory of monopoly, oligopoly, and perfect competition in a unified way. This is what distinguishes his work from practically all the other achievements of his predecessors, who concentrated almost exclusively on perfect markets. Further advances in the analysis of imperfect markets appeared in the work of Hotelling, Chamberlin, and Robinson, nearly a hundred years later. The clarity and conciseness of the exposition is striking. For instance, comparing his analysis of monopoly with that of Alfred Marshall (1890, Ch. 8, Book V), the latter comes across as tedious and laborious, also due to Marshall's outmoded practice of moving diagrams and mathematics to footnotes and appendices. To today's reader,

---

[5]The classic discussion of the *Recherches* is by Irving Fisher (1898). See also Shubik (1998).

[6]Before the *Recherches*, and for a long time after, demand schedules were presented by way of tables or analytically specified mathematical functions, so without generality (see Creedy, 1992, Ch. 2) Cournot's introduction of general mathematical functions was innovative in itself, but also because general functions were hardly used in the natural sciences (Friedman, 1999). Robertson (1949) mentioned the work of Georg von Buquoy as the first thorough attempt to employ general functions to their advantage. The German author is quoted as saying that "it is often sufficient in the study of political economy to use algebraic formulas as sentences expressed in symbols" (Robertson, 1949, p. 529). The functions given by Buquoy (and mentioned in Robertson, 1949) cannot be readily accepted as elementary economic functions.

who is familiar with some degree of abstraction after attending a principles course and is also equipped with some differential calculus from a first-year seminar, Cournot almost speaks in a modern tongue.

Chapter 5 ("Of monopoly") offers a pure theory of monopoly. Cournot first introduced a cost function $\phi(D)$. He then demonstrated which price (rather than quantity) will maximise monopoly profits, by providing the first-order condition:

$$D + \frac{dD}{dp}\left[p - \frac{d[\phi(D)]}{dD}\right] = 0.$$

Hereafter the price charged was studied under decreasing, constant, and increasing marginal costs (though speaking in terms of "function $\phi'(D)$"). Chapter 6 considers the impact of taxation in this market.

Chapter 7 ("Of the competition of producers") turns to a market with two or more producers. Cournot started with a duopoly, writing D_1 and D_2 for the sales of the two producers ($D_1 + D_2 = D$) and ignoring any production cost. Collusion is ruled out: "*each of them independently* will seek to maximize (...) income as large as possible" (Cournot, 1897, p. 79). To express this income or profit, he switched to the inverse notation of market demand, writing $p = f(D)$. Profits of the two producers follow as $D_1 f(D_1 + D_2)$ and $D_2 f(D_1 + D_2)$. This step is crucial because Cournot intends to move here from price-setting under monopoly to quantity-setting under oligopoly. Yet, as Shubik (1998) remarked, Cournot's mathematics do not entirely match his words regarding the identity of the strategic variable:

> Proprietor (1) can have no direct influence on the determination of D_2: all that he can do, when D_2 has been determined by proprietor (2), is to choose for D_1 the value which is best for him. This he will be able to accomplish by properly adjusting his price, except as proprietor (2), who, seeing himself forced to accept this price and this value of D_1, may adopt a new value for D_2, more favourable to his interests than the preceding one. (Cournot, 1897, p. 80)

Cournot continued by formulating the first-order conditions for the two producers with quantity as strategic variable:

$$f(D_1 + D_2) + D_1 f'(D_1 + D_2) = 0$$
$$f(D_1 + D_2) + D_2 f'(D_1 + D_2) = 0.$$

He presented the two conditions (reaction curves) in a diagram and studied the stability of the equilibrium. Next, he proved that the duopoly price is lower than the monopoly price and included a discussion concerning why an equilibrium with the monopoly price would break down. The chapter further contains generalizations to n producers with different cost functions.

Chapter 8 ("Of unlimited competition") deals with perfect competition. Cournot started by pointing out the delicate strategic difference between an oligopolistic market with a large number of producers and a perfectly competitive market. His point of departure is the first-order condition of the kth producer in the former market, which he wrote as

$$D_k + \left[p - \phi_k'(D_k)\right] \frac{dD}{dp} = 0$$

where $\phi_k'(D_k)$ is the marginal cost of the producer.[7] Cournot reasoned that "(t)he effects of competition have reached their limit, when each of the partial productions D_k is *inappreciable*, not only with reference to the total production $D = F(p)$, but also with reference to the derivative $F'(p)$, so that the partial production D_k could be subtracted from D without any appreciable variation resulting in the price of the commodity" (Cournot, 1897, p. 90). It implies that the term D_k can be neglected in the above equation, which yields the familiar price equals marginal cost condition for firms in a competitive market, $p = \phi_k'(D_k)$.[8] In the remainder of the chapter, Cournot identified the aggregate supply curve in a competitive market and the market equilibrium price (also by way of a demand-supply diagram), and showed that an overall increase of the production costs raises the equilibrium price by a smaller amount.

Chapter 9 ("Of the mutual relations of producers") differs from the earlier ones by dealing with a somewhat peculiar market situation. It analyses a setting in which two producers sell different products that are of no use unless combined in a fixed ratio to form some composite good. Cournot emphasised that the two producers are not engaged in competition, since they supply different goods, but have a "mutual relation". He supposed that, just as in his Chapter 5, each producer acts as a price-setting monopolist. A producer's optimal price depends on the price set by the other producer. Thus the model boils down to what we would now call a differentiated duopoly where the two goods are perfect complements and prices are the strategic variable. Remarkably, Sonnenschein (1968) has shown that the duopoly theories of Chapters 7 and 9 are formally identical; one can be obtained from the other by simply replacing prices (quantities) by quantities (prices).

4 Recognition During His Life

The *Recherches* was almost completely ignored when it first appeared. It unsettled Cournot so much that he abandoned economics for 25 years. In the preface of the *Principes*—his second try, now in non-mathematical language—he reflected on its

[7] To see how the equation is derived, begin with $f(D) + D_k f'(D) - \phi_k'(D_k) = 0$. Since $f(D) = p$ and $f'(D) = \frac{dp}{dD}$, this is equivalent to $p + D_k \frac{dp}{dD} - \phi_k'(D_k) = 0$. Next, multiply both sides with $\frac{dD}{dp}$ and rearrange terms.

[8] A simpler way to see this is to write the first order condition of the kth producer as $f(D) + D_k f'(D) \frac{dD}{dD_k} - \phi_k'(D_k) = 0$ and noting that $\frac{dD}{dD_k} = 1$ under oligopoly and vanishes under perfect competition.

reception and, somewhat disillusioned, concluded: “When one wants to go against the accepted habits, either he makes a revolution (...) or he draws no attention at all, which is what happened to me” (Cournot, 1863, p. I, as cited in Theocharis, 1990).[9]

Nearly 20 years passed before the first real appraisal of the *Recherches* was published (Dimand, 1988). John Bradford Cherriman was a Canadian professor in mathematics and philosophy with an interest in economic issues, who had received his training in Cambridge. Perhaps he had come across the *Recherches* through Cournot’s writings on chance and probability. In 1857, Cherriman wrote a thorough review of ten pages full of praise, even valuing the *Recherches* above any other work since Adam Smith. He particularly hailed Cournot’s use of mathematics as a source of clarity and parsimony (“surely no science ever called louder for this aid” (as cited in Dimand, 1988)). Unfortunately, the review was not communicated to Cournot, probably because it appeared in a Canadian journal.

The publication of the *Principes* in 1863 immediately provoked some reaction from French economists. More importantly, it succeeded in drawing attention to Cournot’s earlier work. A young and still unknown Léon Walras wrote a sympathetic review of the book and took the opportunity to laud the attempt of the *Recherches* to apply mathematics to economics (“the first and only one of its kind which has been made, (...) it is of interest to a high degree for the future of Political Economy” (as cited in Theocharis, 1990)). His review failed to raise a stir, however. The most prominent reviews that considered the *Recherches* were highly critical, not only with respect to its mathematical approach, but also with respect to economic content and alleged ideology (see Ekelund and Hébert, 1990, Theocharis, 1990, and Baloglou, 2012). Particularly the criticisms of Roger de Fontenay and Paul-Gustave Fauveau have found a place in the history of economic thought.

Fontenay was an admirer of Frédéric Bastiat (1801–1850), a brilliant advocate of free markets and limited government. He edited the work of Bastiat and also wrote a book on land rents that was greeted by other French economists. Fontenay quickly realized that Cournot’s new book essentially was a dressed-up version of the *Recherches*. In a 20-page review in the *Journal des Économistes* (de Fontenay, 1864), appearing one year after the publication of the *Principes*, Fontenay appreciated Cournot’s mastery of mathematical modelling but regarded his grasp of economics as naive and simplistic. The problem was not the use of mathematics per se,[10] but rather that Cournot had pushed the limits too far. According to Fontenay, submitting behavioural questions to mathematical manipulation inevitably leads to oversimplification and suggests an air of precision that is almost never present in real life. He concluded that “despite all his talent as dialectician and algebraist,

[9]We should not read too much into his words here. Schumpeter was pretty sure that Cournot “thought of this failure as a small disagreeable incident in an otherwise prosperous career. Moreover, he had every reason to congratulate himself on the reception of what he—again, almost certainly—considered his really important works” (Schumpeter, 1954, p. 924–5). Schumpeter referred to Cournot’s work on probability and his contributions to philosophy.

[10]Although Fontenay tended to think Cournot’s mathematics were unnecessary (see note 4).

M. Cournot has only arrived at results which are very mediocre from the point of view of economic interest, and which are sometimes more than questionable, as far as exactitude is concerned" (de Fontenay, 1864, p. 238, as cited in Baloglou, 2012, p. 460). The heaviest blow struck at Cournot came near the end of the review, where he was accused of having socialist leanings. Cournot's mechanistic treatment of the economy implied, according to Fontenay, that he saw the economy as a machine that could be fine-tuned and optimized by human intervention: Clearly he saw a role for government regulation and thus rejected Say's idea of organic harmony (shared by the majority of French economists of his day) in favour of the idea of "mechanical adjustment, even when there is no (apparent) advantage to doing so" (de Fontenay, 1864, p. 249, as cited in Ekelund and Hébert, 1990, p. 144).

Fauveau was a French engineer and mathematician who was probably the first to hint at a basic weakness in Cournot's oligopoly theory, some 16 years before Bertrand aired his critique (see Ekelund and Hébert, 1990; Theocharis, 1990). In 1864 he published a book Fauveau (1864) with a mathematical treatment of tax policy. After approvingly building on Cournot's analysis of taxation in a monopolistic market, Fauveau turned to Cournot's oligopoly theory and questioned one of its key assumptions. "The author assumes that each producer can fix his sales without this changing the sales of the other producers, a thing which is not true. He cannot, in effect, modify his own sales without changing his selling price, which will cause a variation in the sales of his competitors" (Fauveau, 1864, p. 56–7, as cited in Theocharis, 1990, p. 928). In other words, a producer cannot adjust his quantity sold ("sales") without changing the price, and once he changes the price it will affect the quantity sold by the other producers. Fauveau did not, however, elaborate on this point further. Three years later, in an article in the *Journal des Économistes* (Fauveau, 1867), he did return to it when he discussed Cournot's description of the tatonnement process in the *Principes*. Fauveau now proposed that once the other producers have fixed their quantities, a producer simply cannot change the price. "We think that once proprietor N has fixed the quantity n which he intends to deliver to the public, proprietor M can no more fix the selling price, which is found fixed by the force of things" (Fauveau, 1867, p. 35, as cited in Theocharis, 1990, p. 929).[11] Perhaps he had a sort of Stackelberg solution in mind. However, to conclude that Fauveau saw the role of prices as strategic variable, as some authors do (see Ekelund and Hébert, 1990), seems far-fetched.

Cournot hardly responded to his critics. In the preface of the *Revue*, published after his death, he only briefly referred to Fontenay's review (without mentioning his name). To the latter's complaint that he had ignored 25 years of advances in the economic literature, Cournot retorted: "thus the poor author that no one of the

[11]He goes on "Is it not evident, in fact, that if producer M had fixed a selling price, all the sales would be determined by this alone, and that this would forcibly modify the quantity of sales which proprietor N would have been able to assign to himself a priori? Once the price is fixed, proprietor N has an interest to sell whatever is demanded of him, and it would be absurd for him to seek to sell less at the same price" (Fauveau, 1867, p. 35, as cited in Theocharis, 1990, p. 929).

official world of French economists wanted to quote incurred the reproach of not having quoted others enough" (Cournot, 1877, p. iii, as cited in Shubik, 1998). Cournot thus persisted in his independence and, indeed, did little to change the impression of being an outsider and amateur economist. He did not take part in the public policy debates of his day and even did little to defend his ideas and his use of mathematics. According to Ekelund and Hébert, the silence among French academics did not have so much to do with Cournot's use of mathematics but rather with his "ostrich-like behavior" (Ekelund and Hébert, 1990, p. 47).

This may be true but it tells only part of the story. Cournot's reluctance to defend the *Recherches* also stemmed from his growing dissatisfaction with the book. De Villé and Ménard 1989 argue that Cournot increasingly started to doubt the usefulness of mathematics in economics and the relevance of his ideal-typical theories.

At the time of writing the *Recherches*, a positivist-mechanistic ideology prevailed that favoured the application of mathematics to social reality (see de Villé and Ménard, 1989). The recent advances in the natural sciences and Auguste Comte's idea of science as "social physics" must have inspired Cournot to extend the application of mathematics, which had so strongly proved its power in physics, to social disciplines. Moreover, the instability of post-revolutionary France had created an urgent need for discovering and understanding the social and economic regularities that govern society. Social scientists with their new analytical tools could help to regulate the social order, an idea that fitted with the French engineering tradition (see Ekelund and Hébert, 2014, p. 329). Cournot had believed that, although the present showed little continuity and much discontinuity, the progress of civilization would produce a world closer and closer to a mechanistic system. In particular, the increase of markets and competition, the spread of education, and the growing capacity of the state to set the right institutions would, he thought, create a more rational world in which his mathematical method and his theories were in place.

The next 40 years increasingly convinced him that he was wrong (see de Villé and Ménard, 1989). Cournot observed that the extension of markets did not automatically enforce competition but rather appeared to increase the scope for market concentration. The formation of large market-leading firms and coalitions among firms suggested that the coordination problems of a growing economy were not solved by Adam Smith's "invisible hand" but by the powerful interactions between organized forces, including the authority of the state. The social and political turmoil of mid-century France fortified his increasing scepticism. Context was as much part of the explanation as regularity, raising his awareness that rational theory should strike a balance with the analysis of institution-specific and history-specific factors. Mathematics could help but could not be at the heart of the explanation. Gradually Cournot became convinced that economics was closer to biology than to mechanics (de Villé and Ménard, 1989, p. 499). When his admirer Léon Walras published *Éléments d'économie politique pure* in 1874, he criticised the book for placing too much emphasis on perfect competition and mathematical modelling (de Villé and Ménard, 1989, p. 500).

It was only in the latter years of his life that Cournot was able to see the first signs of recognition for his main work. He accepted the praise philosophically in the preface of the *Revue*. Walras gratefully made use of Cournot's mathematical approach and analysis of market types in his *Éléments* and acknowledged his debt to Cournot even more strongly in private correspondence with him. In 1875 an Italian translation of the *Recherches* appeared in a series of economic works edited by Gerolamo Boccardo (the first English translation appeared in 1897). Cournot also knew that the mathematical approach of the *Recherches* was appreciated by Stanley Jevons, then the towering figure among British economists. In the preface of the second edition of his *Theory of political economy*, which appeared about two years after Cournot's death, Jevons concluded that the core of the *Recherches* "contains a wonderful analysis of the laws of supply and demand, and of the relations of prices, production, consumption, expenses and profits" (Jevons, 1957, p. xxx). Regarding its mathematical approach, Jevons felt sure that an adequate treatment of political economy "must be based upon the analysis of Cournot, or at least must follow his general method" (Jevons, 1957, p. xxxi).[12] Thus Cournot's economics finally came in from the cold.

5 Recognition After His Death

Although Cournot's work had now reached a wider public, real success eluded him for a long time. Despite the fact that Alfred Marshall also paid homage to Cournot, after Cournot's death his economic theories underwent a period of criticism and rejection that lasted until midway through the twentieth century. The first major critique in this period was raised by someone who appears to have known nothing about economics, Joseph Bertrand.[13]

Bertrand was a highly respected and well-known French mathematician with a broad interest in the history of science (see Struik, 2008). His many high-ranking academic functions, combined with his literacy, eloquence and charm, had catapulted him to a position of national stature. Although his mathematical work did not dig as deep as that of the great mathematicians of his time, Bertrand's often elegant studies were widely read. Also, he was famous for his entertaining mathematical textbooks written for secondary schools. For 35 years he edited the *Journal des Savants*, the earliest academic journal in Europe. To this he contributed articles of a popular nature, many dealing with the history of science. In the 1883 volume, Bertrand wrote a malicious joint review of Walras's *Théorie mathématique de la richesse sociale*, which had just been published, and the *Recherches*. His

[12] Jevons found the book as early as in 1872, but admitted in the preface of the second edition that he had studied it only recently (Jevons, 1957, p. xxx).

[13] The characterisation is by James Friedman (1999), who attributed to Bertrand a "maximum of arrogance and a minimum of knowledge".

review left a lasting footprint in economic theory, not for its slaying of the mathematical approach of the authors and their mechanistic thinking, but for its attack on Cournot's oligopoly theory. A footprint that deserves to be erased.[14]

The crucial part of the review is about Cournot's duopoly model in Chapter 7, roughly half a page long. According to Bertrand,

> Cournot assumes that one of the proprietors will reduce his price to attract buyers to him, and that the other will in turn reduce his price even more to attract buyers back to him. They will only stop undercutting each other in this way, when either proprietor, even if the other abandoned the struggle, has nothing more to gain from reducing his prices. One major objection to this is that there is no solution under this assumption, in that there is no limit in the downward movement. Indeed, whatever the common price adopted, if one of the proprietors, alone, reduces his price he will (...) attract all the buyers and thus double his revenue if his rival lets him do so. (Bertrand, 1883, p. 503, as cited in Dimand and Dimand, 1996, p. 27)

The quote reveals two mistakes by Bertrand. Perhaps he was deceived by Cournot's dynamic stability analysis and by the passage on the identity of the strategic variable on p. 80 of the *Recherches* (cited above). First, Cournot's assumptions do not allow some sort of price rivalry. Cournot assumed that one of the proprietors may (implicitly) cut his price to attract buyers, but not that the other could respond in a similar way, because he assumed that the other proprietor has to accept this price. Second, even in the case of price rivalry, the downward movement of prices is not without limit, but tends to converge to the competitive-market solution. Hence, there is no indication whatsoever that Bertrand criticised Cournot for using quantity as the firm's strategic variable while he himself proposed price (which underlies the so-called Bertrand model). Similarly, the result that (under certain conditions) price competition leads to an equilibrium with marginal-cost pricing (the so-called Bertrand Paradox) cannot be traced back to this quote. Bertrand actually seemed to believe that an oligopolistic industry is best characterised by the cooperative solution. In the line preceding the quote, he wrote that for firms "it would be in their interest to join forces and fix a common price (...) but this solution is rejected (by Cournot)" (Bertrand, 1883, p. 503, my translation). From what we have noted above, it is tempting to infer that the elderly Cournot might have agreed with Bertrand about the relevance of the cooperative solution.

Therefore, linking Bertrand's name to price competition is unwarranted, just as in the case of Fauveau. The origin of price competition probably lies in the work of Francis Edgeworth. In 1897, he published *La teoria pura del monopolio*,[15] which

[14]Terence Hutchison and Joseph Schumpeter were among the first to observe that Bertrand's review of Cournot's duopoly had been overrated. Hutchison (1953, p. 312) noted that Bertrand's comments "form a very slight foundation indeed for Bertrand's reputation as a modern pioneer of duopoly analysis" and Schumpeter (1954, p. 924) thought that Bertrand's review "received more attention than it deserved". Magnan de Bornier (1992) is generally credited for debunking the myth about Bertrand (see also Friedman, 1999 and Lambertini and Mosca, 2010). Dimand and Dore (1999) discuss the historical roots of the Bertrand myth.

[15]Reprinted in English as "The pure theory of monopoly" in Edgeworth (1993).

gives a fairly complete account of price competition in a duopoly. *La teoria* partly corrects and elaborates on his analysis in *Mathematical psychics* (which appeared as early as in 1881), perhaps in response to Bertrand's review. The result that price competition will generate the competitive-market solution was probably first pointed out by Irving Fisher (1898). Additionally, all this refers to the homogeneous goods case. We should recall that Cournot explicitly recognized the role of prices as strategic variable in his Chapter 9 analysis of a differentiated oligopoly with complementary goods.

After Bertrand, it was Edgeworth who became Cournot's fiercest critic. Though a great fan of Cournot's mathematical approach, referring to him as "the father of Mathematical Economics" (Edgeworth, 1881, p. 83), Edgeworth rejected his analysis of oligopoly. In *La teoria*, he wrote

> Cournot's conclusion has been shown to be erroneous by Bertrand for the case in which there is no cost of production; by Professor Marshall for the case in which the cost follows the law of increasing returns; and by the present writer for the case in which the cost follows the law of diminishing returns. (Edgeworth, 1993, p. 117–8)[16]

And in 1925, when looking back at *La teoria*, Edgeworth handed down his final verdict: "Still in 1897 much of Cournot's construction remained standing (...) Now the demolition of Cournot's theory is generally accepted" (Edgeworth, 1993, p. 111). Thus the many works on imperfect competition that appeared during the late 1920s and 1930s either consigned Cournot's oligopoly theories to the dustbin of past mistakes, or ignored them altogether. For instance, accepting the price-undercutting argument against Cournot, Hotelling (1929) introduced locational product differentiation and Chamberlin (1948) provided a dynamic argument that oligopolists will collude. Robinson (1948) did not even mention Cournot. In the end, the alleged defects of Cournot's oligopoly theories were considered reason enough to dismiss the whole *Recherches*.

6 His Comeback

In the 1950s the tide turned with the advent of game theory and Walrasian economics. The crisp clarity of game theory helped to identify the assumptions Cournot ought to have made to derive the results on oligopoly he was after. John Nash proposed a solution to what constitutes an equilibrium in a non-cooperative

[16]Edgeworth referred to a footnote in the *Principles* where Marshall argued that Cournot's equilibrium would be unstable in the case of increasing returns to scale. "For if any one of the rivals got an advantage, and increased his scale of production, he would thereby gain a further advantage, and soon drive all his rivals out of the field" (Marshall, 1890, p. 485–6). Cournot did not mention whether marginal costs would be increasing, decreasing or constant, though he discussed these possibilities extensively in his earlier Chapter 5 on monopoly. Also note that Edgeworth strongly overrated Bertrand's contribution here—the quote will have reinforced the Bertrand myth (see note 14).

market, and it did not take long for others to recognize that Cournot's equilibrium was essentially a Nash equilibrium. Some even went so far as to label it the Cournot-Nash equilibrium. What looked naive and muddled in Cournot's theory could now be ironed out by seeing his theory as a single-shot game with output as strategic variable. His stability analysis with reaction curves, which had confused Fauveau, Bertrand and many others, could now be dismissed as a poetic flight of fancy. Also, Cournot's mathematical methodology gradually appeared in a new light. The Walrasian research programme on general equilibrium theory got underway, following major advances in fixed point theory in mathematics. In the space of a few decades, it became almost unthinkable to do research in economic theory without using mathematics. As James Friedman noted, "the distinction between the *theorist* and the *mathematical economist* no longer exists" (Friedman, 1999, p. 3). Although the future development of economic ideas will certainly move beyond many current topics and approaches, it is unlikely that the mathematical method will fall out of favour. In view of this, it is probably Cournot's greatest achievement that he was the first to develop economic theory through the use of general mathematical functions, subject to only qualitative restrictions, as exemplified by his demand function $F(p)$.

Acknowledgements The author thanks the editors of this book and Hans-Peter Weikard for their helpful comments on an earlier draft of the article.

References

Baloglou CP (2012) Antoine Augustin Cournot. In: Backhaus JG (ed) Handbook of the history of economic thought. Springer, Berlin, pp 437–463

Bertrand J (1883) (Review of Walras's) Théorie mathématique de la richesse sociale (and Cournot's) Recherches sur les Principes Mathématiques de la Théorie de la Richesses. Journal de Savants 67:499–508. https://cruel.org/econthought/texts/marginal/bertrand83.pdf

Chamberlin EH (1948/1933) The theory of monopolistic competition, 6th edn. Oxford University Press, London

Cournot A (1863) Principes de la Théorie des Richesses. Hachette, Paris

Cournot A (1877) Revue Sommaire des Doctrines Economiques. Hachette, Paris. https://archive.org/details/revuesommairede00courgoog

Cournot A (1897/1838) Researches into the mathematical principles of the theory of wealth (translated by N. T. Bacon). Macmillan, London/New York. https://archive.org/details/researchesintom00fishgoog

Creedy J (1992) Demand and exchange in economic analysis. Edward Elgar, Aldershot

de Fontenay R (1864) Principes de la théorie des richesses, par M. Cournot. Journal des Économistes 15:231–251. http://gallica.bnf.fr/ark:/12148/bpt6k378087.image.swf

de Villé P, Ménard C (1989) An insolent founding father. Eur Econ Rev 33:494–502

Dimand RW (1988) An early Canadian contribution to mathematical economics: J. B. Cherriman's 1857 review of Cournot. Can J Econ 21:610–616

Dimand MA, Dimand RW (1996) The history of game theory, vol I. Routledge, London/New York

Dimand RW, Dore MHI (1999) Cournot, Bertrand, and game theory: a further note. Atl Econ J 27:325–333

Edgeworth FY (1881) Mathematical psychics. C. Kegan Paul, London

Edgeworth FY (1993/1925) The pure theory of monopoly. In: Edgeworth FY (ed) Papers relating to political economy, vol 1. Thoemmes Press, Bristol, pp 111–142
Ekelund RB, Hébert RF (1990) Cournot and his contemporaries: is an obituary the only bad review? South Econ J 57:139–149
Ekelund RB, Hébert RF (2014) A history of economic theory and method, 6th edn. Waveland Press, Long Grove
Fauveau P-G (1864) Considérations Mathématiques sur la Théorie de l'Impôt. Gauthiers-Villars, Paris. http://gallica.bnf.fr/ark:/12148/bpt6k110301q/f55.image
Fauveau P-G (1867) Considérations mathématiques sur la théorie de la valeur. Journal des Économistes 15:31–40. http://gallica.bnf.fr/ark:/12148/bpt6k37818j.image.swf
Fisher I (1898) Cournot and mathematical economics. Q J Econ 12:119–138
Friedman JW (1999) The legacy of Augustin Cournot. Technical report, Department of Economics, University of North Carolina, Chapel Hill
Hotelling H (1929) Stability in competition. Econ J 39(153):41–57
Hutchison TW (1953) A review of economic doctrines 1870–1929. Clarendon Press, Oxford
Jevons WS (1957/1871) The theory of political economy, 5th edn. Kelley and Millman, New York
Keynes JM (1921) A treatise on probability. Macmillan, London
Lambertini L, Mosca M (2010) Give to Caesar what is Caesar's. Or, give to Launhardt what we are used to thinking is Bertrand's. Technical report, Università di Bologna, Università di Lecce
Magnan de Bornier J (1992) The 'Cournot-Bertrand debate': a historical perspective. Hist Pol Econ 24:623–656
Magnan de Bornier J (2007) Cournot as an economist: 200 years of relevance. In: Touffut J-P (ed) Augustin Cournot: modelling economics. Edward Elgar, Cheltenham, pp 8–20
Marshall A (1890) Principles of economics, 1st edn. Macmillan, London/New York
Moore HC (1905) The personality of Antoine Augustin Cournot. Q J Econ 19:370–399
Mounce HO (1998) Cournot, Antoine Augustin (1801–77). In: Craige E (ed) Routledge encyclopedia of philosophy. rep.routledge.com (July 2015)
Nichol AJ (1938) Tragedies in the life of Cournot. Econometrica 6:193–197
Robertson RM (1949) Mathematical economics before Cournot. J Pol Econ 57:523–536
Robinson J (1933/1948) The economics of imperfect competition, 1st edn. Macmillan, London
Schumpeter JA (1954) History of economic analysis. Oxford University Press, New York
Shubik M (1998) Antoine Augustin Cournot. In: Eatwell J, Milgate M, Newman P (eds) The new Palgrave, vol 1. Macmillan, London, pp 708–712
Sonnenschein H (1968) The dual of duopoly is complementary monopoly: or, two of Cournot's theories are one. J Pol Econ 76:316–318
Struik DJ (2008) Bertrand, Joseph Louis François. In: Complete dictionary of scientific biography. Encyclopedia.com (July 2015)
Theocharis RD (1990) A lag in the recognition of Cournot's contribution to economic analysis. Can J Econ 23:923–933
Walras L (1874) Éléments d'Économie Politique Pure, ou Théorie de la Richesse Sociale. Corbaz et al., Lausanne
Walras L, Cournot A, Jevons S, Antonelli E (1935) Léon Walras et sa corréspondence avec Augustin Cournot, Stanley Jevons and Étienne Antonelli. Econometrica 3:119–127

Cournot Tatonnement in Aggregative Games with Monotone Best Responses

Nikolai S. Kukushkin

Abstract This paper establishes the acyclicity of Cournot tatonnement in a strategic game with aggregation and monotone best responses, under the broadest assumptions on aggregation rules allowing the Huang–Dubey–Haimanko–Zapechelnyuk–Jensen trick to work and with minimal topological restrictions.

1 Introduction

The importance of aggregation for the existence of Nash equilibrium was first noticed by Novshek (1985), see also Kukushkin (1994). Kukushkin (2004) showed that monotonicity conditions in games with additive aggregation ensure the acyclicity of Cournot tâtonnement rather than the mere existence of an equilibrium. Dubey et al. (2006), having modified a trick invented by Huang (2002) for different purposes, developed a tool applicable to a broader class of aggregation rules. Kukushkin (2005) and, especially, Jensen (2010) extended its sphere of applicability much further. It looks plausible that the latter paper describes the most general class of aggregation rules for which this approach can still work.

This paper strives to establish the acyclicity of Cournot tâtonnement under the same assumptions on aggregation, but with minimal topological restrictions. The point is that the best response correspondences in the main results of Jensen (2010) were assumed upper hemicontinuous. Although one can plausibly argue that the upper hemiconinuity of the best responses holds in "most" of important economic models, "most" cannot be replaced with "all." Even more importantly, our main theorem implies the existence of an equilibrium where each player uses an arbitrarily fixed monotone selection from the best response correspondence; nothing like that could be derived from the previous literature.

There is an additional reason to look for the weakest possible topological conditions. When viewed as a fixed point theorem, our result occupies a position intermediate between Brouwer's and Tarski's theorems: the former is purely topo-

N.S. Kukushkin
Dorodnicyn Computing Center, 40, Vavilova, Moscow 119333, Russia
e-mail: ququns@inbox.ru

P. von Mouche, F. Quartieri (eds.), *Equilibrium Theory for Cournot Oligopolies and Related Games*, Springer Series in Game Theory,
DOI 10.1007/978-3-319-29254-0_4

logical; the latter, order-theoretical. In our case, some combination of conditions of both kinds seems indispensable; so it is highly desirable to understand exactly what is needed from either side.

In Sect. 2, the most basic definitions are given; in Sect. 3, our (i.e., mostly Jensen's) assumptions, as well as the main theorem, are formulated. Section 4 contains a review of conditions ensuring appropriate monotonicity in strategic games. In Sect. 5, we briefly discuss alternative aggregation rules for which our main findings do, or do not, hold. Section 6 contains the proof of our main theorem and an example of Cournot dynamics when the best responses are not upper hemicontinuous.

2 Preliminaries

In the main theorem, we consider exogenously given best response correspondences rather than games as such. An *abstract game* is defined by a finite set of players N and, for each $i \in N$, a strategy set X_i and the *best response correspondence* $\mathcal{R}_i\colon X_{-i} \to 2^{X_i} \setminus \{\emptyset\}$, where $X_{-i} := \prod_{j \in N \setminus \{i\}} X_j$. We also denote $X_N := \prod_{i \in N} X_i$, the set of *strategy profiles*.

Remark 1 The definition of abstract games belongs to Vives (1990). Under his approach, however, the collection of responses $\mathcal{R}_i$ was immediately replaced with a single correspondence $\mathcal{R}\colon X_N \to X_N$, $\mathcal{R}(x_N) := \prod_{i \in N} \mathcal{R}_i(x_{-i})$, to which Tarski's fixed point theorem could be applied. In our case, the structure of a Cartesian product plays a crucial role and is retained to the end.

An *equilibrium* of an abstract game is $x_N^0 \in X_N$ such that $x_i^0 \in \mathcal{R}_i(x_{-i}^0)$ for each $i \in N$. The first basic question about a particular model is whether it admits an equilibrium. In the case of a positive answer, the next question is whether iteration of the best responses leads to equilibria. In this paper, we derive a positive answer to the first question from the same answer to the second one.

We introduce the *best response improvement* relation on X_N ($i \in N, y_N, x_N \in X_N$):

$$y_N \rhd_i x_N \rightleftharpoons [y_{-i} = x_{-i} \;\&\; x_i \notin \mathcal{R}_i(x_{-i}) \ni y_i]; \tag{1a}$$

$$y_N \rhd x_N \rightleftharpoons \exists i \in N \,[y_N \rhd_i x_N]. \tag{1b}$$

Every equilibrium is a maximizer of $\rhd$. Since $\mathcal{R}_i(x_{-i}) \neq \emptyset$ for all $i \in N$ and $x_{-i} \in X_{-i}$, every maximizer of $\rhd$ is an equilibrium.

A *Cournot path* is a finite or infinite sequence $\langle x_N^k \rangle_{k=0,1,\ldots}$ such that $x_N^{k+1} \rhd x_N^k$ whenever x_N^{k+1} is defined. A *Cournot potential* is an irreflexive and transitive binary relation $\succ$ on X_N such that

$$\forall x_N, y_N \in X_N \left[y_N \rhd x_N \Rightarrow y_N \succ x_N\right]. \tag{2}$$

The existence of a Cournot potential is equivalent to the absence of *Cournot cycles*, i.e., Cournot paths $\langle x_N^0, x_N^1, \ldots, x_N^m \rangle$ such that $m > 0$ and $x_N^0 = x_N^m$. If X_N is finite, this fact implies that every Cournot path, if continued whenever possible, reaches an equilibrium in a finite number of steps. Otherwise, the acyclicity of the best responses does not imply very much by itself. However, if a topological structure is assumed on the strategies and the notion of a Cournot potential is strengthened, conclusions can be drawn, not dissimilar from those in the finite case.

Let X be a metric space. A binary relation $\succ$ on X is *ω-transitive* if it is transitive and

$$\left[x^\omega = \lim_{k\to\infty} x^k \;\&\; \forall k \in \mathbb{N}\, [x^{k+1} \succ x^k]\right] \Rightarrow x^\omega \succ x^0. \tag{3}$$

It is essential that (3) implies $x^\omega \succ x^k$ for all $k = 1, \ldots$ as well.

Remark 2 The property seems to have been considered first by Gillies (1959), and then by Smith (1974), for orderings. The term "ω-transitivity" first appeared in Kukushkin (2003).

Theorem 1 from Kukushkin (2008) implies that an irreflexive and ω-transitive binary relation on a compact set always admits a maximizer. Therefore, the existence of an ω-transitive Cournot potential ensures the existence of an equilibrium. It also ensures the "transfinite convergence" (Kukushkin, 2003, 2010) of all iterations of the best responses to equilibria.

3 Main Result

To derive the existence of an ω-transitive Cournot potential in a class of abstract games, we impose a set of assumptions combining topological and order-theoretical requirements. Quite a number of notations and definitions are needed.

A *chain* is a linearly ordered set. A partially ordered set (*poset*) X is *chain-complete* if every chain $\emptyset \neq C \subseteq X$ admits both $\sup C$ and $\inf C$ in X. A poset X is *chain-complete downwards* (*upwards*) if only the existence of $\inf C$ ($\sup C$) is ensured for every nonempty chain in X. A *well ordered set* is a chain Δ such that every subset $\Delta' \subseteq \Delta$ contains its minimum. Dually, a $*$-*well ordered set* is a chain every subset of which contains its maximum.

Given a poset A and $b \in B \subseteq A$, we denote $B^{\rightarrow}(b) := \{a \in B \mid a > b\}$ and $B^{\leftarrow}(b) := \{a \in B \mid a < b\}$. Given a metric space A and $B \subseteq A$, $\operatorname{cl} B$ denotes the topological closure of B in A while $\operatorname{Int} B$ denotes its interior.

We assume throughout that a "universal" separable metric space A is given, which is simultaneously a poset. The order on A is consistent with the topology in the sense that:

$$\forall x \in A \left[\{y \in A \mid y \geq x\} = \operatorname{cl}(\operatorname{Int} A^{\rightarrow}(x))\right]; \tag{4a}$$

$$\forall x \in A \left[\{y \in A \mid y \leq x\} = \operatorname{cl}(\operatorname{Int} A^{\leftarrow}(x))\right]. \tag{4b}$$

These conditions imply that all upper sets $\{y \in A \mid y \geq x\}$ and all lower sets $\{y \in A \mid y \leq x\}$ $(x \in A)$ are closed. A good example of such A is $\mathbb{R}^m$; a more general example is the space of continuous functions on a compact space C with point-wise order and the metric $d(f, g) := \max_{x\in C} |f(x) - g(x)|$. Both examples are also *lattices*, which fact comes in handy in Sect. 4.

Lemma 1 *If* $x^k \to x^\omega$ *and* $x^{k+1} \geq x^k$ *for all* $k \in \mathbb{N}$, *then* $x^\omega = \sup_k x^k$.

Proof First, we denote $X^k := \{y \in A \mid y \geq x^k\}$ for each $k \in \mathbb{N}$; clearly, $x^h \in X^k$ whenever $h \geq k$. Since X^k is closed by (4a), we have $x^\omega \in X^k$, i.e., $x^\omega \geq x^k$, for each k. Assuming $y \geq x^k$ for each k, we denote $Y := \{x \in A \mid y \geq x\}$; Y is closed by (4b). Since $x^k \in Y$ for each k, we have $x^\omega \in Y$ as well; hence $x^\omega \leq y$. Thus, $x^\omega = \sup_k x^k$ indeed. □

Lemma 2 *If* $x^k \to x^\omega$ *and* $x^{k+1} \leq x^k$ *for all* $k \in \mathbb{N}$, *then* $x^\omega = \inf_k x^k$.

The proof is dual to that of Lemma 1. □

Henceforth, we assume that each X_i is a compact subset of the universal set A, and endow X_N with, say, the maximum metric. By Lemmas 1 and 2, each X_i is also chain-complete.

Further, we assume that there are continuous mappings $\sigma_i : X_{-i} \to \mathbb{R}$ $(i \in N)$, *aggregation rules*. Denoting $S_i := \sigma_i(X_{-i}) \subset \mathbb{R}$, $s_i^- := \min S_i$ and $s_i^+ := \max S_i$, we assume the existence of correspondences $R_i : S_i \to 2^{X_i} \setminus \{\emptyset\}$ and continuous mappings $g : X_N \to \mathbb{R}$, $F_i : [s_i^-, s_i^+] \times X_i \to \mathbb{R}$ and $v_i : X_{-i} \to \mathbb{R}$ $(i \in N)$ such that

$$\mathcal{R}_i(x_{-i}) = R_i(\sigma_i(x_{-i})) \tag{5}$$

and

$$g(x_N) = F_i(\sigma_i(x_{-i}), x_i) + v_i(x_{-i}) \tag{6}$$

for all $i \in N$, $x_{-i} \in X_{-i}$, and $x_N \in X_N$. Additionally, we assume that each F_i has a continuous derivative w.r.t. its first argument,

$$D_i(s_i, x_i) := \frac{\partial F_i}{\partial s_i}(s_i, x_i), \tag{7}$$

on $[s_i^-, s_i^+] \times X_i$.

Finally, we impose monotonicity assumptions:

$$\forall i \in N\, \forall s_i', s_i \in S_i \left[[s_i' > s_i \;\&\; x_i' \in R_i(s_i') \;\&\; x_i \in R_i(s_i)] \Rightarrow x_i' \geq x_i \right]; \tag{8}$$

$$\forall i \in N\, \forall s_i \in S_i\, \forall x_i', x_i \in X_i \left[x_i' > x_i \Rightarrow D_i(s_i, x_i') > D_i(s_i, x_i) \right]. \tag{9}$$

Theorem 1 *An abstract game satisfying all the above assumptions admits an* ω*-transitive Cournot potential (hence admits an equilibrium as well).*

The proof is deferred to Sect. 6.

Our conditions (6), (7), and (9) are exactly the same as in Jensen (2010). As to (8), Jensen preferred to consider *decreasing* best responses; the difference is of no significance since one can always replace σ_i with $-\sigma_i$. There was no need for explicit conditions like (4) in Jensen (2010) since all strategy sets were assumed to be subsets of $\mathbb{R}^m$.

Thus, if we assumed that each $\mathcal{R}_i$ is upper hemicontinuous, the difference between our Theorem 1 and the main result of Jensen (2010) would be quite minor. However, we do *not* impose that assumption.

The implications of Theorem 1 for the existence of equilibria in strategic games are considered in the next section. We follow the same logic as in Kukushkin (2005) and Jensen (2010): each $\mathcal{R}_i$ may be perceived either as the total best response correspondence or as an increasing selection from it.

4 Monotonicity Conditions in Strategic Games

The difference between a *strategic game* and an abstract game is that the best responses in the former model are not exogenous, but generated by the maximization of (ordinal) utility functions $u_i: X_N \to \mathbb{R}$,

$$\mathcal{R}_i(x_{-i}) := \operatorname*{Argmax}_{x_i \in X_i} u_i(x_i, x_{-i}) \tag{10}$$

for each player $i \in N$ and every $x_{-i} \in X_{-i}$.

Our Theorem 1 can be applied directly to the best response correspondences in a strategic game if each X_i is a compact subset of an appropriate universal set A, while $\mathcal{R}_i$'s defined by (10) satisfy all those assumptions. Let there be continuous aggregation rules $\sigma_i: X_{-i} \to \mathbb{R}$ $(i \in N)$ such that

$$u_i(x_N) = U_i(\sigma_i(x_{-i}), x_i) \tag{11}$$

for all $i \in N$ and $x_N \in X_N$. For each $i \in N$, we denote $S_i := \sigma_i(X_{-i}) \subset \mathbb{R}$, and redefine the best response correspondence:

$$R_i(s_i) := \operatorname*{Argmax}_{x_i \in X_i} U_i(s_i, x_i).$$

Our assumption $\mathcal{R}_i(x_{-i}) \neq \emptyset$ is equivalent to $R_i(s_i) \neq \emptyset$ for each $s_i \in S_i$.

Remark 3 In principle, equality (5) may hold for all $i \in N$ and $x_{-i} \in X_{-i}$ without (11) holding for all $i \in N$ and $x_N \in X_N$. However, no natural class of strategic games where this happens is known.

Our assumptions on σ_i, i.e., (6), (7), and (9), do not depend on whether it is about an abstract game or a strategic game. Jensen (2010) provides quite a list of game models where the assumptions imposed on aggregation rules in Sect. 3 are satisfied.

As to assumption (8), the well-known studies of monotone comparative statics (Topkis, 1978; Vives, 1990; Veinott AF Jr (1992) Lattice programming. Qualitative optimization and equilibria. Seminar notes; Milgrom and Shannon, 1994) allow us to derive it from assumptions imposed directly on utility functions. We have to start with definitions.

A utility function exhibits the *single crossing* property (Milgrom and Shannon, 1994) if these conditions hold for all $i \in N$, $y_i, x_i \in X_i$, and $s'_i, s_i \in S_i$:

$$[y_i > x_i \ \& \ s'_i > s_i \ \& \ U_i(s_i, y_i) > U_i(s_i, x_i)] \Rightarrow U_i(s'_i, y_i) > U_i(s'_i, x_i); \tag{12a}$$

$$[y_i > x_i \ \& \ s'_i > s_i \ \& \ U_i(s_i, y_i) \geq U_i(s_i, x_i)] \Rightarrow U_i(s'_i, y_i) \geq U_i(s'_i, x_i). \tag{12b}$$

Also important are the *strict single crossing* property (Milgrom and Shannon, 1994):

$$[y_i > x_i \ \& \ s'_i > s_i \ \& \ U_i(s_i, y_i) \geq U_i(s_i, x_i)] \Rightarrow U_i(s'_i, y_i) > U_i(s'_i, x_i); \tag{13}$$

and the *weak single crossing* property (Shannon, 1995):

$$[y_i > x_i \ \& \ s'_i > s_i \ \& \ U_i(s_i, y_i) > U_i(s_i, x_i)] \Rightarrow U_i(s'_i, y_i) \geq U_i(s'_i, x_i). \tag{14}$$

When strategies are scalar, conditions like (12), (13) or (14) are sufficient for monotone comparative statics results. When X_i's are lattices, some versions of *quasisupermodularity* (Milgrom and Shannon, 1994; see also LiCalzi and Veinott 1992), are needed too. We reproduce a few of them here.

First, four "quarters" of quasisupermodularity proper:

$$U_i(s_i, x_i) \geq U_i(s_i, y_i \wedge x_i) \Rightarrow [U_i(s_i, y_i \vee x_i) \geq U_i(s_i, x_i) \ \text{ or } \ U_i(s_i, y_i \vee x_i) \geq U_i(s_i, y_i)]; \tag{15a}$$

$$U_i(s_i, y_i) \geq U_i(s_i, y_i \vee x_i) \Rightarrow [U_i(s_i, y_i \wedge x_i) \geq U_i(s_i, x_i) \ \text{ or } \ U_i(s_i, y_i \wedge x_i) \geq U_i(s_i, y_i)]; \tag{15b}$$

$$U_i(s_i, x_i) > U_i(s_i, y_i \wedge x_i) \Rightarrow [U_i(s_i, y_i \vee x_i) > U_i(s_i, x_i) \ \text{ or } \ U_i(s_i, y_i \vee x_i) > U_i(s_i, y_i)]; \tag{15c}$$

$$U_i(s_i, y_i) > U_i(s_i, y_i \vee x_i) \Rightarrow [U_i(s_i, y_i \wedge x_i) > U_i(s_i, x_i) \ \text{ or } \ U_i(s_i, y_i \wedge x_i) > U_i(s_i, y_i)]. \tag{15d}$$

Then, two "halves" of *weak quasisupermodularity*:

$$U_i(s_i, x_i) > U_i(s_i, y_i \wedge x_i) \Rightarrow [U_i(s_i, y_i \vee x_i) \geq U_i(s_i, x_i) \text{ or } U_i(s_i, y_i \vee x_i) \geq U_i(s_i, y_i)]; \tag{16a}$$

$$U_i(s_i, y_i) > U_i(s_i, y_i \vee x_i) \Rightarrow [U_i(s_i, y_i \wedge x_i) \geq U_i(s_i, x_i) \text{ or } U_i(s_i, y_i \wedge x_i) \geq U_i(s_i, y_i)]. \tag{16b}$$

Remark 4 Each of conditions (15) and (16) holds trivially when $x_i \geq y_i$ or $x_i \leq y_i$.

Proposition 22 from Kukushkin (2013b) immediately implies this result:

Proposition 1 *Let, in a strategic game satisfying* (11)*, X_i be a lattice and the utility function $U_i(s_i, x_i)$ satisfy these assumptions: the strict single crossing condition* (13) *holds for every $x_i, y_i \in X_i$ and $s_i, s_i' \in S_i$; there is $s_i^* \in S_i$ such that* (15a) *holds for all $x_i, y_i \in X_i$ and $s_i < s_i^*$ while* (15b) *holds for all $x_i, y_i \in X_i$ and $s_i > s_i^*$. Then the best response correspondence R_i is increasing in the sense of* (8).

If X_i is a *semilattice*, i.e., only the meet $y_i \wedge x_i$ is guaranteed to exist for every $x_i, y_i \in X_i$ (e.g., a budget set), then conditions (15) and (16) make no sense. However, a similar role is played by a condition that could be called *semiquasisupermodularity*:

$$U_i(s_i, y_i) > U_i(s_i, y_i \wedge x_i) \Rightarrow U_i(s_i, y_i \wedge x_i) \geq U_i(s_i, x_i). \tag{17}$$

Proposition 2 *Let, in a strategic game satisfying* (11)*, X_i be a semilattice, the strict single crossing condition* (13) *hold for every $x_i, y_i \in X_i$ and $s_i, s_i' \in S_i$, and* (17) *hold for every $x_i, y_i \in X_i$ and $s_i \in S_i$. Then the best response correspondence R_i is increasing in the sense of* (8).

Proof Let $s_i' > s_i$, $x_i \in R_i(s_i)$ and $y_i \in R_i(s_i')$; we have to show $y_i \geq x_i$. Otherwise, we would have $x_i > y_i \wedge x_i$. Since $U_i(s_i, x_i) \geq U_i(s_i, y_i \wedge x_i)$, we have $U_i(s_i', x_i) > U_i(s_i', y_i \wedge x_i)$ by (13); hence $U_i(s_i', y_i \wedge x_i) \geq U_i(s_i', y_i)$ by (17). Thus, $U_i(s_i', x_i) > U_i(s_i', y_i)$, which contradicts $y_i \in R_i(s_i')$. □

Remark 5 Undoubtedly, condition (17) is much more demanding than (15) or (16); however, it may hold in a non-trivial way. Essentially, it shows that U_i may increase in x_i only along a "fixed path."

When the best response correspondences in a strategic game with appropriate aggregation do *not* satisfy (8), our Theorem 1 may still be applicable in a more sophisticated way. If there are increasing selections from the best response correspondences, we may consider the abstract game defined by those selections rather than total best responses. We cannot obtain any statement about *all* Cournot dynamics in this way, but the existence of an equilibrium, at least, will be ensured. In the case when those selections are "natural" in one sense or another, the acyclicity of the corresponding dynamics may also be of interest. Finally, the existence of an equilibrium where each player uses a fixed selection from the best response correspondence may be important in some contexts.

Combining Proposition 26 from Kukushkin (2013b) and Theorem 2.2 from Kukushkin (2013a), we obtain this result:

Proposition 3 *Let, in a strategic game satisfying* (11), X_i *be a lattice and the utility function* $U_i(s_i, x_i)$ *satisfy these conditions: the weak single crossing condition* (14) *holds for every* $x_i, y_i \in X_i$ *and* $s_i, s_i' \in S_i$*; there is* $s_i^* \in S_i$ *such that* (15c) *holds for all* $x_i, y_i \in X_i$ *and* $s_i < s_i^*$ *while* (15d) *holds for all* $x_i, y_i \in X_i$ *and* $s_i > s_i^*$*; every* $R_i(s_i)$ *is nonempty and chain-complete. Then there exists an increasing selection* r_i *from the best response correspondence* R_i*.*

Invoking Proposition 28 from Kukushkin (2013b) and the same Theorem 2.2 from Kukushkin (2013a), we obtain this result:

Proposition 4 *Let, in a strategic game satisfying* (11), X_i *be a lattice and the utility function* $U_i(s_i, x_i)$ *satisfy these conditions: the single crossing conditions* (12) *hold for every* $x_i, y_i \in X_i$ *and* $s_i, s_i' \in S_i$*; there is* $s_i^* \in S_i$ *such that* (16a) *holds for all* $x_i, y_i \in X_i$ *and* $s_i < s_i^*$ *while* (16b) *holds for all* $x_i, y_i \in X_i$ *and* $s_i > s_i^*$*; every* $R_i(s_i)$ *is nonempty and chain-complete. Then there exists an increasing selection* r_i *from the best response correspondence* R_i*.*

A similar statement is valid for semilattices too.

Proposition 5 *Let, in a strategic game satisfying* (11), X_i *be a semilattice and the utility function* $U_i(s_i, x_i)$ *satisfy conditions* (12a) *and* (17) *for every* $x_i, y_i \in X_i$ *and* $s_i, s_i' \in S_i$*, while every* $R_i(s_i)$ *be nonempty and chain-complete downwards. Then there exists an increasing selection* r_i *from the best response correspondence* R_i*.*

Proof For every $s_i \in S_i$, we denote $R_i^-(s_i)$ the "lower frontier" of $R_i(s_i)$, i.e., $R_i^-(s_i) \rightleftharpoons \{x_i \in R_i(s_i) \mid \nexists y_i \in R_i(s_i)\, [y_i < x_i]\}$. Since $R_i(s_i)$ is nonempty and chain-complete downwards, Zorn's Lemma implies that $R_i^-(s_i) \neq \emptyset$. Let us show that an arbitrary selection $r_i: S_i \to X_i$ from $R_i^-(s_i)$ is increasing.

Let $s_i' > s_i$, $x_i = r_i(s_i)$ and $y_i = r_i(s_i')$; we have to show $y_i \geq x_i$. Otherwise, we would have $x_i > y_i \wedge x_i$; hence $U_i(s_i, x_i) > U_i(s_i, y_i \wedge x_i)$ because $x_i \in R_i^-(s_i)$. Now we have $U_i(s_i', x_i) > U_i(s_i', y_i \wedge x_i)$ by (12a); hence $U_i(s_i', y_i \wedge x_i) \geq U_i(s_i', y_i)$ by (17). Thus, $U_i(s_i', x_i) > U_i(s_i', y_i)$, which contradicts $y_i \in R_i(s_i')$. □

5 On Other Aggregation Rules

It is important to note that Theorem 1, i.e., the Huang–Dubey–Haimanko–Zapechelnyuk–Jensen trick, does not cover *all* nice aggregation rules. For instance, abstract games satisfying (5) and (8) with $X_i \subset \mathbb{R}$ and $\sigma_i(x_{-i}) = \min_{j \neq i} x_j$ for all $i \in N$ or $\sigma_i(x_{-i}) = -\min_{j \neq i} x_j$ for all $i \in N$ also admit ω-transitive Cournot potentials. Moreover, given a subset $I(i) \subseteq N \setminus \{i\}$ for each $i \in N$ such that $j \in I(i) \iff i \in I(j)$, the aggregation rules $\sigma_i(x_{-i}) = \min_{j \in I(i)} x_j$ or $\sigma_i(x_{-i}) = -\min_{j \in I(i)} x_j$ are also acceptable (Kukushkin, 2003, Theorems 7 and 8). It goes without saying that the minimum can be replaced with the maximum. Apparently, functions g and F_i satisfying (6) and (9) cannot exist in this case.

Aggregation rules mapping X_{-i} into chains "longer" than $\mathbb{R}$ (lexicographies) may also be nice although there are very few established facts so far. On the other hand, *partially* ordered aggregates seem hopeless. For instance, additive aggregation makes sense for $X_i \subset \mathbb{R}^m$ with $m > 1$ as well, in which case $S_i \subset \mathbb{R}^m$ too. Moreover, in the case of $\sigma_i(x_{-i}) = \sum_{j \neq i} x_j$, condition (8) implies that the best responses are increasing, and hence the existence of equilibria is looked after by the Tarski Theorem (provided the strategy sets remain lattices). Nonetheless, there may be Cournot cycles in such games. When $\sigma_i(x_{-i}) = -\sum_{j \neq i} x_j$, even the mere existence of an equilibrium is not guaranteed.

Example 1 Consider an abstract game where $N := \{1, 2, 3\}$, $X_1 := \{(0,0,0), (1,0,0)\}$, $X_2 := \{(0,0,0), (0,1,0)\}$, $X_3 := \{(0,0,0), (0,0,1)\}$, $\sigma_i(x_{-i}) := -\sum_{j \neq i} x_j$, and

$$R_1(s_1) := \begin{cases} \{(0,0,0)\} & \text{if } s_1 \leq (0,-1,0); \\ \{(1,0,0)\} & \text{otherwise;} \end{cases}$$

$$R_2(s_2) := \begin{cases} \{(0,0,0)\} & \text{if } s_2 \leq (0,0,-1); \\ \{(0,1,0)\} & \text{otherwise;} \end{cases}$$

$$R_3(s_3) := \begin{cases} \{(0,0,0)\} & \text{if } s_3 \leq (-1,0,0); \\ \{(0,0,1)\} & \text{otherwise.} \end{cases}$$

Condition (8) is obvious; nonetheless, there is no equilibrium.

6 Proof

For every $s_i \in S_i$, we define $\bar{R}_i(s_i) := \{x_i \in X_i \mid (s_i, x_i) \in \operatorname{cl}(\operatorname{graph} R_i)\}$; clearly, (8) holds for $\bar{R}_i$ as well. For each $i \in N$, we define $X_i^0 := \bigcup_{s_i \in S_i} \bar{R}_i(s_i)$. The compactness of S_i and upper hemicontinuity of $\bar{R}_i$ imply that X_i^0 is closed in X_i; hence it is compact too. For every $x_N \in X_N$, we set $N^0(x_N) := \{i \in N \mid x_i \in X_i^0\}$.

For each $i \in N$, we pick an arbitrary selection $r_i : S_i \to X_i$ from R_i; by (8), r_i is increasing in the sense of $s_i' \geq s_i \Rightarrow r_i(s_i') \geq r_i(s_i)$. For every $s_i \in S_i$, we denote $S_i^{\to}(s_i) := \{s_i' \in S_i \mid s_i' > s_i\}$ and $S_i^{\leftarrow}(s_i) := \{s_i' \in S_i \mid s_i' < s_i\}$, and then $S_i^{\to} := \{s_i \in S_i \mid s_i = \inf S_i^{\to}(s_i)\}$ and $S_i^{\leftarrow} := \{s_i \in S_i \mid s_i = \sup S_i^{\leftarrow}(s_i)\}$. For $s_i \in S_i^{\to}$, we set $r_i^+(s_i) := \inf r_i(S_i^{\to}(s_i))$; for $s_i \in S_i^{\leftarrow}$, $r_i^-(s_i) := \sup r_i(S_i^{\leftarrow}(s_i))$.

Assumption (8) immediately implies that

$$\forall s_i \in S_i^{\rightarrow} \cap S_i^{\leftarrow} \, [r_i^+(s_i) \geq r_i^-(s_i)];$$
$$\forall s_i \in S_i^{\rightarrow} \, \forall s_i' > s_i \, \forall x_i \in \bar{R}_i(s_i') \, [x_i \geq r_i^+(s_i)];$$
$$\forall s_i \in S_i^{\leftarrow} \, \forall s_i' < s_i \, \forall x_i \in \bar{R}_i(s_i') \, [x_i \leq r_i^-(s_i)];$$

in the following equality, if $r_i^+(s_i)$ and/or $r_i^-(s_i)$ are not defined, the corresponding term(s) should be ignored:

$$\forall s_i \in S_i \, [\bar{R}_i(s_i) = \mathrm{cl}(R_i(s_i)) \cup \{r_i^-(s_i)\} \cup \{r_i^+(s_i)\}];$$
$$\forall s_i \in S_i^{\rightarrow} \cap S_i^{\leftarrow} \, [r_i^+(s_i) = r_i^-(s_i) \Rightarrow \bar{R}_i(s_i) = R_i(s_i) = \{r_i(s_i)\}].$$

Step 6.1 *If $s_i \in S_i^{\leftarrow}$ and $r_i(s_i) = r_i^-(s_i)$, then r_i is left continuous at s_i.*

Proof Let $\langle s_i^k \rangle_k$ be a strictly increasing sequence such that $s_i^k \to s_i$; then $s_i = \sup_k s_i^k$, and hence $r_i(s_i) = r_i^-(s_i) = \sup_k r_i(s_i^k)$. Since X_i is compact, we may assume $r_i(s_i^k) \to x_i^* \in X_i$; by Lemma 1, we have $x_i^* = \sup_k r_i(s_i^k) = r_i(s_i)$; hence $r_i(s_i^k) \to r_i(s_i)$. □

Step 6.2 *If $s_i \in S_i^{\rightarrow}$ and $r_i(s_i) = r_i^+(s_i)$, then r_i is right continuous at s_i.*

The proof is dual to that of Step 6.1. □

Now we extend r_i to the whole $[s_i^-, s_i^+]$ with the following construction. For every $s_i \in [s_i^-, s_i^+]$ we define $\xi_i^+(s_i) = \min\{\xi_i \in S_i \mid \xi_i \geq s_i\}$ and $\xi_i^-(s_i) = \max\{\xi_i \in S_i \mid \xi_i \leq s_i\}$. Obviously, $\xi_i^+(s_i) = \xi_i^-(s_i) = s_i$ if and only if $s_i \in S_i$; otherwise, $\xi_i^-(s_i) < s_i < \xi_i^+(s_i)$. Now for every $s_i \in [s_i^-, s_i^+] \setminus S_i$ we define $r_i(s_i) = r_i(\xi_i^-(s_i))$ if $s_i - \xi_i^-(s_i) \leq \xi_i^+(s_i) - s_i$, and $r_i(s_i) = r_i(\xi_i^+(s_i))$ otherwise.

Step 6.3 *r_i is continuous at every $s_i \in [s_i^-, s_i^+]$ except for a countable subset.*

Proof S_i being compact, its complement, $[s_i^-, s_i^+] \setminus S_i$, consists of a countable number of disjoint open intervals. $S_i \setminus (S_i^{\rightarrow} \cap S_i^{\leftarrow})$ consists of the end points of those same intervals. The way we extended r_i beyond S_i ensures continuity everywhere with the possible exception of the end points and the middle of each interval. By Steps 6.1 and 6.2, r_i is continuous wherever $r_i^-(s_i) = r_i^+(s_i)$.

Therefore, we only have to prove that the set $\{s_i \in S_i^{\rightarrow} \cap S_i^{\leftarrow} \mid r_i^+(s_i) > r_i^-(s_i)\}$ is countable. We pick a countable and dense subset of A and denote it Z. Given $s_i \in S_i^{\rightarrow} \cap S_i^{\leftarrow}$ such that $r_i^+(s_i) > r_i^-(s_i)$, we denote $U' := \mathrm{Int}\, A^{\leftarrow}(r_i^-(s_i))$ and $U'' := \mathrm{Int}\, A^{\leftarrow}(r_i^+(s_i))$. By (4b), we have $r_i^+(s_i) \notin \mathrm{cl}\, U'$; hence there is an open set $U \subset A$ such that $r_i^+(s_i) \in U$ and $U \cap \mathrm{cl}\, U' = \emptyset$. Since $r_i^+(s_i) \in \mathrm{cl}\, U''$, we have $U \cap U'' \neq \emptyset$. Now we pick $z(s_i) \in Z \cap U \cap U''$ arbitrarily; this is possible since Z is dense in A. Since $z(s_i) \notin \mathrm{cl}\, U'$, we have $r_i^-(s_i) \ngtr z(s_i)$; since $z(s_i) \in U''$, we have $r_i^+(s_i) > z(s_i)$. Finally, $z(s_i') \neq z(s_i)$ whenever $s_i' > s_i$, because $r_i^-(s_i') \geq r_i^+(s_i) > z(s_i)$ while $r_i^-(s_i') \ngtr z(s_i')$. □

For every $x_N \in X_N$, we define a function

$$H(x_N) := g(x_N) + \sum_{i \in N} \Big[-F_i(s_i^+, x_i) + \int_{s_i^-}^{s_i^+} \min\{D_i(s_i, x_i), D_i(s_i, r_i(s_i))\} \, ds_i \Big]. \tag{18}$$

In light of Step 6.3, the integral in (18) exists even in the Riemann sense. Let $i \in N$, $x_N \in X_N$, and $x_i \in \bar{R}_i(s_i^*)$ for $s_i^* \in S_i$. The key role in the Huang–Dubey–Haimanko–Zapechelnyuk–Jensen trick is played by this equality, easily following from (6) and the monotonicity (9) of $D_i(s_i, x_i)$:

$$H(x_N) = F_i(\sigma_i(x_{-i}), x_i) - F_i(s_i^*, x_i) + \int_{s_i^-}^{s_i^*} D_i(s_i, r_i(s_i)) \, ds_i + C(x_{-i}). \tag{19}$$

It will be easily derived from (19) that $H(y_N) = H(x_N)$ whenever $x_{-i} = y_{-i}$ and $x_i \in \bar{R}_i(\sigma_i(x_{-i})) \ni y_i$. Not so easy, but also straightforward is the derivation from (19) that $H(y_N) > H(x_N)$ whenever $x_{-i} = y_{-i}$ and $x_i \notin \bar{R}_i(\sigma_i(x_{-i})) \ni y_i$. The most cumbersome part of the whole construction is the definition of the order in the sense of which the current strategy profile goes upwards when $x_i \in \bar{R}_i(\sigma_i(x_{-i})) \setminus R_i(\sigma_i(x_{-i}))$ is replaced with $y_i \in R_i(\sigma_i(x_{-i}))$. We generally follow Kukushkin (2005), but additional subtleties are needed because X_i's are no longer chains. This whole part becomes superfluous if every R_i is upper hemicontinuous, so $R_i = \bar{R}_i$.

For every $i \in N$, we define binary relations on X_i:

$$y_i \rhd\!\!\rhd_i x_i \rightleftharpoons \exists \bar{s}_i \in S_i \, [y_i \in R_i(\bar{s}_i) \,\&\, x_i \in \bar{R}_i(\bar{s}_i) \setminus R_i(\bar{s}_i)] \tag{20}$$

(in the following, we say "$y_i \rhd\!\!\rhd_i x_i$ holds at $\bar{s}_i$");

$$y_i \rhd\!\!\rhd_i^+ x_i \rightleftharpoons [y_i \rhd\!\!\rhd_i x_i \,\&\, y_i > x_i];$$

$$y_i \rhd\!\!\rhd_i^- x_i \rightleftharpoons [y_i \rhd\!\!\rhd_i x_i \,\&\, y_i < x_i].$$

An *i-singular upward chain* is a countable well ordered subset $\Delta \subseteq X_i$ such that (i) $y_i \rhd\!\!\rhd_i^+ x_i$ whenever $y_i \in \Delta$ and $x_i = \max \Delta^{\leftarrow}(y_i)$ [in which case $y_i = \min \Delta^{\rightarrow}(x_i)$], and (ii) $y_i = \sup \Delta^{\leftarrow}(y_i)$ whenever $y_i \in \Delta$ and $\max \Delta^{\leftarrow}(y_i)$ does not exist. We set $y_i \ggcurly_i^+ x_i$ iff $y_i > x_i$ and there is an i-singular upward chain $\Delta \subseteq X_i$ such that $y_i = \max \Delta$ and $x_i = \min \Delta$.

An *i-singular downward chain* is defined dually as a countable $*$-well ordered subset $\Delta \subseteq X_i$ such that (i) $y_i \rhd\!\!\rhd_i^- x_i$ whenever $y_i \in \Delta$ and $x_i = \min \Delta^{\rightarrow}(y_i)$ [in which case $y_i = \max \Delta^{\leftarrow}(x_i)$], and (ii) $y_i = \inf \Delta^{\rightarrow}(y_i)$ whenever $y_i \in \Delta$ and $\min \Delta^{\rightarrow}(y_i)$ does not exist. We set $y_i \ggcurly_i^- x_i$ iff $y_i < x_i$ and there is an i-singular downward chain $\Delta \subseteq X_i$ such that $y_i = \min \Delta$ and $x_i = \max \Delta$.

Then, we define

$$y_i \ggcurly_i x_i \rightleftharpoons [y_i \rhd\!\!\rhd_i x_i \ \text{or} \ y_i \ggcurly_i^+ x_i \ \text{or} \ y_i \ggcurly_i^- x_i]. \tag{21}$$

Remark 6 Obviously, $y_i \succ\!\!\succ_i^+ x_i$ if $y_i \rhd\!\!\rhd_i^+ x_i$, while $y_i \succ\!\!\succ_i^- x_i$ if $y_i \rhd\!\!\rhd_i^- x_i$. However, $y_i \rhd\!\!\rhd_i x_i$ may hold when y_i and x_i are incomparable in the order on X_i.

Now, we are ready to define our potential, a binary relation on X_N:

$$y_N \succ x_N \rightleftharpoons \big[N^0(y_N) \supset N^0(x_N) \text{ or } [N^0(y_N) = N^0(x_N) \,\&\, H(y_N) > H(x_N)] \text{ or } \big(N^0(y_N) = N^0(x_N) \,\&\, H(y_N) = H(x_N) \,\& \; \forall i \in N\,[y_i = x_i \text{ or } y_i \succ\!\!\succ_i x_i] \,\&\, \exists i \in N\,[y_i \succ\!\!\succ_i x_i]\big)\big]. \quad (22)$$

Obviously, $\succ$ is irreflexive. Checking its ω-transitivity and (2) needs quite some effort.

Step 6.4 *Both relations $\succ\!\!\succ_i^+$ and $\succ\!\!\succ_i^-$ are ω-transitive.*

Proof It is sufficient to consider one of the relations, say, $\succ\!\!\succ_i^+$. Let $z_i \succ\!\!\succ_i^+ y_i \succ\!\!\succ_i^+ x_i$. By definition, there are two i-singular upward chains, Δ' and Δ'', such that $\min \Delta' = x_i$, $\max \Delta' = y_i = \min \Delta''$, and $\max \Delta'' = z_i$. Defining $\Delta = \Delta' \cup \Delta''$, we see that Δ is an i-singular upward chain—when checking each condition in the definition, we will find ourselves either totally inside Δ' or totally inside Δ''. Since $x_i = \min \Delta$ and $z_i = \max \Delta$, we have $z_i \succ\!\!\succ_i^+ x_i$.

The proof of (3) is quite similar. Let $x_i^k \to x_i^\omega$ and $x_i^{k+1} \succ\!\!\succ_i^+ x_i^k$ for all k; let Δ^k ($k = 0, 1, \dots$) be an i-singular upward chain such that $x_i^k = \min \Delta^k$ and $x_i^{k+1} = \max \Delta^k$. Denoting $\Delta = \{x_i^\omega\} \cup \bigcup_{k\in\mathbb{N}} \Delta^k$, we again obtain that Δ is an i-singular upward chain (the condition $x_i^\omega = \sup_{k\in\mathbb{N}} x_i^k$ is essential here), $x_i^0 = \min \Delta$ and $x_i^\omega = \max \Delta$. □

Step 6.5 *If $z_i \rhd\!\!\rhd_i y_i \rhd\!\!\rhd_i x_i$, then either $z_i \rhd\!\!\rhd_i^+ y_i \rhd\!\!\rhd_i^+ x_i$ or $z_i \rhd\!\!\rhd_i^- y_i \rhd\!\!\rhd_i^- x_i$.*

Proof Let $y_i \rhd\!\!\rhd_i x_i$ at $s_i' \in S_i$ and $z_i \rhd\!\!\rhd_i y_i$ at $s_i'' \in S_i$. Since $y_i \in R_i(s_i') \setminus R_i(s_i'')$, $s_i'' \neq s_i'$.

Suppose that $s_i'' > s_i'$. If $y_i < x_i$ or y_i and x_i are incomparable in the order on X_i, then $y_i' > y_i$ for every $y_i' \in \bar{R}_i(s_i'')$; hence $y_i \in \bar{R}_i(s_i'')$ is impossible, and hence $z_i \rhd\!\!\rhd_i y_i$ at s_i'' is impossible too. If $y_i > x_i$, then $y_i \rhd\!\!\rhd_i^+ x_i$. Since $y_i' \geq y_i$ for every $y_i' \in R_i(s_i'')$, $z_i \rhd\!\!\rhd_i^+ y_i$ too.

The case of $s_i'' < s_i'$ is treated dually. □

Step 6.6 *If $z_i \succ\!\!\succ_i y_i \succ\!\!\succ_i x_i$, then either $z_i \succ\!\!\succ_i^+ y_i \succ\!\!\succ_i^+ x_i$ or $z_i \succ\!\!\succ_i^- y_i \succ\!\!\succ_i^- x_i$.*

Proof First, if we suppose that $z_i \rhd\!\!\rhd_i y_i$ does not hold, then either $z_i \succ\!\!\succ_i^+ y_i$ or $z_i \succ\!\!\succ_i^- y_i$. In the first case, there is an i-singular upward chain Δ such that $z_i = \max \Delta$ and $y_i = \min \Delta$. Setting $z_i' := \min(\Delta \setminus \{y_i\})$, we obtain $z_i' \rhd\!\!\rhd_i^+ y_i$. Dually, if $z_i \succ\!\!\succ_i^- y_i$, we pick an i-singular downward chain Δ such that $z_i = \min \Delta$ and $y_i = \max \Delta$. Setting $z_i' := \max(\Delta \setminus \{y_i\})$, we obtain $z_i' \rhd\!\!\rhd_i^- y_i$. Obviously, if $z_i \rhd\!\!\rhd_i y_i$ does hold, then we may set $z_i' := z_i$ and have $z_i' \rhd\!\!\rhd_i y_i$ again. Let $z_i' \rhd\!\!\rhd_i y_i$ at $\bar{s}_i \in S_i$.

Turning to the second relation, we see that Step 6.5 applies if $y_i \rhd\!\!\rhd_i x_i$; hence either $z_i' \rhd\!\!\rhd_i^+ y_i \rhd\!\!\rhd_i^+ x_i$ or $z_i' \rhd\!\!\rhd_i^- y_i \rhd\!\!\rhd_i^- x_i$; hence either $z_i \succ\!\!\succ_i^+ y_i \succ\!\!\succ_i^+ x_i$ or $z_i \succ\!\!\succ_i^- y_i \succ\!\!\succ_i^- x_i$ indeed.

Let $y_i \ggcurly_i^+ x_i$ while $y_i \not\trianglerighteq\!\!\triangleright_i x_i$. Then there is an i-singular upward chain Δ such that $x_i = \min \Delta$ and $y_i = \sup(\Delta \setminus \{y_i\})$. By definition, for every $x_i' \in \Delta \setminus \{y_i\}$, there is $y_i' \in \Delta \setminus \{y_i\}$ such that $y_i' \triangleright\!\!\triangleright_i^+ x_i'$; hence $y_i', x_i' \in \bar{R}(s_i)$ for some $s_i \in S_i$. Since $y_i', x_i' < y_i$, we must have $s_i < \bar{s}_i$; hence $y_i', x_i' \leq z_i'$. Clearly, $y_i = \sup(\Delta \setminus \{y_i\})$ is only possible if $z_i' \geq y_i$; hence $z_i' \triangleright\!\!\triangleright_i^+ y_i$, and hence $z_i \ggcurly_i^+ y_i \ggcurly_i^+ x_i$.

The case of $y_i \ggcurly_i^- x_i$ is treated dually. □

Step 6.7 *The relation $\ggcurly_i$ is ω-transitive.*

Proof The statement immediately follows from Steps 6.6 and 6.4. □

Step 6.8 *The relation $\succ$ is irreflexive and ω-transitive.*

Proof The irreflexivity of $\succ$ is obvious; checking transitivity is very easy. Let us check (3). The situation $N^0(x_N^{k+1}) \supset N^0(x_N^k)$ can only happen for a finite number of k; without restricting generality, $N^0(x_N^{k+1}) = N^0(x_N^k)$ for all k, and hence $N^0(x_N^\omega) \supseteq N^0(x_N^0)$ since each X_i^0 is closed. If $H(x_N^{k+1}) > H(x_N^k)$ for a single k, then $H(x_N^\omega) > H(x_N^0)$ since H is continuous, and we are home. Finally, let $N^0(x_N^{k+1}) = N_N^0(x^k)$ and $H(x_N^{k+1}) = H(x_N^k)$ for all k. Then, for each $i \in N$, either $x_i^{k+1} \ggcurly_i x_i^k$ for some k, or $x_i^{k+1} = x_i^k$ for all k. In the first case, Step 6.7 applies, producing $x_i^\omega \ggcurly_i x_i^0$; in the second, $x_i^\omega = x_i^0$. In either case, we have $x_N^\omega \succ x_N^0$. □

Step 6.9 *If $y_N \triangleright x_N$, then $y_N \succ x_N$.*

Proof Let $y_N \triangleright_i x_N$ and $\bar{s}_i := \sigma_i(x_{-i})$. We have $y_i \in R_i(\bar{s}_i)$ by definition; hence $y_i \in X_i^0$ and $N^0(y_N) \supseteq N^0(x_N)$. If the inclusion is strict, we are home.

Let us assume $N^0(y_N) = N^0(x_N)$, i.e., $x_i \in \bar{R}_i(s_i^*)$ for $s_i^* \in S_i$. Invoking (19) separately for x_N and y_N, we obtain

$$\begin{aligned} H(y_N) - H(x_N) \\ &= \int_{s_i^-}^{\bar{s}_i} D_i(s_i, r_i(s_i))\, ds_i - F_i(\bar{s}_i, x_i) + F_i(s_i^*, x_i) - \int_{s_i^-}^{s_i^*} D_i(s_i, r_i(s_i))\, ds_i \\ &= \int_{\bar{s}_i}^{s_i^*} [D_i(s_i, x_i) - D_i(s_i, r_i(s_i))]\, ds_i. \end{aligned} \tag{23}$$

If $s_i^* = \bar{s}_i$, i.e., $x_i \in \bar{R}_i(\bar{s}_i)$, then $H(y_N) = H(x_N)$. Since $x_i \in \bar{R}_i(\bar{s}_i) \setminus R_i(\bar{s}_i)$ and $y_i \in R_i(\bar{s}_i)$, we have $y_i \triangleright\!\!\triangleright_i x_i$ by (20) and hence $y_i \ggcurly_i x_i$ by (21). Therefore, $y_N \succ x_N$ by the third term in (22).

Finally, let $x_i \notin \bar{R}_i(\bar{s}_i)$; hence $s_i^* \neq \bar{s}_i$. If $s_i^* > \bar{s}_i$, then the integrand in (23) is nonnegative on the whole interval because $x_i \geq r_i(s_i)$; since $x_i \notin \bar{R}_i(\bar{s}_i)$ and the graph of $\bar{R}_i$ is closed, $x_i > r_i(s_i)$ in an open neighborhood of $\bar{s}_i$ and hence the integrand is strictly positive by (9). If $s_i^* < \bar{s}_i$, then the integrand is nonpositive on the whole interval and strictly negative in an open neighborhood of $\bar{s}_i$, but $ds_i < 0$ (the lower limit is greater than the upper one). In either case, $H(y_N) > H(x_N)$, hence $y_N \succ x_N$ by the second term in (22). □

Thus, $\succ$ is an ω-transitive Cournot potential and Theorem 1 is proved.

Example 2 Let us consider a game where $N := \{1, 2\}$, $X_i := [0, 2]$, and the utility functions are "isomorphic":

$$u_i(x_N) := \begin{cases} \min\{4x_i + 4x_{-i}, 7x_{-i} - 2x_i + 6\}, & \forall j \in N\, [x_j \geq 1]; \\ \min\{4x_i + 4x_{-i}, 7x_{-i} - 2x_i + 3\}, & \text{otherwise.} \end{cases}$$

Obviously, both u_i are upper semicontinuous, but not continuous. The unique best responses are easy to compute:

$$\mathscr{R}_i(x_{-i}) = \begin{cases} \{x_{-i}/2 + 1\}, & x_{-i} \geq 1; \\ \{x_{-i}/2 + 1/2\}, & x_{-i} < 1. \end{cases}$$

There is a unique Nash equilibrium, $(2, 2)$.

The monotonicity condition (8) being obvious, this game belongs to the class covered by Theorem 1 with $\sigma_i(x_{-i}) := x_{-i}$. The function $H(x_N)$ defined by (18) looks as follows:

$$H(x_N) = x_1 x_2 + \psi(x_1) + \psi(x_2),$$

where

$$\psi(x) := \begin{cases} 2x - x^2 - 3/2, & 3/2 \leq x \leq 2; \\ 3/4 - x, & 1 \leq x \leq 3/2; \\ x - x^2 - 1/4, & 1/2 \leq x \leq 1; \\ 0, & x \leq 1/2. \end{cases}$$

Every Cournot path started from x_N^0 with $x_i^0 < 1$ for both i converges to $x_N^\omega = (1, 1)$, which is not an equilibrium; $H(x_N^k)$ strictly increases at each step. When, say, player 1 makes a best response improvement by replacing $x_1^\omega = 1$ with $x_1^{\omega+1} = 3/2$, H remains the same, but $3/2 \succ\!\!\succ_1^+ 1$; hence the strategy profile goes upwards in the sense of $\succ$ defined by (22). After that, a unique Cournot path converges to $(2, 2)$, which *is* an equilibrium; $H(x_N^{\omega+k})$ again strictly increases at each step $k > 1$.

This example demonstrates why Jensen's construction (18) alone is insufficient when the best responses need not be upper hemicontinuous.

Actually, this game admits a simpler Cournot potential, represented by an upper semicontinuous real-valued function

$$P(x_N) := \min_i u_i(x_N).$$

It is easy to check that $P(y_N) > P(x_N)$ whenever $y_N \rhd x_N$. On the other hand, this simpler potential hinges on specifics of this particular example, whereas (18) and (22) work for every game from the class.

Remark 7 Kukushkin (2005) contains a more elaborate example where the function H remains constant along a "double infinite" Cournot path (i.e., up to the second limit). Similarly, an example can be produced where all assumptions of Theorem 1 are satisfied, while an arbitrary (countable) transfinite number of best response improvements may be needed to reach an equilibrium, with H being constant all the way. Thus, the behavior of Cournot dynamics when the best responses are not upper hemicontinuous may be much more complicated than in Jensen (2010, Theorem 2).

Acknowledgements Financial support from the Russian Foundation for Basic Research (project 14-07-00075) is acknowledged. I thank an anonymous referee for helpful comment on the original version.

References

Dubey P, Haimanko O, Zapechelnyuk A (2006) Strategic complements and substitutes, and potential games. Games Econ Behav 54:77–94

Gillies DB (1959) Solutions to general non-zero-sum games. In: Tucker AW, Luce RD (eds) Contributions to the theory of games 4. Annals of mathematical studies 40. Princeton University Press, Princeton NJ, pp 47–85

Huang Z (2002) Fictitious play in games with a continuum of strategies. Ph.D. thesis, State University of New York at Stony Brook, Department of Economics

Jensen MK (2010) Aggregative games and best-reply potentials. Econ Theory 43:45–66

Kukushkin NS (1994) A fixed-point theorem for decreasing mappings. Econ Lett 46:23–26

Kukushkin NS (2003) Acyclicity of monotonic endomorphisms. Russian Academy of Sciences, Dorodnicyn Computing Center, Moscow. http://www.ccas.ru/mmes/mmeda/ququ/MonoEndo.pdf

Kukushkin NS (2004) Best response dynamics in finite games with additive aggregation. Games Econ Behav 48:94–110

Kukushkin NS (2005) "Strategic supplements" in games with polylinear interactions. EconWPA paper. http://ideas.repec.org/p/wpa/wuwpga/0411008.html

Kukushkin NS (2008) Maximizing an interval order on compact subsets of its domain. Math Soc Sci 56:195–206

Kukushkin NS (2010) Improvement paths in strategic games: a topological approach. Russian Academy of Sciences, Dorodnicyn Computing Center, Moscow. http://www.ccas.ru/mmes/mmeda/ququ/ImPath.pdf

Kukushkin NS (2013a) Increasing selections from increasing multifunctions. Order 30:541–555

Kukushkin NS (2013b) Monotone comparative statics: changes in preferences versus changes in the feasible set. Econ Theory 52:1039–1060

LiCalzi M, Veinott AF Jr (1992) Subextremal functions and lattice programming. EconWPA Paper. http://ideas.repec.org/p/wpa/wuwpge/0509001.html

Milgrom P, Shannon C (1994) Monotone comparative statics. Econometrica 62:157–180

Novshek W (1985) On the existence of Cournot equilibrium. Rev Econ Stud 52:85–98

Shannon C (1995) Weak and strong monotone comparative statics. Econ Theory 5:209–227

Smith TE (1974) On the existence of most-preferred alternatives. Int Econ Rev 15:184–194

Topkis DM (1978) Minimizing a submodular function on a lattice. Oper Res 26:305–321

Vives X (1990) Nash equilibrium with strategic complementarities. J Math Econ 19:305–321

Existence and Uniqueness of Nash Equilibrium in Aggregative Games: An Expository Treatment

Richard Cornes and Takashi Sato

Abstract In this study, we present an elementary treatment of Cournot's oligopolistic competition model in which an industry's demand function has a constant elasticity and technology has a convex cost function. Our treatment uses the 'share function', which exploits the aggregative structure of the resulting game. In this setting, the best response functions are not monotonic, ruling out the use of techniques previously applied to analyze submodular and supermodular games. Share functions, which model each firm's most preferred share of total output as a proportion of total output, allow for a direct and transparent method to derive the properties of a Nash equilibrium. To illustrate this, we establish the existence of a unique equilibrium and examine its response to exogenous shocks.

1 Introduction

Many treatments of the static Cournot's oligopoly model adopt assumptions that imply downward-sloping best response functions. However, such assumptions are unnecessarily restrictive—simple examples of games can be found which do not possess this property, but are known to be well-behaved in the sense of possessing a unique Nash equilibrium in pure strategies. This study aims to exploit the aggregative structure inherent in the standard oligopoly models in the manner first suggested by Reinhard Selten (1970), Ferenc Szidarovszky (1970), and Szidarovszky and

T. Sato (✉)
Shimonoseki City University, Shimonoseki, Japan
e-mail: sato@shimonoseki-cu.ac.jp

P. von Mouche, F. Quartieri (eds.), *Equilibrium Theory for Cournot Oligopolies and Related Games*, Springer Series in Game Theory,
DOI 10.1007/978-3-319-29254-0_5

Yakowitz (1977)[1] and the more recently developed methods of Cornes and Hartley (2003, 2005, 2007, 2012a,b,c)[2] to provide a simple and direct treatment of a family of static oligopoly games which are neither sub- nor supermodular[3] but which are well-behaved in the sense of possessing a unique equilibrium. Our analysis deals with existence, uniqueness, and certain comparative static properties, which we aim to present in an easily comprehensible manner.[4]

2 Analyzing Aggregative Games: Brief Exposition

Cournot oligopoly is possibly the most fundamental concept known to all researchers studying the economic application of noncooperative game theory. In Nash equilibrium each firm chooses an output level that is a best response to the choices made by all others. This naturally seems to suggest the notion of a best response function. However, although this may be a natural way of viewing the game from a participant's viewpoint, it is not necessarily the most useful tool for an outsider—the economist or game theorist who wants to analyze its properties.

We start in familiar territory, with the standard best response function. This expresses a player's most preferred output choice as a function of the sum of all other players' output levels. Cournot's oligopoly theory typically concentrates on the 2-firm example, and makes simple assumptions on the firms' cost functions and the industry demand function that are (much more than) enough to guarantee that both firms' best response functions are everywhere nonincreasing, and that an increase of one unit by one firm reduces the other's best response by less than one unit. In a standard diagram, one best response function everywhere has a slope less than -1, while the other has a slope everywhere greater than -1. It then becomes intuitively clear that, at most, a single equilibrium exists.

We denote player i's best response function as $\hat{q}_i = b_i(Q_{-i})$, where Q_{-i} denotes the sum of all firms' output levels, except that of firm i.[5] Note that each firm has a different argument for its best function. Suppose that, given an arbitrary level

[1] Historically, Selten (1970) and Szidarovszky (1970) independently developed the idea of transforming the problem of establishing an equilibrium to identifying the fixed point of the joint best reply correspondence. Okuguchi (1993, 1998), Okuguchi and Szidarovszky (1999), and Szidarovszky and Okuguchi (1997) applied this to oligopoly models. Other studies include Corchon (2001), Novshek (1985), Quartieri and Okuguchi (2008), Vives (2001), and Yamazaki (2008, 2009).

[2] Also see Vasin et al. (2007) and von Mouche et al. (2012).

[3] See Amir (1996), Amir and Lambson (2000), and Vives (2001) for supermodular games of Cournot oligopoly.

[4] See Corchon (2001), Friedman (1983), Okuguchi and Szidarovszky (1999), and Vives (2001) for constructive discussions on oligopoly models.

[5] More accurately, $\hat{q}_i = b_i(q_1, q_2, \cdot\cdot, q_{i-1}, q_{i+1}, \cdot\cdot, q_n)$ is the best response function, depending on $n-1$ variables, whereas $\hat{q}_i = b_i(Q_{-i})$ may be called the 'reduced best response function' depending on one variable (i.e., the aggregate actions of the other firms).

of industry output level Q, each player has a unique best response satisfying the requirement $\hat{q}_i = b_i(Q_{-i}) = b_i(Q - \hat{q}_i)$. There is no general presumption that such an arbitrarily chosen Q is consistent with the allocation being a Nash equilibrium. A Nash equilibrium is an allocation that satisfies an additional consistency requirement each and every player must be choosing her best response to the choices of all others. In a game with n players, the best response function approach expresses this consistency requirement in the form of n equations in n unknowns—the n best response functions.

To be sure that at least one equilibrium—that is, a solution to this system of equations—exists, certain continuity properties must be possessed by each individual's best response function. However, we will not dwell on this issue—the examples that interest us enjoy such properties. Our principal focus is not on existence, but on conditions that guarantee its uniqueness. Whether a system has a unique equilibrium or not is significant because of its implications for the comparative static properties of the model.

The development and serious application of this approach encounters two significant limitations. The first becomes evident when we consider adding extra firms to the industry. The addition of a firm to the story requires the addition of an extra dimension—equilibrium in a 3-firm oligopoly is the intersection of three planes in a diagram with three axes, and so on. It may still be shown that, if the other assumptions are maintained, uniqueness of equilibrium survives the addition of extra firms.[6] However, this is far from intuitively, or geometrically, obvious, and deeper mathematics has to be invoked in order to analyze such issues.

The second issue is that it is simple to write down a Cournot model in which firms do not have monotonic best responses. To do so, one can follow Cournot's example and let production—in his example the bottling of spring water—be zero. Then assume that the demand function facing the industry has unit elasticity $-Q_D = 1/P$. We leave the reader to confirm that the individual firm in such an industry has a nonmonotonic best response function.

The Nobel laureate Reinhard Selten noted 40 years ago that oligopoly models have special structure that permits alternative ways of describing the game, and he suggested one such alternative. His suggestion, which appeared in a book written in German, seems to have gone largely overlooked, except amongst a small group of industrial economics researchers. More recently, it has been further developed by others in a variety of applications, notably public good models and contest theory. However, this approach and its more recent developments remain, in our view, insufficiently appreciated and exploited. The best response function remains the standard tool, not only in the classroom, but also in the research literature.

Selten and Szidarovszky observed that the Cournot model involves a game in which the players cared, not about an arbitrary vector of strategic choices by their fellow players, but about a very particular aggregate of those choices. In an industry

---

[6]See Quartieri and Okuguchi (2008) and von Mouche and Quartieri (2013) for recent studies on a unique Cournot equilibrium.

of many firms, each producing essentially the same good, it is the unweighted sum of outputs of all those firms that determines the price at which each unit of output can be sold, and it is therefore this sum that is relevant for each firm.

The approach to modeling that we want to expound here extends Selten and Szidarovszky's suggested alternative approach in order to relax both of these limitations. Selten and Szidarovszky's own suggested function what we have elsewhere called the replacement function exploits the models structure to permit a straightforward analysis of a noncooperative game in which there may be large numbers of firms, each with its own idiosyncratic cost function. We then introduce a further modification that permits a clearer analysis of the properties of models in which the monotonicity property mentioned above is violated.

3 Cournot with Constant Demand Elasticity: Simple Example

3.1 Replacement Function

Suppose that the inverse demand function facing a single-product industry takes the unit elasticity form

$$P(Q) = Q^{-1} \ (Q \neq 0).$$

The inverse demand function is unbounded at 0, such as $P(Q) = 1/Q \ (Q \neq 0)$ (and, for example, $P(0) = 137$).[7] For the purposes of this example, we allow each profit-maximizing firm to enjoy idiosyncratic but constant unit costs. Firm i's unit cost is c_i. The number of firms, n, is given as $i = 1, \cdots, n$. In static Cournot equilibrium, firm i chooses its quantity, q_i, to maximize $\pi_i(q_i, Q) = (Q^{-1} - c_i)q_i \ (Q \neq 0)$. The first-order condition for an interior profit-maximizing choice, $\hat{q}_i$, requires that

$$\frac{\partial \pi_i}{\partial q_i} + \frac{\partial \pi_i}{\partial Q} = Q^{-1} - c_i - \hat{q}_i Q^{-2} = 0 \ (Q \neq 0),$$

or

$$\hat{q}_i = Q(1 - Qc_i) \ (Q \neq 0).$$

[7] Clearly 137 is assigned arbitrarily. See von Mouche and Quartieri (2013, footnote 5 of p. 709).

We must also allow for the possibility of a corner solution at which $q_i = 0$. Taking this into consideration, player i's best response can be modeled by the following relationship:

$$\hat{q}_i = r_i(Q) = \max\{Q(1 - Qc_i), 0\} \quad (Q \neq 0).$$

An observant reader may also worry that, for certain assumed values of the parameters, a player's best response as indicated by this function may exceed the industry output itself, which in the current context would be an economically meaningless situation. We will see later, however, that this issue need not worry us at any Nash equilibrium.

The replacement function of a firm is essentially an individual consistency condition. It uniquely determines i's most preferred output level as a function of industry output Q (the best reply of the firm is compatible with industry output Q), given the values of the exogenous cost parameters. It tells us that, if the level of industry output is to be a Nash equilibrium, then we must require, at the very least, that every player's best response be consistent with that level that is, it must be the value indicated by that player's replacement function. The function $r_i(Q)$, which Cornes and Hartley call the "replacement function", is clearly not monotonic as the value of Q increases from zero, that of $\hat{q}_i$ at first also increases, before subsequently falling, eventually becoming zero again at $Q = c_i^{-1}$.

3.2 *Share Function*

It is certainly possible to analyze the present model using the replacement function. But we will find it more convenient to describe each player's most preferred behavior using the share function, as suggested by Cornes and Hartley. For reasons that will become very clear shortly, this offers a particularly simple way of modeling a Nash equilibrium and of confirming that, for the present model, there can only exist one such equilibrium.

Consider again the replacement function derived above. We are allowed to divide both sides by the quantity Q, as long as it is strictly positive. Doing so generates what Cornes and Hartley call the player's "share function". In the present example, this takes the following form:

$$\hat{s}_i = \frac{r_i(Q)}{Q} = \max\{1 - Qc_i, 0\} \quad (Q \neq 0).$$

By contrast with the replacement function, this is everywhere strictly decreasing in Q.

3.3 Nash Equilibrium

The replacement and share functions describe the behavior of individual players. They each describe that behavior by imposing the condition that the player's best response be consistent with the observed aggregate. In order for an allocation to be a Nash equilibrium, we must add to these n individual consistency conditions a collective consistency requirement. We must require that the sum of all players' best responses should be consistent with more precisely, should equal the industry total. In terms of the replacement function, we require that

$$R(Q) \equiv \sum_{i=1}^{n} r_i(Q) = Q.$$

In terms of the share function, we require that

$$S(Q) \equiv \sum_{i=1}^{n} s_i(Q) = 1.$$

The functions $R(\cdot)$ and $S(\cdot)$ can be labelled respectively as the aggregate replacement and aggregate share functions. Note that, whichever of these formulations we choose to adopt, a Nash equilibrium is described by a single collective consistency requirement. This contrasts with the best response function, which describes a Nash equilibrium of an n-player game using n consistency requirements, the best response functions.

Stated informally, the issue of whether a Nash equilibrium exists is equivalent to that of whether there exists a value of Q at which the sum of all players' share values is unity. The uniqueness issue is equivalent to that of whether there exists just one such value.

4 Existence of a Unique Cournot-Nash Equilibrium

4.1 Properties of Share Functions

We now slightly generalize the model introduced above and show conditions under which an equilibrium exists in pure strategies. We extend our earlier example to permit constant, but not necessarily unit elasticity, and also more general forms for the individual firms' cost function. The number of firms, n, is given as $j = 1, \cdots, n$.

A.1: The cost function of firm j, $C_j(q_j)$, is everywhere twice differentiable and has the following properties: $C_j(0) = 0, C_j'(q_j) > 0, C_j''(q_j) \geq 0$, for all q_j.
In short, positive output requires incurring positive costs, and both total and marginal costs increase with output.

A.2: The (inverse) demand function facing all firms in the industry is $P(Q) = Q^{-\alpha}(Q \neq 0)$, where $Q \equiv \sum_{j=1}^{n} q_j$ and the parameter $\alpha(\alpha > 0)$ are exogenously determined.

The inverse demand function is also unbounded at 0, such as $P(Q) = 1/Q$ $(Q \neq 0)$ (e.g., $P(0) = 137$).

The profit function of firm j takes the form

$$\pi_j(q_j, Q) = P(Q)q_j - C_j(Q_j) = Q^{-\alpha}q_j - C_j(q_j) \ (Q \neq 0).$$

The first order condition associated with firm j's profit-maximizing response requires that

$$\frac{\partial \pi_j(\hat{q}_j, Q)}{\partial q_j} + \frac{\partial \pi_j(\hat{q}_j, Q)}{\partial Q} = Q^{-\alpha}\left(1 - \frac{\alpha \hat{q}_j}{Q}\right) - C'(\hat{q}_j) \leq 0 \ (Q \neq 0) \quad (1)$$

where $\hat{q}_j$ denotes the firm's best response.[8]

Equation (1) implicitly determines the best response of firm j associated with any given value of the aggregate Q. It is convenient to rewrite this first-order condition using, not the best response itself, but the best response expressed as a share of the industry output:

$$Q^{-\alpha}\left(1 - \alpha\hat{\sigma}_j\right) - C_j'(\hat{\sigma}_j Q) \leq 0 \ (Q \neq 0) \quad (2)$$

where $\hat{\sigma}_j \equiv \hat{q}_j/Q$ $(Q \neq 0)$ is firm j's most preferred share of total industry output.

Equation (2) implicitly defines the firm's most preferred share of industry output in terms of the total industry output. This is the relationship that we will be exploiting on subsequent analysis. That analysis proceeds first by confirming that there exists a share function for the player. Then we tease out some relevant properties of the share function. These properties will then allow us to establish that an equilibrium must exist.

To establish that the relationship we seek is indeed a single-valued share function, not a potentially many-valued correspondence, we must establish that, for any value of Q, there is at most a single value of $\hat{\sigma}_j$ that satisfies (2). Consider an allocation at which Eq. (2) is satisfied at some strictly positive level of $\hat{\sigma}_j$. At that allocation, we know that

$$Q^{-\alpha}\left(1 - \alpha\hat{\sigma}_j\right) - C_j'(\hat{\sigma}_j Q) = 0 \ (Q \neq 0) \quad (3)$$

[8] If we have **A1** and **A2**, the share function is strictly decreasing (Proposition 2). This implies that the second order condition is satisfied. See the Appendix for details.

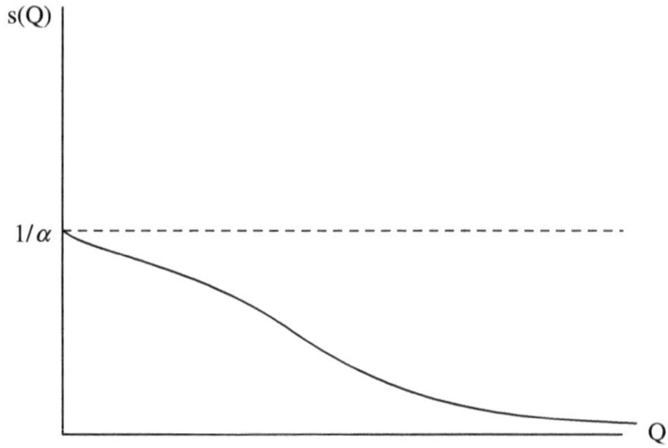

Fig. 1 Share function

Holding Q constant, it is clear that as the value of the firm's share rise, that of the left hand side must everywhere fall. Therefore, there cannot exist other values of $\hat{\sigma}_j$ that satisfy the first order condition for profit maximization. We may therefore describe the relationship between $\hat{\sigma}_j$ and Q in explicit form: $\hat{\sigma}_j = s_j(Q)$ (Fig. 1). More formally, we have established.

Proposition 1 *Assumptions* ***A.1*** *and* ***A.2*** *imply that there exists a share function,* $s_j(Q)$ $(Q \neq 0)$, *for every player j.*

Having established the existence of a share function for every player, we now tease out their properties by examining the implicit function presented in (3).

Proposition 2 *Assumptions* ***A.1*** *and* ***A.2*** *imply that*

(i) $s_j(Q)(Q \neq 0)$ *is continuous and strictly decreasing where positive,*
(ii) $\lim_{Q \to 0} s_j(Q) = \frac{1}{\alpha}$, *and*
(iii) $\lim_{Q \to \infty} s_j(Q) = 0$.

Continuity of the share function follows directly from the implicit function theorem. Inspection of Eq. (3) confirms that a higher value of Q must be accompanied by a lower value of the firm's share if the first-order condition for an interior profit-maximizing allocation for that is to be maintained. It is worth noting that details of cost conditions are not relevant in determining the limiting values of individual shares as Q becomes either very large or very small.[9]

[9]See the Appendix for a detailed proof of Proposition 2.

4.2 Existence of Equilibrium

Armed with our definition of a Nash equilibrium in terms of share functions, and the properties possessed by those share functions, we can establish conditions under which a Nash equilibrium exists, given our assumptions, in an industry with n firms, where n is exogenously determined.

Recall that a Nash equilibrium is characterized by the consistency requirement that the sum of all firms' share values equal unity. Since each share function is continuous, so is their sum—the operation of addition preserves continuity.

However, continuity is not the sole relevant property for the existence question. As the aggregate Q varies over all conceivable values, we require the range of the "aggregate share value" to include the value unity. Suppose, for the sake of argument, that $\alpha = 5/2$. Then there cannot exist a Nash equilibrium if there are assumed to be only two active firms. The reason is, quite simply, that at all values of Q, the sum of share values falls short of unity. Figure 2 shows the graph of an individual share function as the thin curve. The graph of the aggregate share function for a two-firm example the lower of the two thick curves—nowhere reaches the value unity, reflecting the nonexistence of a Nash equilibrium. However, an equilibrium in the presence of three firms exists, and is shown as the higher of the two thick curves.

The special case of this result is commonly observed in the field of industrial organization, which commonly reports that when the demand function has unit elasticity, there is no well-defined monopolistic equilibrium. This, of course, corresponds to the case of $n = 1$. If, for example, $\alpha = 1 + \varepsilon$, where ε takes on a small positive value, it is easy to confirm that, at any positive level of Q, the monopolist can increase its revenue ($Q^{-\varepsilon}$), and thus, profits, by further reducing its output.

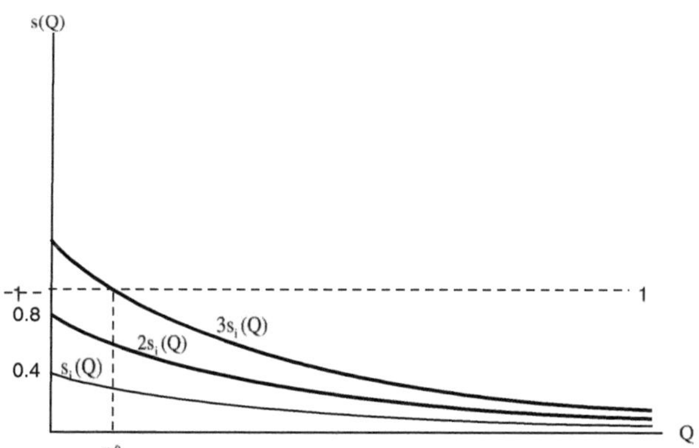

Fig. 2 Cournot-Nash equilibrium in the aggregate share function

4.3 Uniqueness of Nash Equilibrium

Given that at least one equilibrium exists, the fact that there is at most a single equilibrium follows in a straightforward way from the fact that each player's share function is nonincreasing in Q, and is strictly decreasing wherever that share value is strictly positive. From Propositions 1 and 2 we have Theorem 1 as follows.[10,11]

Theorem 1 *If* **A.1** *and* **A.2** *hold for firm* j *and* $n \geq \alpha$, *there exists a unique Cournot-Nash equilibrium other than* $n = \alpha = 1$, $(j = 1, \cdots, n)$.

5 Comparative Static Properties of Nash Equilibrium

5.1 Firm Entry

The comparative static properties are, for the most part, straightforward. But there is one result which may be unfamiliar and worth noting.[12]

Suppose that there is a Nash equilibrium with n firms, and consider the consequences of an extra firm joining the industry. Before the entry of this firm, the sum of share values of existing firms at the initial equilibrium total output level equals unity. If the new firm is allowed to enter the industry, there are two possible equilibrium responses. If the extra firm would not choose to produce positive output at the prevailing total output level, nothing changes the extra firm remains inactive and the industry output level remains unchanged. If the extra firm would choose to produce positive output at the prevailing equilibrium output level, its entry raises the sum of all share values at that level. The aggregate share graph shifts upward in the neighbourhood of the initial equilibrium and, since the aggregate share function is downward-sloping, the new equilibrium total output level exceeds the initial level.

The observation that the entry of an additional firm will either leave the equilibrium total output of the industry unchanged or will expand it is neither new nor surprising. However, the non-monotonicity of individual firms' replacement functions suggests a more interesting possibility concerning the responses of levels of individual existing firms' outputs. We demonstrate this with a simple numerical example.

---

[10] See the Appendix for a proof of Theorem 1.

[11] Note that the constant elasticity of the demand function does not satisfy Hahn or Novshek's condition, because $P'(Q) + P''(Q)q_i = \alpha Q^{-\alpha-1}[-1 + (\alpha + 1)\sigma_i] > 0$, if $\sigma_i > 1/(\alpha + 1)$ is satisfied. Our share function approach is much weaker than Hahn or Novshek's condition. See Novshek (1985).

[12] See Okuguchi's classical paper Okuguchi (1973) for quasi-competitiveness of Cournot oligopoly.

Example: An industry contains two firms, 1 and 2. The demand function facing the industry has unit elasticity, and the two firms' total cost functions are, respectively,

$$C_1 = q_1 + (q_1)^2 \text{ and } C_2 = 2q_2.$$

Clearly, the two firms have very different cost functions, reflecting their adoption of different technologies. Firm 1 faces lower marginal costs at low output levels ($0 < q_1 < 1$), but as its output level rises, its marginal cost rises steeply. By contrast firm 2 is a higher cost producer at low output levels ($0 < q_2 < 1$), but has constant marginal and unit cost at all output levels. It is easily confirmed that the firms' share functions are

$$s_1(Q) = \frac{1 - Q}{2Q^2 + 1} \text{ and } s_2(Q) = 1 - 2Q.$$

The unique initial Nash equilibrium is at the nonnegative solution of the condition

$$s_1(Q) + s_2(Q) = \frac{1 - Q}{2Q^2 + 1} + 1 - 2Q = 1,$$

which is

$$Q = 0.29804.$$

At this allocation, the equilibrium shares and output levels are

$$(s_1, s_2, q_1, q_2) = (0.59607, 0.40392, 0.17765, 0.12038).$$

We now allow a third firm to enter the industry. Suppose it has the same cost function as firm 2. Then it will, indeed, choose to produce a positive quantity, and at the new equilibrium the following consistency requirement holds:

$$s_1(Q) + s_2(Q) + s_3(Q) = \frac{1 - Q}{2Q^2 + 1} + 1 - 2Q + 1 - 2Q = 1,$$

which implies that, at the new equilibrium,

$$Q = 0.37272$$

and

$$(s_1, s_2, q_1, q_2, q_3) = (0.49089, 0.25456, 0.25456, 0.18296, 0.09488, 0.09488)$$

Fig. 3 Effects of entry in Cournot market

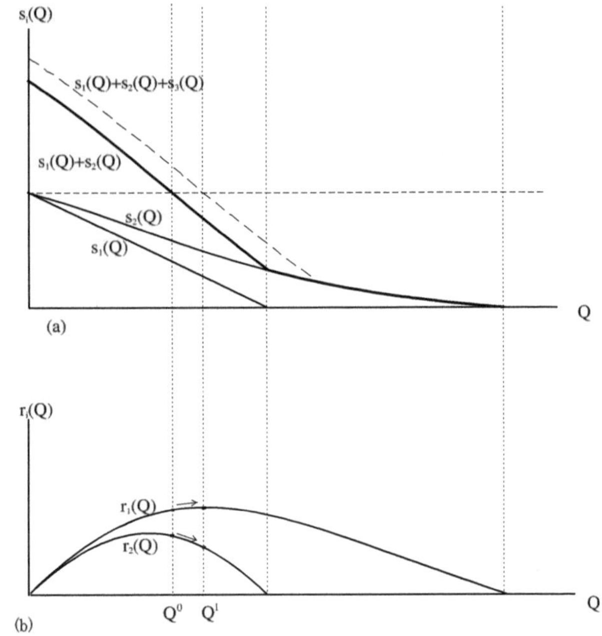

Industry output has risen, and the shares of firms 1 and 2 have fallen. None of that is surprising. Of more interest, however, is the observation that, in response to the entry of firm 3, firm 1's output has risen, but firm 2's output has fallen. Figure 3 summarizes the situation. Figure (a) shows the graphs of the share functions of firms 1 and 2. The thick continuous line graphs the 'aggregate share function' and shows the initial equilibrium, determined by the intersection of this graph with the dashed line along which aggregate shares equal unity. The entry of firm 3 shifts part of the aggregate graph upwards it is shown by the thick dashed line in 3(a). Panel 3(b) shows explicitly the responses of firms 1 and 2 to the resulting expansion in industry output.

In the present example, the difference between the two initial firms' cost functions implies that, over an interval, one is on an upward-sloping part of the replacement function, while the other is on a downward-sloping part. In the simpler examples of textbooks, where the demand curve is typically assumed to be piecewise linear, the firms' replacement functions are everywhere downward-sloping. In the present example, we can investigate the potentially significant change in the pattern of the industry in terms of the fortunes of firms with 'old' and 'new' technologies.

5.2 *Other Shocks*

Maybe we should explicitly look at the consequences of an exogenous change in one firm's cost function. This allows us to examine whether, for example, a lowering of a single firm's costs increases all firms' profits or not.

Also, perhaps a change in a tax or subsidy on the industry's output. Old literature by Dixit (1986), Sead (1985) and Stern (1987) suggests that imposing taxes on an oligopoly can increase firms' profits.

5.3 *Relaxing Cost Assumptions*

This remains to be done. If we can find an interesting example—either where a share function exists but is not monotonic, or where the share correspondence is multi-valued—would be worthwhile.

6 Concluding Remarks

This study aimed to show how readily simple oligopolistic games can be analyzed by exploiting their aggregative structure. The family of models analyzed here—those with constant elasticity industry demand function do not appear to have received the same systematic attention as the classroom models with linear demand. We hope that this paper persuades others that there is an interesting family of models in which best response functions are not monotonic, but which nevertheless are susceptible to straightforward analysis.

Appendix

Proof of Proposition 2

(i) Substituting the share function $\sigma_i = s_i(Q)$ into (3) and differentiating (3) with respect to Q yields, if $\sigma_i > 0$

$$\frac{ds_i(Q)}{dQ} = \frac{P'(Q)(1-\alpha\sigma_i) - C_i''(\sigma_i Q)\sigma_i}{\alpha P(Q) + C_i''(\sigma_i Q)Q}, (i = 1, \cdots, n). \tag{4}$$

Assumptions **A1** and **A2** imply that the share function $s_i(Q)$ is continuous by the implicit function theorem and is strictly decreasing from (4) where positive.[13]

(ii) From (3) we have the Lerner condition, which is

$$\sigma_i = \frac{P(Q) - C_i'(\sigma_i Q)}{\alpha P(Q)} \equiv \frac{L_i(\sigma_i Q, Q)}{\alpha}, (i = 1, \cdots, n). \tag{5}$$

The left-hand side of (5) is the share value, while the right-hand side is the *Lerner index* $L_i(= (P(Q) - C_i'(q_i))/P(Q)), (i = 1, \cdots, n)$ of the monopoly power of firm i and the constant elasticity of demand $(1/\alpha)$. Since we have $P(Q)(= Q^{-\alpha}) \to \infty$ as $Q \to 0$ and $C_i'(\sigma_i Q) = C_i'(q_i) < \infty$, (5) implies that

$$\lim_{Q\to 0} s_i(Q) = \lim_{Q\to 0}\left[\frac{1}{\alpha}\left(1 - \frac{C_i'(\sigma_i Q)}{P(Q)}\right)\right] = \frac{1}{\alpha}.$$

(iii) Since we have $P(Q)(= Q^{-\alpha}) \to 0$ as $Q \to \infty$, $C_i'(\sigma_i Q) = C_i'(q_i) < \infty$ and $s_i(Q) \geq 0$, (5) implies that

$$\lim_{Q\to \infty} s_i(Q) = \lim_{Q\to \infty}\left[\frac{1}{\alpha}\left(1 - \frac{C_i'(\sigma_i Q)}{P(Q)}\right)\right] = 0$$

Proof of Theorem 1

From Proposition 2, if $n(1/\alpha) \geq 1$, the aggregate share function equals or exceeds one for small Q and is less than one for large Q. By the continuity of the share function, we see that Nash equilibrium values, $Q^*$of Q exist when the aggregate share function equals unity.

$$\sum_{j=1}^{n} s_j(Q^*) = \sum_{j=1}^{n} \frac{L_j(s_j(Q^*)Q^*, Q^*)}{\alpha} = 1.$$

Furthermore, since the aggregate share function is strictly decreasing, we have the unique equilibrium.

Proof of the second order condition

From (4), we see that the share function is decreasing, which implies that

$$\begin{aligned} d^2\pi_i/dq_i^2 &= P'(Q) - C_i''(q_i) + P''(Q)q_i + P'(Q) \\ &= Q^{-1}\left\{(Q - q_i)[P'(Q) - C_i''(q_i)] + q_i[P'(Q) - C_i''(q_i)] + Q[P''(Q)q_i + P'(Q)]\right\} \\ &= Q^{-1}\left\{-(Q - q_i)Q^{-1}[\alpha P(Q) + C_i''(q_i)Q] + Q[P'(Q)(1 - \alpha\sigma_i) - C_i''(q_i)\sigma_i]\right\} < 0. \end{aligned}$$

[13]From (3), we have $1 - \alpha\sigma_i \geq 0$.

References

Amir R (1996) Cournot oligopoly and the theory of supermodular games. Games Econ Behav 15:132–148
Amir R, Lambson E (2000) On the effects of entry in Cournot markets. Rev Econ Stud 67:235–254
Corchon L (2001) Theoy of imperfectly competitive markets, 2nd edn. Springer, Berlin
Cornes R, Hartley R (2003) Risk aversion, heterogeneity and contests. Public Choice 117:1–25
Cornes R, Hartley R (2005) Asymmetric contests with general technologies. Econ Theory 26:923–946
Cornes R, Hartley R (2007) Aggregative public good games. J Public Econ Theory 9:201–219
Cornes R, Hartley R (2012a) Risk aversion in symmetric and asymmetric contests. Econ Theory 51:247–275
Cornes R, Hartley R (2012b) Fully aggregative games. Econ Lett 116:631–633
Cornes R, Hartley R (2012c) Well-behaved aggregative games. Working paper, Australian National University
Dixit A (1986) Comparative statics for oligopoly. Int Econ Rev 27:107–122
Friedman JW (1983) Oligopoly theory. Cambridge University Press, Cambridge
Novshek W (1985) On the existence of Cournot equilibrium. Rev Econ Stud 52:85–98
Okuguchi K (1973) Quasi-competitiveness and Cournot oligopoly. Rev Econ Stud 40:145–148
Okuguchi K (1993) Unified approach to Cournot models, oligopoly, taxation, and aggregate provision of a pure public good. Eur J Pol Econ 9:233–245
Okuguchi K (1998) Existence of equilibrium for Cournot oligopoly-oligopsony. Keio Econ Stud 35:45–53
Okuguchi K, Szidarovszky F (1999) The theory of oligopoly with multi-product firms, 2nd revised and enlarged edn. Springer, Berlin
Quartieri F (2008) Necessary and sufficient conditions for the existence of a unique Cournot equilibrium. Ph.D. thesis, Seina-Universita di Siena, Italy
Sead J (1985) Profitable cost increases and the shifting of taxation: equilibrium responses of markets in oligopoly. Warwick Economic Research Papers No 260. The University of Warwick
Selten R (1970) Preispolitik der Mehrproduktunternehmung in der Statischen Theorie. Springer, Berlin
Stern N (1987) The effects of taxation, price control and government contracts in oligopoly and monopolistic competition. J Public Econ 32:133–158
Szidarovszky F (1970) On the oligopoly game. Technical report, Karl Marx University of Economics, Budapest
Szidarovszky F, Okuguchi K (1997) On the existence and uniqueness of pure nash equilibrium in rent-seeking games. Games Econ Behav 18:135–140
Szidarovszky F, Yakowitz S (1977) A new proof of the existence and uniqueness of the Cournot equilibrium. Int Econ Rev 18:787–789
Vasin A, Vasina P, Ruleva T (2007) On organization of markets of homogeneous goods. J Comput Syst Sci Int 46:93–106
Vives X (2001) Oligopoly pricing: old ideas and new tools. MIT, Cambridge
von Mouche P, Quartieri F (2013) On the uniqueness of Cournot equilibrium in case of concave integrated price flexibility. J Glob Optim 57:707–718
von Mouche P, Quartieri F, Szidarovszky F (2012) On a fixed point problem transformation method. In: Proceedings of the 10th IC-FPTA, Cluj-Napoca, pp 179–190
Yamazaki T (2008) On the existence and uniqueness of pure-strategy Nash equilibrium in asymmetric rent-seeking contests. J Public Econ Theory 10:317–327
Yamazaki T (2009) The uniqueness of pure-strategy Nash equilibrium in rent-seeking games with risk-averse players. Public Choice 139:335–342

References

On the Geometric Structure of the Cournot Equilibrium Set: The Case of Concave Industry Revenue and Convex Costs

Pierre von Mouche

Abstract The recent results in von Mouche and Quartieri (Econ Bull 35(2):1299–1305, 2015) on equilibrium (semi-)uniqueness for homogeneous Cournot oligopolies with concave industry revenue and convex costs are refined and conceptualised. For this class of oligopolies also new results concerning the geometric structure of the equilibrium set E are provided. In particular, a subclass is identified for which E is a non-empty polytope on which the aggregator is constant and a subclass for which E is a 1-dimensional polytope on which the aggregator is injective.

1 Introduction

The first results on equilibria for homogeneous Cournot oligopolies[1] appeared some decades ago; concerning equilibrium existence I mention here Burger (1959), Frank and Quandt (1963) and concerning equilibrium uniqueness Okuguchi and Suzumura (1971). Up to now the dependence of the equilibrium set on its primitive objects, i.e. the price function and the cost functions, has been the object of research. These results quite exclusively concern equilibrium existence, equilibrium uniqueness and equilibrium semi-uniqueness.[2]

Although there are various results dealing with equilibrium sets of cardinality at most one, recent results being von Mouche and Quartieri (2013), Ewerhart (2014), and von Mouche and Quartieri (2015), there are only a few results for the geometric structure of the equilibrium set in the case it has a higher cardinality. Ignoring

[1]Terminology and notations for oligopolies are provided in Sect. 2.

[2]With E the equilibrium set, equilibrium uniqueness boils down to equilibrium existence, i.e. $\#E \geq 1$, together with equilibrium semi-uniqueness, i.e. $\#E \leq 1$.

P. von Mouche
Wageningen UR, Postbus 6700 EW, Wageningen, The Netherlands
e-mail: pvmouche@deds.nl

P. von Mouche, F. Quartieri (eds.), *Equilibrium Theory for Cournot Oligopolies and Related Games*, Springer Series in Game Theory,
DOI 10.1007/978-3-319-29254-0_6

the results for Cournot oligopolies which are super modular,[3] I am only aware of Szidarovszky and Yakowitz (1982) where sufficient conditions for a Cournot oligopoly with a concave price function and convex cost functions are given for the equilibrium set to be polyhedral and of a generalisation of this result in von Mouche (2011) to aggregative games with concave conditional payoff functions and constant equilibrium aggregator. Based on this generalisation, recently results in Folmer and von Mouche (2015) on the geometric structure of the equilibrium set of transboundary pollution games were established. In the present article I will provide such a result for Cournot oligopolies with concave industry revenue and convex cost functions. In addition for such oligopolies, I will provide sufficient conditions for the equilibrium set to be a 1-dimensional polytope.

Starting point for this contribution to the Festschrift is the following result in von Mouche and Quartieri (2015).

Theorem 1 *Consider a Cournot oligopoly with concave industry revenue function r_p and convex cost functions c_i. In addition suppose that at least one of the following three conditions holds:*

I. r_p is strictly concave.
II. Every c_i is strictly convex.
III. r_p is decreasing on the interior of its domain and every c_i is strictly increasing.

If there exists at least one equilibrium, then there is

1. *a unique equilibrium aggregate, say $y_\star$;*
2. *a unique equilibrium if the price function p is differentiable at $y_\star$;*
3. *a unique equilibrium if p is differentiable on the interior of its domain.* ◇

Theorem 1 is a strongest result for this class of oligopolies and improves upon a result in Murphy et al. (1982); for related results see the references in von Mouche and Quartieri (2015). One important improvement is that the industry revenue function may be discontinuous at 0; for example the proper price function $p(y) = 1/y$ is allowed for. Also note that cost functions are not supposed to be differentiable and that Theorem 1 is closely related to Theorem 2 of Szidarovszky and Yakowitz (1982) dealing with the more simple traditional case of a decreasing concave p.[4]

[3] As proved in Zhou (1994), the equilibrium set of such a game is a complete lattice.

[4] The difficulties for concave industry revenue are related to the fact that for this case best replies may not have decreasing single-valued selections nor increasing single-valued selections. This can be easily verified for the following Cournot oligopoly taken from von Mouche and Quartieri (2012): $n = 2,\ X_1 = X_2 = \mathbb{R}_+,\ c_1 = c_2 = \mathrm{Id}/100,\ p(y) = 1/(y+1)$. Indeed, for the best reply correspondence $R_1 : \mathbb{R}_+ \multimap \mathbb{R}_+$ of firm 1 one has: $R_1(3) = \{16\},\ R_1(8) = \{21\},\ R_1(24) = \{25\},\ R_1(63) = \{16\}$.

Equilibrium existence is not guaranteed under the general conditions of Theorem 1. However, under usual further conditions, equilibrium existence is no issue. Indeed: as for each firm all conditional profit functions[5] are concave, the Nikaido-Isoda theorem Nikaido and Isoda (1955) guarantees that it is sufficient for existence in Theorem 1 that, in addition, an appropriate compactness condition is imposed (like a binding capacity constraint for each firm[6]) and price and cost functions are continuous. So, properly speaking, Theorem 1 is a result about equilibrium semi-uniqueness.

The proof of Theorem 1 uses new ideas from Quartieri (2008); von Mouche and Quartieri (2013, 2015) for Cournot oligopolies with convex cost functions and does not rely on concave programming techniques as used in Murphy et al. (1982), but just on an analysis of first order conditions. Up to now in the literature these new ideas have hardly been commented on; in von Mouche and Quartieri (2015), for example, the emphasis was on just providing a simple short proof. In the present article in Sects. 3 and 4 I take the opportunity to rectify this shortcoming. New results can be found in Sects. 5 and 6. These results concern the geometric structure of the equilibrium set in the case where its cardinality is at least two.

The organisation of the present article is as follows. Section 2 fixes the oligopolistic setting and various notations. Section 3 makes some observations on Theorem 1 and its proof; in particular the importance of the equilibrium aggregator is explained here. In Sect. 4 four topics are identified that play a role in the analysis of the first order conditions: activeness of equilibria, sufficient conditions for the injectivity of the equilibrium aggregator, a crucial inequality for price functions, and sufficient conditions for the equilibrium aggregator to be constant. The following two sections merely deal with the class of Cournot oligopolies with concave industry revenue and convex cost functions. Section 5 identifies with Theorem 2 a subclass for which the equilibrium set is a 1-dimensional polytope and Sect. 6 with Theorem 3 a subclass for which the equilibrium set is a non-empty polytope. Section 7 illustrates with examples and Sect. 8 concludes. There is an appendix dealing with effective capacity constraints and with some convex analysis related to price functions.

2 Setting and Notations

A *(homogeneous) Cournot oligopoly* is a game in strategic form with a player set $N := \{1, \ldots, n\}$ whose elements are called *firms*. I assume that each firm i has a strategy set X_i that is a proper interval of $\mathbb{R}_+$ containing 0. (More generally, one could deal with strategy sets which are non-empty subsets of $\mathbb{R}_+$.) This implies that $X_i = [0, m_i]$ or $X_i = [0, m_i[$ (with $m_i > 0$) or $X_i = \mathbb{R}_+$. X_i also is called *production level set* and its elements *production levels*; those of (the Minkowski-

[5] I.e. the payoff of the firm as a function of its own production level.

[6] Also see Sect. 3.3.

sum) $Y := \sum_{l \in N} X_l$ are called *industry production levels*. With $\mathbf{X} := \mathsf{X}_{i=1}^n X_i$, each firm i has a payoff function, called *profit function*, $u_i : \mathbf{X} \to \mathbb{R}$ given by[7]

$$u_i(\mathbf{x}) := p(x_N)x_i - c_i(x_i).$$

Here $p : Y \to \mathbb{R}$ is called *price function* (also called *inverse demand function*) and $c_i : X_i \to \mathbb{R}$ is called *firm i's (net) cost function*. A Nash equilibrium of a Cournot oligopoly is called *(Cournot) equilibrium*.

In the case where $X_i = [0, m_i]$ or $X_i = [0, m_i[$ firm i is said to have (m_i as) a *capacity constraint*. When $X_i = [0, m_i]$, then I shall speak of a *binding* capacity constraint. Also Y is a proper interval of $\mathbb{R}$ with $0 \in Y \subseteq \mathbb{R}_+$ and Y is compact if and only if $X_i = [0, m_i]$ for each i; in this case $Y = [0, m_N]$. I say that $\overline{x}_i \in X_i$ is an *effective capacity constraint* of firm i if every $x_i \in X_i$ with $x_i > \overline{x}_i$ is strongly dominated by some element in $[0, \overline{x}_i]$. So in the case $X_i = [0, m_i]$, the (binding) capacity constraint m_i also is an effective capacity constraint.

Note that the profit functions (and therefore equilibria too) do not depend on the value of p at 0. Thus, with $Y^{\oplus} := Y \setminus \{0\}$, only the (restricted) price function[8]

$$\tilde{p} := p \upharpoonright Y^{\oplus}$$

matters. I call $\tilde{p}$ the *proper price function*.

The *industry revenue function* (associated with p) is the function $r_p : Y \to \mathbb{R}$ defined by

$$r_p := p \cdot \mathrm{Id}.$$

So $r_p(y) = p(y)y$.

With E I denote the set of (Nash) equilibria. An equilibrium is called *active* if it is not $\mathbf{0}$. An *equilibrium aggregate* is an element of $\{e_N \mid \mathbf{e} \in E\}$. The elements of $\{p(e_N) \mid \mathbf{e} \in E\}$ are called *equilibrium prices*.

I define $\alpha : \mathbf{X} \to \mathbb{R}$ by

$$\alpha(\mathbf{x}) := x_N.$$

I refer to α as the *aggregator* and to $\alpha \upharpoonright_E$ as the *equilibrium aggregator*.

For $i \in N$ and $\mathbf{z}$ a production level profile of the other firms, the *conditional profit function* $u_i^{(\mathbf{z})} : X_i \to \mathbb{R}$ of firm i is defined by $u_i^{(\mathbf{z})}(x_i) := u_i(z_1, \ldots, z_{i-1}, x_i, , z_{i+1}, \ldots, z_n)$.

Further, if not said otherwise, I always consider a Cournot oligopoly.

---

[7]For all $\mathbf{x} \in \mathbf{X}$ and $A \subseteq N$, I shall denote $\sum_{i \in A} x_i$ by x_A.

[8]I use the following notation for a subset J of $\mathbb{R}$: $l(J) := \{\min(J)\}$ if $\min(J)$ exists and $l(J) := \emptyset$ otherwise; $r(J) := \{\max(J)\}$ if $\max(J)$ exists and $r(J) := \emptyset$ otherwise. Moreover, $\mathrm{Int}(J)$ denotes the topological interior of J, $J^{\ominus} := \mathrm{Int}(J) \cup l(J)$ and $J^{\oplus} := \mathrm{Int}(J) \cup r(J)$.

3 Theorem 1 Reconsidered

Below I reconsider Theorem 1. In doing so I shall refer a few times to results in the following sections.

3.1 *Reformulation*

Note that in Theorem 1(1) it holds that $y_\star \in \mathrm{Int}(Y)$, i.e. that the equilibrium aggregate $y_\star$ is interior. In the case there exist at least two equilibria. Therefore part 3 of this theorem follows from its part 2. With this it is easily seen that under the general conditions of Theorem 1 the following holds:

If there exist at least two equilibria, then

- there is a unique equilibrium aggregate, say $y_\star$, and $y_\star \neq 0$ holds;
- the proper price function is not differentiable at $y_\star$.

In fact this concerns a reformulation of Theorem 1. The merit of this reformulation is that it makes more clear the importance of differentiability of the proper price function for equilibrium semi-uniqueness.

3.2 *Implicit Properties*

In Theorem 1 the concavity of r_p and the convexity of c_i imply that all conditional profit functions of firm i are concave.[9] In the cases I and II all conditional profit functions are even strictly concave. Below I shall mention other implicit properties.

I denote by

$$\mathscr{S}$$

the set of (price) functions p with the following properties: p is semi-differentiable[10] at each interior point of its domain, and when m is a positive non-interior industry production level, then the (left) derivative of p at m exists as an element of $\mathbb{R} \cup \{-\infty\}$. And I denote by

$$\mathscr{S}_\star$$

[9] In order to see this (well-known) result, Proposition 8(1) in the appendix is very useful.

[10] I.e. left and right differentiable.

the set of (price) functions p with the following properties: p is semi-differentiable at each interior equilibrium aggregate and when m is a positive non-interior equilibrium aggregate, then the (left) derivative of p at m exists as an element of $\mathbb{R} \cup \{-\infty\}$.

Although the price and cost functions in Theorem 1 may not be differentiable, the concavity of r_p and the convexity of c_i imply that the following smoothness properties hold:

- $p \in \mathscr{S}$ with $D^+p \leq D^-p \leq 0$ on the interior of the domain and when m is a positive non-interior industry production level, then $D^-p(m) \leq 0$. If r_p is strictly concave, then even $D^+p \leq D^-p < 0$ on the interior of the domain and $D^-p(m) < 0$.[11]
- c_i is semi-differentiable at each interior point of its domain, the right derivative of c_i at 0 exists as an element of $\mathbb{R} \cup \{-\infty\}$ and, if $X_i = [0, m_i]$, its left derivative at m_i exists as an element of $\mathbb{R} \cup \{+\infty\}$.

Although in Theorem 1 there are no explicit monotonicity properties for the price function, the following holds.[12]

- If r_p is concave, then $\tilde{p}$ is decreasing.
- If r_p is strictly concave, then $\tilde{p}$ is strictly decreasing; even $D^-\tilde{p} < 0$.
- If r_p is concave and r_p is decreasing on the interior of its domain, then for each positive industry production level y with $p(y) > 0$ it holds that p is strictly decreasing on $]0, y]$ and $D^-p < 0$ on $]0, y]$.

3.3 *Little Improvement of Theorem 1*

Here I present, with results A–C below, a little improvement of Theorem 1.

Suppose that the industry revenue function is concave, that each cost function is convex and that at least one of the conditions I–III of Theorem 1 holds. Then the following three statements hold.

A. If $\mathbf{a}$ and $\mathbf{b}$ are equilibria, then for every $i \in N$: $a_i > 0 \Leftrightarrow b_i > 0$.
B. If $n = 2$ and $\#E \geq 2$, then in each equilibrium all firms are active.
C. If the industry revenue function and the cost functions are continuous and each firm i has an effective capacity constraint $\overline{x}_i$, then the equilibrium set E is a non-empty compact subset of $\mathbb{R}^n$ with $E \subseteq \mathsf{X}_{i=1}^n [0, \overline{x}_i]$.

Proof of A: first note that in case III, as the cost functions are strictly increasing, p is positive at each positive equilibrium aggregate. This implies (with the implicit

[11]If wished, see Lemma 5 in the appendix.

[12]That the first two statements are true is guaranteed by Lemma 5(4) in the appendix. Concerning the third one, note that (as r_p is concave) $r_{\tilde{p}}$ is decreasing and apply Lemma 6.

properties from Sect. 3.2) that Proposition 1(1) below applies. As $a_N = b_N$ by Theorem 1(1), even $a_i > 0 \Rightarrow b_i > 0$ follows. Proof of B: by A and the fact that the equilibrium aggregator is constant. Proof of C: by Theorem 4 in the appendix.

Finally, I mention here the following corollary by Theorem 1 together with the above result C and Proposition 7 in the appendix: each Cournot oligopoly with a strictly decreasing affine price function and linear cost functions has an equilibrium.[13]

3.4 On the Proof of Theorem 1

The proof of Theorem 1 in von Mouche and Quartieri (2015) relies on an analysis of first order conditions. This proof uses in an implicit way the following simple fundamental observation for a game in strategic form with subsets of $\mathbb{R}$ as strategy sets:

Sufficient and necessary for equilibrium semi-uniqueness is that the equilibrium aggregator α is injective and constant.

The analysis of the first order conditions, and in fact the whole proof, only uses elementary mathematics. In the rest of this subsection I explain the idea of this analysis. The topics involved therein are treated in the next section.

Consider a Cournot oligopoly where $p \in \mathscr{S}_\star$. Concerning the cost functions suppose, with $\hat{X}_i := \{e_i \mid \mathbf{e} \in E\}$, that every c_i is semi-differentiable at each interior production level x_i with $x_i \in \hat{X}_i$ and further that, if $0 \in \hat{X}_i$, then the right derivative of c_i at 0 exists as an element of $\mathbb{R} \cup \{-\infty\}$ and, if $X_i = [0, m_i]$ and $m_i \in \hat{X}_i$, then the left derivative of c_i at m_i exists as an element of $\mathbb{R} \cup \{+\infty\}$.[14]

Notation: for $\mathbf{a}, \mathbf{b} \in \mathbf{X}$ let

$$J_{\mathbf{a};\mathbf{b}} = \{i \in N \mid b_i > a_i\}.$$

I further also just will write J instead of $J_{\mathbf{a};\mathbf{b}}$. It holds that $a_J \leq a_N$, $b_J \leq b_N$ and

$$a_N - a_J \geq b_N - b_J, \tag{1}$$

$$[\mathbf{a} \neq \mathbf{b} \wedge b_N \geq a_N] \Rightarrow [\#J \geq 1 \wedge a_J < b_J]. \tag{2}$$

Now suppose $\mathbf{a}, \mathbf{b} \in E$ and that in the case $a_N = 0$, the limit $\lim_{y \downarrow 0} \tilde{p}(y) \in \mathbb{R} \cup \{-\infty\}$ is well-defined. Then for every $i \in J_{\mathbf{a};\mathbf{b}}$, first order conditions imply

$$D^- u_i(\mathbf{b}) \geq 0 \geq D^+ u_i(\mathbf{a}). \tag{3}$$

[13]Of course, this result is well-known, but follows here without explicit calculations.

[14]We already have seen in Sect. 3.2 that sufficient for all this to hold is that the industry revenue function is concave and the cost functions are convex.

In the case where $a_N \neq 0$, these inequalities read in terms of the price function and the cost functions

$$D^- p(b_N) b_i + p(b_N) - D^- c_i(b_i) \geq 0 \geq D^+ p(a_N) a_i + p(a_N) - D^+ c_i(a_i). \tag{4}$$

In the case further assumptions on the price function and the cost functions are made, like in Theorem 1, certain inequalities concerning $a_i, b_i, D^+ p(a_N), D^- p(b_N), p(a_N), p(b_N), D^+ c_i(a_i)$ and $D^- c_i(b_i)$ hold. The further analysis now concerns the comparison of these inequalities with (4). The first analysis of this kind can be found in Corchón (1996). Even the analysis there holds for aggregative games for situations with one-dimensional strategy sets, differentiable conditional payoff functions and interior equilibria. In Folmer and von Mouche (2004) this analysis for one-dimensional strategy sets was extended to situations with semi-differentiable conditional payoff functions and non-interior equilibria and in Finus et al. (2014) to higher dimensional strategy sets. In the thesis Quartieri (2008) it was shown, among other things, how this way of analysis can handle (the difficult case of) Cournot oligopolies with concave industry revenue and convex costs.

In the case of the further assumptions as in Theorem 1, these "certain inequalities" concerning a (strictly) convex cost function c_i for firm $i \in J_{\mathbf{a};\mathbf{b}}$ are

$$D^+ c_i(a_i) \ (<) \leq D^- c_i(b_i). \tag{5}$$

The inequalities concerning the price function are more subtle and are presented in Sect. 4.3.

4 Analysis of the First Order Conditions

The proof of Theorem 1 in von Mouche and Quartieri (2015) is along the same lines as Theorem 3 in von Mouche and Quartieri (2013) for Cournot oligopolies with convex cost functions and a price function with decreasing price flexibility.[15] In this section I identify and study the topics involved in these proofs. This section in fact contains straightforward generalisations of results in von Mouche and Quartieri (2013) and von Mouche and Quartieri (2015).

[15] In the meantime I became aware of the fact that Vasin et al. (2007) deals with a similar result like Theorem 3 in von Mouche and Quartieri (2013), but with a completely different proof. This result also can be found (as Proposition 1) in the contribution Vasin and Kartunova (2016) in this Festschrift.

4.1 Active Equilibria

I refer to the following property

$$\#E \geq 2 \;\Rightarrow\; \text{each equilibrium is active,}$$

as the *activity property*. The activity property is equivalent with: if $\mathbf{0}$ is an equilibrium, then it is a unique equilibrium. Of course, sufficient for the activity property to hold is that the equilibrium aggregator is constant. The next proposition gives less obvious sufficient conditions.

Proposition 1 *Suppose each cost function is convex. Also suppose that*

- $p \in \mathscr{S}_{\star}$;
- *for each positive equilibrium aggregate* $y_{\star}$ *it holds that* p *is decreasing on* $]0, y_{\star}]$;
- *at least one of the following two conditions holds*

  I. *for each positive equilibrium aggregate* $y_{\star}$ *it holds that* $D^- p < 0$ *on* $]0, y_{\star}]$,
  II. *each cost function is strictly convex.*

Then:

1. *if* $i \in N$ *and* $\mathbf{a}, \mathbf{b} \in E$ *with* $b_N \geq a_{N\setminus\{i\}}$, *then* $b_i > 0 \;\Rightarrow\; a_i > 0$;
2. *the activity property holds.* ⋄

Proof I may assume that $\#E \geq 2$.

1. By contradiction suppose $i \in N$, $\mathbf{a}, \mathbf{b} \in E$ with $b_N \geq a_{N\setminus\{i\}}$ and $a_i = 0 < b_i$. So $b_N > 0$. As $\tilde{p}$ is decreasing on $]0, b_N]$, $\overline{p}(0) := \lim_{y \downarrow 0} \tilde{p}(y) \in \mathbb{R} \cup \{+\infty\}$ is well-defined. It follows that

$$D^+ u_i(\mathbf{a}) = \begin{cases} p(a_{N\setminus\{i\}}) - D^+ c_i(0) \geq p(b_N) - D^+ c_i(0) \text{ if } \sum_{l \neq i} a_l \neq 0, \\ \overline{p}(0) - D^+ c_i(0) \geq p(b_N) - D^+ c_i(0) \text{ if } \sum_{l \neq i} a_l = 0. \end{cases}$$

   Thus

$$D^+ u_i(\mathbf{a}) \geq p(b_N) - D^+ c_i(0).$$

   As $D^- p(b_N) \leq 0$ (< 0 if *I* holds) and $D^- c_i(b_i) \geq D^+ c_i(0)$ (> 0 if *II* holds), I have $D^- p(b_N) b_i + p(b_N) - D^- c_i(b_i) < p(b_N) - D^+ c_i(0)$. Thus

$$D^+ u_i(\mathbf{a}) > D^- p(b_N) b_i + p(b_N) - D^- c_i(b_i) = D^- u_i(\mathbf{b}).$$

   But, as $\mathbf{a}, \mathbf{b} \in E$ and $a_i < b_i$, this is a contradiction with (3).
2. Suppose $\mathbf{0} \in E$. Part 1 implies that $\mathbf{0}$ is the unique equilibrium. □

Sufficient for each equilibrium to be active is that at least one firm i with $c_i(0) \geq 0$ can make a monopoly profit (i.e. $p(x_i)x_i - c_i(x_i) > 0$ for some $x_i \in X_i$).[16] Here is a related result:

Proposition 2 *Suppose $n \geq 2$, each conditional profit function $u_i^{(\mathbf{0})}$ is strictly decreasing on $X_i \setminus \{0\}$ and $\lim_{x_i \downarrow 0} u_i^{(\mathbf{0})}(x_i) > u_i^{(\mathbf{0})}(0)$. Then each equilibrium* **e** *is active. Even at each equilibrium at least two firms are active.* ◇

Proof By contradiction suppose $\#\{j \in N \mid e_j \neq 0\} \leq 1$. As $n \geq 2$, I can fix $i \in N$ such that $e_j = 0$ $(j \neq i)$. If $e_i = 0$, then by assumption $\lim_{x_i \downarrow 0} u_i^{(\mathbf{0})}(x_i) > u_i^{(\mathbf{0})}(e_i)$. If $e_i \neq 0$, then as $u_i^{(\mathbf{0})}$ is strictly decreasing on $X_i \setminus \{0\}$, it also follows that $\lim_{x_i \downarrow 0} u_i^{(\mathbf{0})}(x_i) > u_i^{(\mathbf{0})}(e_i)$. Thus $\lim_{x_i \downarrow 0} u_i^{(\mathbf{0})}(x_i) > u_i^{(\mathbf{0})}(e_i)$ holds which is in contradiction with $\mathbf{e} \in E$. □

4.2 Injective Equilibrium Aggregator

Proposition 3 *Suppose each cost function is convex and p is semi-differentiable at each interior equilibrium aggregate. Each of the following conditions is sufficient for the equilibrium aggregator to be injective:*

I. $D^-p \leq D^+p < 0$ at each interior equilibrium aggregate.
II. $D^-p \leq D^+p$ at each interior equilibrium aggregate and every c_i is strictly convex. ◇

Proof By way of contradiction, suppose $\mathbf{a}, \mathbf{b} \in E$ with $\mathbf{a} \neq \mathbf{b}$ and $a_N = b_N =: y$. This implies that y is an interior equilibrium aggregate. Fix $i \in J_{\mathbf{a};\mathbf{b}}$. As $\mathbf{a}, \mathbf{b}$ are equilibria and $a_i < b_i$, I have (4). So the inequality

$$D^-p(y)b_i - D^+p(y)a_i \geq D^-c_i(b_i) - D^+c_i(a_i)$$

holds. If condition I holds, then with (5) I see that the left hand side of the last inequality is an element of $]-\infty, 0[$ and that the right hand side is an element of $[0, +\infty]$; if II holds, then the left hand side of the last inequality is an element of $]-\infty, 0]$ and the right hand side is an element of $]0, +\infty]$. Contradiction. □

4.3 A Crucial Inequality

The next definition provides the in Sect. 3.4 mentioned "certain inequalities".

[16] Indeed, then $\mathbf{0}$ is not an equilibrium as $u_i^{(\mathbf{0})}(x_i) = p(x_i)x_i - c_i(x_i) > 0 \geq -c_i(0) = u_i^{(\mathbf{0})}(0)$.

Definition 1 Let p be a price function. For $\mathbf{a}, \mathbf{b} \in \mathbf{X}$ with $0 < a_N < b_N$,

- inequality W($\mathbf{a}, \mathbf{b}$) is said to hold if $p \in \mathscr{S}$ and

$$D^+p(a_N)a_J + \#Jp(a_N) \geq D^-p(b_N)b_J + \#Jp(b_N).$$

- inequality S($\mathbf{a}, \mathbf{b}$) is said to hold if $p \in \mathscr{S}$ and

$$D^+p(a_N)a_J + \#Jp(a_N) > D^-p(b_N)b_J + \#Jp(b_N). \ \diamond$$

Lemma 1 *Suppose industry revenue r_p is concave. Let $\mathbf{a}, \mathbf{b} \in \mathbf{X}$ with $0 < a_N < b_N$.*

1. *Inequality W($\mathbf{a}, \mathbf{b}$) holds.*
2. *Each of the following three assumptions separately is sufficient for inequality S($\mathbf{a}, \mathbf{b}$) to hold.*

   a. *r_p is strictly concave;*
   b. *$a_J < a_N$ and $D^-p(b_N) < 0$;*
   c. *$p(a_N) > p(b_N)$ and $\#J \geq 2$.* $\diamond$

Proof Remember that $\#J \geq 1$ and $b_J > a_J$, that $\tilde{p}$ is decreasing and in part 2a even strictly decreasing with $D^-p < 0$. Clearly the desired results hold if $D^-p(b_N) = -\infty$. Henceforth suppose that $D^-p(b_N) \neq -\infty$.

As r_p is concave (and strictly concave in part 2a), the function $r_p^{(a_N - a_J)}$ defined by (8) in the appendix is by Proposition 8 concave (and strictly concave in parts 2a and 2b). Together with $a_J < a_J + b_N - a_N$ this implies $D^+r_p^{(a_N-a_J)}(a_J) \geq D^-r_p^{(a_N-a_J)}(a_J + b_N - a_N)$ ($>$ in parts 2a and 2b), i.e.,

$$D^+p(a_N)a_J + p(a_N) \geq D^-p(b_N)(a_J + b_N - a_N) + p(b_N) \ (> \text{ in parts 2a and 2b}).$$

As, by (1), $a_J + b_N - a_N \leq b_J$ and $D^-p(b_N) \leq 0$ ($<$ in parts 2a and 2b), I obtain

$$D^+p(a_N)a_J + p(a_N) \geq D^-p(b_N)b_J + p(b_N) \ (> \text{ in parts 2a and 2b}). \tag{6}$$

As $p(a_N) \geq p(b_N)$ ($>$ in parts 2a and 2c) and $\#J \geq 1$ ($\#J > 1$ in part 2c), I have

$$(\#J - 1)p(a_N) \geq (\#J - 1)p(b_N) \ (> \text{ in part 2c}). \tag{7}$$

Adding (6) and (7) gives $D^+p(a_N)a_J + \#Jp(a_N) > D^-p(b_N)b_J + \#Jp(b_N)$. $\square$

4.4 Constant Equilibrium Aggregator

With the next proposition I present sufficient conditions for the equilibrium aggregator to be constant. (Note that a constant equilibrium aggregator implies that all equilibrium prices are the same.) This proposition is a variant of an equilibrium uniqueness argument in (Quartieri, 2008, p. 33) which has been subsequently employed—for essentially less general classes of oligopolies—also in von Mouche and Quartieri (2013) and von Mouche and Quartieri (2015).

Proposition 4 *Suppose the activity property holds. Suppose that cost functions are convex and that for all equilibria* $\mathbf{a}, \mathbf{b}$ *with* $0 < a_N < b_N$ *inequality* $W(\mathbf{a}, \mathbf{b})$ *holds. Then each of the following two additional assumptions separately is sufficient for the equilibrium aggregator to be constant.*

I. For all equilibria $\mathbf{a}, \mathbf{b}$ *with* $0 < a_N < b_N$ *inequality* $S(\mathbf{a}, \mathbf{b})$ *holds.*
II. Each cost function is strictly convex. ◇

Proof By contradiction, suppose that $\mathbf{a}, \mathbf{b} \in E$ with $a_N \neq b_N$. I may suppose that $a_N < b_N$. As the activity property holds, I have $a_N > 0$. By assumption, inequality $\mathrm{W}(\mathbf{a}, \mathbf{b})$ holds and in case I even $\mathrm{S}(\mathbf{a}, \mathbf{b})$. So

$$D^+ p(a_N) a_J + s p(a_N) \geq D^- p(b_N) b_J + s p(b_N) \ (> \text{ in case I}).$$

As each c_i is convex (and strictly convex in case II), I obtain with (5)

$$D^+ p(a_N) a_J + s p(a_N) - \sum_{i \in J} D^+ c_i(a_i) > D^- p(b_N) b_J + s p(b_N) - \sum_{i \in J} D^- c_i(b_i)$$

and hence that $D^+ p(a_N) a_i + p(a_N) - D^+ c_i(a_i) > D^- p(b_N) b_i + p(b_N) - D^- c_i(b_i)$ for some $i \in J$, in contradiction with (4). □

4.5 Proof of Theorem 1

Lemma 2 *For* $\mathbf{a}, \mathbf{b} \in \mathbf{X}$ *with* $a_N < b_N$ *and* $\#\{i \in N \mid b_i \neq 0\} \geq 2$ *it holds that* $a_J < a_N$ *or* $\#J \geq 2$. ◇

Proof By contradiction suppose $a_J \geq a_N$ and $\#J \leq 1$. By (2) this implies $a_J = a_N$ and $\#J = 1$. Let $m \in N$ be such that $J = \{m\}$. This in turn implies that $b_m > a_m$ and $a_i = b_i = 0 \ (i \neq m)$, which is a contradiction with $\#\{i \in N \mid b_i \neq 0\} \geq 2$. □

Having Proposition 4 and the above lemma, part 1 of Theorem 1, reformulated here in the next proposition, follows in a comfortable way:

Proposition 5 *Suppose the industry revenue function* r_p *is concave and each cost function* c_i *is convex. Each of the following three conditions separately is sufficient for the equilibrium aggregator to be constant.*

I. r_p *is strictly concave.*
II. Every c_i *is strictly convex.*
III. r_p *is decreasing on* Int(Y) *and every* c_i *is strictly increasing.* ⋄

Proof By Proposition 1(2), in each of the three cases I, II, III the activity property holds.[17] Also by Lemma 1(1), for all equilibria $\mathbf{a}, \mathbf{b}$ with $0 < a_N < b_N$ inequality W($\mathbf{a}, \mathbf{b}$) holds. Lemma 1(2a) shows that even inequality S($\mathbf{a}, \mathbf{b}$) holds in the case I. So by Proposition 4 the proof is complete in the cases I and II.

Now further consider the case III. Well, it is easy to see that in the case $n = 1$ there is at most one equilibrium. And also that $\mathbf{0}$ is the unique equilibrium in the case $r_p \leq 0$. So if $n = 1$ or $r_p \leq 0$, then the equilibrium aggregator is constant. Now further suppose $n \geq 2$ and that r_p is in at least one point positive. Suppose $\mathbf{a}, \mathbf{b}$ with $0 < a_N < b_N$; I show that again inequality S($\mathbf{a}, \mathbf{b}$) holds. Well, as each function $u_i^{(\mathbf{0})} = r_p - c_i$ is strictly decreasing on $X_i \setminus \{0\}$, Proposition 2 applies and so $\#\{i \in N \mid b_i \neq 0\} \geq 2$. Lemma 2 applies and so $a_J < a_N$ or $\#J \geq 2$. So, by Lemma 1(2b,2c) inequality S($\mathbf{a}, \mathbf{b}$) holds. □

Having Proposition 5, now also part 2 of Theorem 1 follows: for the unique equilibrium aggregate $y_\star$ I have $y_\star \in \text{Int}(Y)$ and $D^- p(y_\star) = D^+ p(y_\star)$. As also $D^+ p(y_\star) = D^- p(y_\star) < 0$ or each c_i is strictly convex, Proposition 3 implies that the equilibrium aggregator also is injective and thus that there is a unique equilibrium.

5 One-Dimensional Equilibrium Set

First I provide a result concerning an implication injectivity of the equilibrium aggregator has for the equilibrium set in the case of concave industry revenue and convex cost functions.

Lemma 3 *Suppose* $\#E \geq 2$, *the industry revenue function is concave, each cost function is convex and for each positive equilibrium aggregate* $y_\star$ *it holds that* $D^- p < 0$ *on* $]0, y_\star]$.

1. *For all* $\mathbf{a}, \mathbf{b} \in E$ *with* $a_N \neq b_N$ *there exists a unique* $m \in N$ *such that* $a_i = b_i$ $(i \neq m)$.
2. *If the equilibrium aggregator is injective, then there exists a unique* $m_\star \in N$ *such that* $a_i = b_i$ $(i \neq m_\star)$ *for all* $\mathbf{a}, \mathbf{b} \in E$ *with* $\mathbf{a} \neq \mathbf{b}$.
3. *If the equilibrium aggregator is injective, then there exists a unique* $m_\star \in N$ *such that* $a_i = 0$ $(i \neq m_\star)$ *for every* $\mathbf{a} \in E$. ⋄

[17] Remember that in Sect. 3.2 I mentioned implicit monotonicity properties and also that p is positive at each positive equilibrium aggregate in the case cost functions are strictly increasing.

Proof

1. Suppose $\mathbf{a}, \mathbf{b} \in E$ with $a_N \neq b_N$. I may suppose that $b_N > a_N$. Let $J = J_{\mathbf{a};\mathbf{b}}$. With (1) and (2) I see that $b_N > 0,\ \#J \geq 1,\ a_J \leq a_N,\ b_J \leq b_N,\ a_J < b_J$ and $a_N - a_J \geq b_N - b_J$. And Proposition 1(I,2) implies $a_N \neq 0$.

   Now I prove by contradiction that $a_N - a_J = b_N - b_J$ and $\#J = 1$. So suppose

$$a_N - a_J > b_N - b_J \text{ (Case A) or } \#J \geq 2 \text{ (Case B).}$$

   In case A, I have $a_J < a_N$ and $D^- p(b_N) < 0$. In case B, I have $\#J \geq 2$ and $p(a_N) > p(b_N)$. Lemma 1(2b,2c) implies that inequality S($\mathbf{a}, \mathbf{b}$) holds, i.e. that

$$D^+ p(a_N) a_J + \#J p(a_N) > D^- p(b_N) b_J + \#J p(b_N).$$

   As each c_i is convex, I obtain

$$D^+ p(a_N) a_J + \#J p(a_N) - \sum_{i \in J} D^+ c_i(a_i) > D^- p(b_N) b_J + \#J p(b_N) - \sum_{i \in J} D^- c_i(b_i)$$

   which in turn implies the existence of $i \in J$ such that

$$D^+ p(a_N) a_i + p(a_N) - D^+ c_i(a_i) > D^- p(b_N) b_i + p(b_N) - D^- c_i(b_i),$$

   i.e. a contradiction with (4).

   Now let m be the unique element of J. I have $b_m > a_m$ and $a_i \geq b_i\ (i \neq m)$. So $b_N - b_J = a_N - a_J = \sum_{i \neq m} a_l \geq \sum_{i \neq m} b_l = b_N - b_J$. So $\sum_{i \neq m} a_i = \sum_{i \neq m} b_i$. It follows that $a_i = b_i\ (i \neq m)$.
2. If I prove that there exists an $m_\star \in N$ such that $a_i = b_i\ (i \neq m_\star)$ for all $\mathbf{a}, \mathbf{b} \in E$ with $\mathbf{a} \neq \mathbf{b}$, then $m_\star$ is automatically unique. Having part 1, I denote for $\mathbf{a}, \mathbf{b} \in E$ with $\mathbf{a} \neq \mathbf{b}$ (and then $a_N \neq b_N$) the unique $m \in N$ such that $a_i = b_i\ (i \neq m)$ and $a_m \neq b_m$ by

$$m_{\mathbf{a};\mathbf{b}}.$$

   I now finish the proof in 4 steps.

   Step 1. If $\mathbf{v}, \mathbf{w}, \mathbf{a} \in E$ with $\#\{\mathbf{v}, \mathbf{w}, \mathbf{a}\} = 3$, then $m_{\mathbf{a};\mathbf{v}} = m_{\mathbf{v};\mathbf{w}}$ or $m_{\mathbf{a};\mathbf{w}} = m_{\mathbf{v};\mathbf{w}}$.
   Proof: write

$$m^0 := m_{\mathbf{v};\mathbf{w}},\ m^1 := m_{\mathbf{a};\mathbf{v}},\ m^2 := m_{\mathbf{a};\mathbf{w}}.$$

   By contradiction suppose $m^1 \neq m^0$ and $m^2 \neq m^0$. This implies $a_{m^0} = v_{m^0}$ and $a_{m^0} = w_{m^0}$. Thus $v_{m^0} = w_{m^0}$, a contradiction.
   Step 2. If $\mathbf{v}, \mathbf{w}, \mathbf{a} \in E$ with $\#\{\mathbf{v}, \mathbf{w}, \mathbf{a}\} = 3$, then $m_{\mathbf{a};\mathbf{v}} = m_{\mathbf{a};\mathbf{w}} = m_{\mathbf{v};\mathbf{w}}$.
   Proof: let m^0, m^1 and m^2 be as in Step 1. By this step 1 I have $m^1 = m^0$ or $m^2 = m^0$. Suppose $m^1 = m^0$; I shall prove that $m^2 = m^0$. (The other case is

proved in the same way.) Well, I have

$$a_l = v_l \ (l \neq m^0), \ v_l = w_l \ (l \neq m^0).$$

This implies $a_l = w_l \ (l \neq m_0)$ and therefore also $a_{m^0} \neq w_{m^0}$. As also $a_{m^2} \neq w_{m^2}$ it follows that $m^0 = m^2$.
Step 3. As $\#E \geq 2$, I can fix $\mathbf{v}, \mathbf{w} \in E$ with $\mathbf{v} \neq \mathbf{w}$. Let $m_\star := m_{\mathbf{v};\mathbf{w}}$.
Step 4. For all $\mathbf{a}, \mathbf{b} \in E$ with $\mathbf{a} \neq \mathbf{b}$ it holds that $m_{\mathbf{a};\mathbf{b}} = m_\star$.
Proof: Case where $\mathbf{a}, \mathbf{b}, \mathbf{v}, \mathbf{w}$ are 4 different equilibria. As $\#\{\mathbf{a}, \mathbf{b}, \mathbf{v}\} = 3$, I have by Step 2,

$$m_{\mathbf{a};\mathbf{b}} = m_{\mathbf{a};\mathbf{v}}.$$

As $\#\{\mathbf{a}, \mathbf{w}, \mathbf{w}\} = 3$, different, I have by Step 2,

$$m_{\mathbf{v};\mathbf{w}} = m_{\mathbf{a};\mathbf{v}}.$$

So $m_{\mathbf{a};\mathbf{b}} = m_{\mathbf{v};\mathbf{w}} = m_\star$.
Case where $\mathbf{a} = \mathbf{v}$ or $\mathbf{a} = \mathbf{w}$ or $\mathbf{b} = \mathbf{v}$ or $\mathbf{b} = \mathbf{w}$. Suppose $\mathbf{a} = \mathbf{v}$. (The other possibilities are handled analogously.) If $\mathbf{b} = \mathbf{w}$, then $m_{\mathbf{a};\mathbf{b}} = m_{\mathbf{v};\mathbf{w}} = m_\star$. If $\mathbf{b} \neq \mathbf{w}$, then by Step 2 , as $\#\{\mathbf{a}, \mathbf{b}, \mathbf{w}\} = 3$, I have $m_{\mathbf{a};\mathbf{b}} = m_{\mathbf{a};\mathbf{w}} = m_{\mathbf{v};\mathbf{w}} = m_\star$.

3. I shall prove that for $m_\star$ as in part 2 it holds that $a_i = 0 \ (i \neq m_\star)$ for every $\mathbf{a} \in E$; then $m_\star$ automatically is unique and the proof is complete.
   Fix $\mathbf{a} \in E$. As $\#E \geq 2$, I also can fix $\mathbf{b} \in E$ with $\mathbf{b} \neq \mathbf{a}$. As the equilibrium aggregator is injective I have $a_N \neq b_N$. I may suppose that $a_N < b_N$. Now I have

$$a_i = b_i \ (i \neq m_\star).$$

Finally, I prove by contradiction that $a_i = 0 \ (i \neq m_\star)$. So suppose there exists $j \neq m_\star$ with

$$a_j \neq 0.$$

So $a_j = b_j \neq 0$. Note that $a_{m_\star} < a_{m_\star} + a_j \leq a_N$; so $a_{m_\star} < a_N$. Lemma 1(2b) implies

$$D^+p(a_N)a_{m_\star} + p(a_N) > D^-p(b_N)b_{m_\star} + p(b_N).$$

As $c_{m_\star}$ is convex, it follows that

$$D^+p(a_N)a_{m_\star} + p(a_N) - D^+c_{m_\star}(a_{m_\star}) > D^-p(b_N)b_{m_\star} + p(b_N) - D^-c_{m_\star}(b_{m_\star}),$$

which is a contradiction with (4). □

Proposition 6 *Suppose the industry revenue function is concave and each cost function is convex. Suppose that the equilibrium aggregator is injective and that for each positive equilibrium aggregate $y_\star$ it holds that $D^-p < 0$ on $]0, y_\star]$.*

1. *The equilibrium set E is convex and at most* 1*-dimensional.*
2. *If* $\#E \geq 2$, *then there exists* $m_\star \in N$ *and a proper interval* I *of* $\mathbb{R}$ *with* $0 \notin I \subseteq X_{m_\star}$ *such that* $E = \{(0, \ldots, 0, x_{m_\star}, 0, \ldots, 0) \mid x_{m_\star} \in I\}$. $\diamond$

Proof First I note that, as $\tilde{p}$ is continuous on $\mathrm{Int}(Y)$, it follows for each positive equilibrium aggregate $y_\star$ that p is decreasing on $]0, y_\star]$.

As part 2 implies part 1, it is sufficient to prove part 2.

Lemma 3(3) guarantees the existence of $m_\star \in I$ and a subset I of $\mathbb{R}$ with $I \subseteq X_{m_\star}$ such that $E = \{(0, \ldots, 0, x_{m_\star}, 0, \ldots, 0) \mid x_{m_\star} \in I\}$. It remains to be proved that I is an interval; as $\#E \geq 2$ it automatically is proper. Without loss of generality I may suppose that $m_\star = n$. So

$$E = \{(0, \ldots, 0, x_n) \mid x_n \in I\}.$$

Proposition 1(I,2) applies implies $\mathbf{0} \notin E$. So

$$0 \notin I.$$

Define the functions $g_n^+ : \mathrm{Int}(X_n) \to \mathbb{R}$, $g_i^+ : X_n \setminus \{0\} \to \mathbb{R}$ $(i \neq n)$, $g_n^- : X_n \setminus \{0\} \to \mathbb{R}$ by

$$g_n^+(x_n) := D_n^+ u_n^{(\mathbf{0})}(x_n) = D^+p(x_n)x_n + p(x_n) - D^+c_n(x_n),$$

$$g_i^+(x_n) := D_i^+ u_i^{(0,\ldots,0,x_n)}(0) = p(x_n) - D^+c_i(0),$$

$$g_n^-(x_n) := D_n^- u_n^{(\mathbf{0})}(x_n) = D^-p(x_n)x_n + p(x_n) - D^-c_n(x_n).$$

Let

$$K_n^+ := \{x_n \in \mathrm{Int}(X_n) \mid g_n^+(x_n) \leq 0\},$$

$$K_i^+ := \{x_n \in X_n \setminus \{0\} \mid g_i^+(x_n) \leq 0\} \ (i \neq n),$$

$$K_n^- := \{x_n \in X_n \setminus \{0\}) \mid g_n^-(x_n) \geq 0\}.$$

As the conditional payoff functions are concave, I have for $(0, \ldots, 0, x_n)$ with $x_n \in X_n \setminus \{0\}$ that $(0, \ldots, 0, x_n) \in E$ if and only if

$$[x_n \in \mathrm{Int}(X_n) \Rightarrow g_n^+(x_n) \leq 0] \wedge [g_i^+(x_n) \leq 0 \ (i \neq n)] \wedge [g_n^-(x_n) \geq 0].$$

With

$$I' := \Big((\cap_{i=1}^{n-1} K_i^+) \cap (K_n^+ \cap K_n^-) \Big) \cup \Big((\cap_{i=1}^{n-1} K_i^+) \cap \{x_n \in X_n \setminus X_n^\ominus \mid g_n^-(x_n) \geq 0\} \Big),$$

this implies for $x_n \in X_n \setminus \{0\}$:

$$(0, \ldots, 0, x_n) \in E \;\Leftrightarrow\; x_n \in I'.$$

So $I' = I$.

Finally, I prove that I is an interval. So suppose $a, b \in I$ with $a < b$ and fix $c \in]a, b[$. Now, $a, c \in \mathrm{Int}(X_n)$ and $b \in X_n^\oplus$. It follows that $g_n^+(a) \leq 0 \leq g_n^-(a)$, $g_i^+(a) \leq 0$ $(i \neq n)$, $g_i^+(b) \leq 0$ $(i \neq n)$, $g_n^-(b) \geq 0$. As $\tilde{p}$ is decreasing and $u_i^{(\mathbf{0})}$ is concave, the functions g_n^+, g_i^+, g_n^- are decreasing. So I obtain

$$g_n^+(c) \leq g_n^+(a) \leq 0,\; g_i^+(c) \leq g_i^+(a) \leq 0,\; g_n^-(c) \geq g_n^-(b \geq 0.$$

So $c \in K_n^+$, $c \in K_i^+$, $c \in K_n^-$. Thus $c \in I$. □

Theorem 2 *Consider a Cournot oligopoly with a concave industry revenue function and convex cost functions. Suppose that the proper price function is differentiable with negative derivative. Also suppose that* $\#E \geq 2$.

1. *E is a* 1*-dimensional convex set, even there exists* $m_\star \in N$ *and a proper interval* I *of* $\mathbb{R}$ *with* $0 \notin I \subseteq X_{m_\star}$ *such that* $E = \{(0, \ldots, 0, x_{m_\star}, 0, \ldots, 0) \mid x_{m_\star} \in I\}$.
2. *Suppose the industry revenue function and the cost functions are continuous.*
   a. *If each production level set is closed, then I in part 1 is closed and so E is polyhedral.*
   b. *If each firm has an effective capacity constraint, then I in part 1 is compact and so E is a non-empty polytope.* ◇

Proof

1. This follows from Proposition 6 together with Proposition 3.
2. Now the payoff functions are continuous. This implies that E is closed in $\mathbf{X}$.
   a. Now $\mathbf{X}$ is closed in $\mathbb{R}^n$ and therefore E is a closed subset of $\mathbb{R}^n$. This implies that the interval I in part 1 is closed.
   b. By Theorem 4(2) the equilibrium set is non-empty and compact. This implies that I in part 1 is compact. □

6 Polyhedral Structure

In this section I study the polyhedral structure of Cournot oligopolies with convex cost functions and concave industry revenue. Having Theorem 1(1), the main result, Theorem 3, essentially follows from a general result for aggregative games in von Mouche (2011). However, as I here deal with Cournot oligopolies, the proof of the result in von Mouche (2011) can be simplified considerably.

I start with a lemma from convex analysis.

Lemma 4 *Suppose that I is an interval of $\mathbb{R}^n$ (i.e. a Cartesian product of n real intervals), A is an affine subset of $\mathbb{R}^n$ and $I \cap A$ is closed. Then $I \cap A$ is polyhedral.* $\diamond$

Proof I may assume that $I \cap A \neq \emptyset$. I will prove that, denoting by $\overline{I}$ the topological closure of I, $I \cap A = \overline{I} \cap A$. Then the proof is complete because $\overline{I} \cap A$ is polyhedral as intersection of polyhedral sets. $I \cap A \subseteq \overline{I} \cap A$ is evident. Now I prove by contradiction that $\overline{I} \cap A \subseteq I \cap A$. So suppose that

$$\mathbf{a} \in (\overline{I} \cap A) \setminus (I \cap A).$$

With ∂I the topological boundary of I, I have $\mathbf{a} \in \overline{I} \setminus I \subseteq \partial I$. Also $\mathbf{a} \in A$.

Let $(\mathbf{a}_n)$ be a sequence in A with limit $\mathbf{a}$. For n large enough I have $\mathbf{a}_n \notin I$. (Indeed: if not, then there exists a sub sequence $(\mathbf{b}_n)$ of $(\mathbf{a}_n)$ with $\mathbf{b}_n \in A \cap I$. As $I \cap A$ is closed, it follows that $\mathbf{a} \in I \cap A$, which is a contradiction.) It follows that there exists a neighbourhood U of $\mathbf{a}$ with $U \cap A \subseteq \mathbb{R}^n \setminus I$. So there exists a neighbourhood of $\mathbf{a}$ in A that does not intersect I. Take $\mathbf{y} \in I \cap A$. As A is convex, I have $\{\lambda \mathbf{a} + (1-\lambda)\mathbf{y} \mid \lambda \in [0,1]\} \subseteq A$. Also $\mathbf{a} \in \partial I$ and $\mathbf{y} \in I$. As I is an interval, this implies $\{\lambda \mathbf{a} + (1-\lambda)\mathbf{y} \mid \lambda \in [0,1[\,\} \subseteq I$. But, there is a neighbourhood of $\mathbf{a}$ in A that does not intersect I. Contradiction. □

Theorem 3 *Consider a Cournot oligopoly with a concave industry revenue function and convex cost functions. In addition suppose that at least one of the following three conditions holds:*

I. r_p is strictly concave.
II. Every c_i is strictly convex.
III. r_p is decreasing on the interior of its domain and every c_i is strictly increasing.

Then the following results hold.

1. *The equilibrium aggregator is constant.*
2. *E is a convex subset of $\mathbb{R}^n$. Even: E is the intersection of an interval of $\mathbb{R}^n$ and an affine subset of $\mathbb{R}^n$.*
3. *Suppose the industry revenue function and the cost functions are continuous.*
   a. *E is closed in $\mathbf{X}$.*
   b. *If each production level set is closed, then E is polyhedral.*
   c. *If each firm has an effective capacity constraint, then E is a non-empty polytope.* $\diamond$

Proof

1. By Proposition 5.

2. I may assume that $\#E \geq 2$. By part 1 the aggregator is constant on E, say $y_\star$. As $\#E \geq 2$ it follows that $y_\star \in \mathrm{Int}(Y)$. Now I provide the proofs in 5 steps.

   Step 0. The functions $t_i^- : X_i^\oplus \times Y^\oplus \to \mathbb{R} \cup \{-\infty\}$ and $t_i^+ : X_i^\ominus \times \mathrm{Int}(Y) \to \mathbb{R} \cup \{+\infty\}$ are well-defined by

$$t_i^\pm(x_i, y) := D^\pm p(y) x_i + p(y) - D^\pm c_i(x_i).$$

   Step 1. The functions t_i^+ and t_i^- are decreasing in their first variable. Indeed: $\tilde{p}$ is decreasing, so $D^\pm p(y) \leq 0$. Also, $D^\pm c_i$ are increasing as c_i is convex.
   Step 2. With, for $i \in N$,

$$\begin{aligned} L_i &:= \{x_i \in \mathrm{Int}(X_i) \mid t_i^+(x_i, y_\star) \leq 0 \leq t_i^-(x_i, y_\star)\} \\ &\cup \{x_i \in X_i \setminus X_i^\ominus \mid t_i^-(x_i, y_\star) \geq 0\} \cup \{x_i \in X_i \setminus X_i^\oplus \mid t_i^+(x_i, y_\star) \leq 0\} \end{aligned}$$

   it holds that

$$L_i \text{ is an interval of } \mathbb{R}.$$

   Indeed: suppose $a, b \in L_i$ with $a < b$ and fix $c \in]a, b[$. Now, $c \in \mathrm{Int}(X_i)$. Noting that $a \in X_i^\ominus$ and $b \in X_i^\oplus$, I have $t_i^+(a, y_\star) \leq 0 \leq t_i^-(b, y_\star)$. As, by Step 1, t_i^+ and t_i^- are decreasing in their first variable it follows that $t_i^+(c, y_\star) \leq t_i^+(a, y_\star) \leq 0 \leq t_i^-(b, y_\star) \leq t_i^-(c, y_\star)$. So $c \in L_i$.
   Step 3. $E = (L_1 \times \cdots \times L_n) \cap \alpha^{<-1>}(y_\star)$.
   Proof of '$\subseteq$': suppose $\mathbf{e} \in E$. As $\alpha(\mathbf{e}) = y_\star$, I have $\mathbf{e} \in \alpha^{<-1>}(y_\star)$. Now fix $i \in N$. In the case $e_i \in \mathrm{Int}(X_i)$ I have

$$t_i^+(e_i, y_\star) = D^+ u_i^{(\mathbf{e}_{\hat{\imath}})}(e_i) \leq 0 \leq D^- u_i^{(\mathbf{e}_{\hat{\imath}})}(e_i) = t_i^-(e_i, y_\star).$$

   In the case $e_i = 0$ I have

$$t_i^+(e_i, y_\star) = D^+ u_i^{(\mathbf{e}_{\hat{\imath}})}(e_i) \leq 0$$

   and in the case ($X_i = [0, m_i]$ and) $e_i = m_i$ I have

$$t_i^-(e_i, y_\star) = D^- u_i^{(\mathbf{e}_{\hat{\imath}})}(e_i) \geq 0.$$

   Thus, as desired, $e_i \in L_i$.
   Proof of '$\supseteq$': let $\mathbf{e} \in (L_1 \times \cdots \times L_n) \cap \alpha^{<-1>}(y_\star)$. Now $e_i \in L_i$ $(i \in N)$ and $\alpha(\mathbf{e}) = y_\star$. If $e_i \in \mathrm{Int}(X_i)$, then

$$D^+ u_i^{(\mathbf{e}_{\hat{\imath}})}(e_i) = t_i^+(e_i, y_\star) \leq 0 \leq t_i^-(e_i, y_\star) = D^- u_i^{(\mathbf{e}_{\hat{\imath}})}(e_i).$$

In the case $e_i = 0$

$$D^+ u_i^{(\mathbf{e}_{\hat{i}})}(e_i) = t_i^+(e_i, y_\star) \leq 0$$

and in the case ($X_i = [0, m_i]$ and) $e_i = m_i$

$$D^- u_i^{(\mathbf{e}_{\hat{i}})}(e_i) = t_i^-(e_i, y_\star) \geq 0.$$

As the conditional profit function $u_i^{(\mathbf{e}_{\hat{i}})}$ is concave, the above signs of the left and right derivatives imply that e_i is a maximiser of $u_i^{(\mathbf{e}_{\hat{i}})}$. Thus $\mathbf{e} \in E$.
Step 4. By Steps 2 and 3, E is the intersection of an interval of $\mathbb{R}^n$ and an affine subset of $\mathbb{R}^n$. Thus part 2 is proven.

3. Now the payoff functions are continuous. This implies that E is closed in $\mathbf{X}$.
   a. Now $\mathbf{X}$ is closed in $\mathbb{R}^n$. With part 2 it follows that E is a closed subset of $\mathbb{R}^n$ that is an intersection of an interval of $\mathbb{R}^n$ and an affine subset of $\mathbb{R}^n$. By Lemma refLeme:polywp such a subset is polyhedral.
   b. By Theorem 4(2) the equilibrium set is non-empty and compact. With part 2 I see that E is a bounded polyhedron and therefore a polytope. □

7 Examples

Consider the Cournot oligopoly specified by

$$n = 2, \; X_1 = X_2 = [0, 5/4], \; c_1 = c_2 = \mathrm{Id}/2,$$

$$p(y) = \begin{cases} 7/4 - y/2 \text{ if } 0 \leq y \leq 3/2, \\ 5/2 - y \text{ if } 3/2 \leq y \leq 5/2. \end{cases}$$

This oligopoly is taken from Szidarovszky and Yakowitz (1982). Its industry revenue function is strictly concave. Theorem 3(I) implies that the equilibrium set is a non-empty polytope and the equilibrium aggregator is constant, say $y_\star$, on this polytope. Now even without determining the equilibrium set E, I have

$$\#E \geq 2 \Rightarrow y_\star = 3/2.$$

Indeed, as $3/2$ is the unique point where the (proper) price function is not differentiable, the reformulation of Theorem 1 in Sect. 3.1 shows that $y_\star = 3/2$. Finally, it is an easy exercise to determine E:

$$E = \{(x_1, x_2) \in \mathbb{R}^2 \mid 1/2 \leq x_1 \leq 1, \; 1/2 \leq x_2 \leq 2, \; x_1 + x_2 = 3/2\}.$$

In accordance with result B in Sect. 3.3 I see that at each equilibrium each firm is active.

Now consider the following Cournot oligopoly specified by

$$n = 2,\ X_1 = X_2 = \mathbb{R}_+,\ c_1 = 1000\,\mathrm{Id},\ c_2 = \mathrm{Id}/4,$$

$$p(y) := \begin{cases} \frac{81}{16y+16} - \frac{5}{16} & \text{if } y \in [0, 2], \\ \frac{9}{4y} + \frac{1}{4} & \text{if } y \in [2, 3], \\ \frac{4}{y+1} & \text{if } y \in [3, \infty[. \end{cases}$$

This oligopoly is taken from von Mouche and Quartieri (2015) (and served there as a counter example for a statement in Murphy et al. (1982)). It is straightforward to verify that the industry revenue function is concave and that the price function p is (continuously) differentiable with $Dp < 0$. In fact Theorem 2 applies in case $\#E \geq 2$. So, even without determining the equilibrium set E, it follows that

$$\#E \geq 2 \ \Rightarrow \ \text{there exists a proper real interval } I \text{ such that } E = I \times \{0\} \text{ or } E = \{0\} \times I.$$

In fact, a straightforward determination of the equilibrium set gives

$$E = \{0\} \times [2, 3].$$

8 Concluding Remarks

1. The article deals with old and new results for homogeneous Cournot oligopolies with a concave industry revenue function (that may be discontinuous at 0) and convex cost functions. An appendix provides a little toolbox and in particular an equilibrium existence result that can handle non-compact production level sets.
2. The main new results are Theorems 2 and 3 on the geometric structure of the equilibrium set. In Theorem 2 the equilibrium aggregator is injective and in Theorem 3 the equilibrium aggregator is constant.
3. The method of analysis is a local one: just analysing first order conditions. An important role is played here by the specific inequalities in Definition 1.
4. It would be very interesting if one could show that the uniqueness results in Ewerhart (2014) for biconcave price functions and not necessarily convex cost functions can also be proved by a local analysis.[18]

[18]The method used in Ewerhart (2014) is similar to that in Gaudet and Salant (1991) and concerns a global analysis of the aggregate backward reply function (see for this function also the contribution Cornes and Sato (2016) in this Festschrift).

5. In none of the lemma's, propositions and theorems is differentiability of the proper price function assumed (but semi-differentiability assumptions are made). The only exception is Theorem 1(3) dealing with equilibrium semi-uniqueness.
6. A discrete variant of Theorem 1 can be found in the contribution Iimura and Watanabe (2016) in this Festschrift.

Appendix

This appendix contains some results, which are not claimed to be new, but for which it is difficult to give a proper reference.

Effective Capacity Constraints

For equilibrium existence compact production level sets are very welcome. However, effective capacity constraints (see Sect. 2) also will do the job as Theorem 4 below shows.

Proposition 7 *Sufficient for firm i to have an effective capacity constraint* $\overline{x}_i$ *is that* $(r_p - c_i)(x_i) < -c_i(0)$ *for every* $x_i \in X_i$ *with* $x_i > \overline{x}_i$ *and that p is decreasing on* $\{y \in Y \mid y \geq \overline{x}_i\}$. $\diamond$

Proof Suppose $x_i \in X_i$ with $x_i > \overline{x}_i$. I show that x_i is strongly dominated by 0 ($\in [0, \overline{x}_i]$). So fix $\mathbf{z} \in \mathbf{X}_{\hat{\imath}}$. Now, as desired, $u_i^{(\mathbf{z})}(0) = -c_i(0) > (r_p - c_i)(x_i) = p(x_i)x_i - c_i(x_i) \geq p(x_i + z_{\hat{\imath}})x_i - c_i(x_i) = u_i^{(\mathbf{z})}(x_i)$. □

The condition in Proposition 7 has the following economic interpretation: the variable monopoly profit becomes negative.

Theorem 4 *Consider a Cournot oligopoly* Γ *where each firm i has an effective capacity constraint* $\overline{x}_i$.

1. *Let* $K_i := [0, \overline{x}_i]$ $(i \in N)$ *and let* Γ' *be the game in strategic form with for firm i as production level set* K_i *and as payoff function* $u_i \upharpoonright K_i$. *Then the equilibrium sets of* Γ *and* Γ' *are the same.*
2. *If, in* Γ, *each profit function is continuous and each conditional profit function is quasi-concave, then the equilibrium set E of* Γ *is a non-empty compact subset of* $\mathbb{R}^n$ *with* $E \subseteq \mathsf{X}_{i=1}^n [0, \overline{x}_i]$. $\diamond$

Proof

1. It is straightforward to verify that $\operatorname{argmax} u_i^{(\mathbf{z})} \upharpoonright K_i = \operatorname{argmax} u_i^{(\mathbf{z})}$ for every $i \in N$ and $\mathbf{z} \in \mathbf{X}_{\hat{\imath}}$. This implies that Γ and Γ' have the same equilibrium set.
2. The Nikaido-Isoda theorem guarantees that the equilibrium set of Γ' is non-empty. So by part 1, Γ has an equilibrium and $E \subseteq \mathsf{X}_{i=1}^n [0, \overline{x}_i]$. As the profit

functions of Γ' are continuous, the equilibrium set of Γ' is is closed in $\mathsf{X}_{i=1}^{n}[0,\overline{x}_i]$ and therefore also in $\mathbb{R}^n$. So E is a closed bounded subset of $\mathbb{R}^n$ and therefore compact. □

Some Convex Analysis Related to Price Functions

First, I introduce some useful notations. For a proper real interval with $J \subseteq \mathbb{R}_+$ and $k \in \mathbb{R}_+$, let

$$J_k := (J - \{k\}) \cap \mathbb{R}_+.$$

So $J_0 = J$ and also J_k is an interval; for $k \in J^{\ominus}$, the interval J_k is proper. And for a function $g : J \to \mathbb{R}$ with J and k as above, the function $g^{(k)} : J_k \to \mathbb{R}$ is defined by

$$g^{(k)}(x) := g(x+k) \tag{8}$$

and the function $r_g : J \to \mathbb{R}$ is defined by

$$r_g := g \cdot \mathrm{Id}.$$

In the rest of this appendix, K is a proper real interval with $0 \in K \subseteq \mathbb{R}_+$, $f : K \to \mathbb{R}$ is a function and

$$\tilde{f} := f \restriction K^{\oplus}.$$

Lemma 5 *Suppose r_f is (strictly) concave.*

1. *f is semi-differentiable at each interior point of its domain and in the case $K = [0,m]$, the (left) derivative $D^-f(m)$ of f at m exists as element of $\mathbb{R} \cup \{-\infty\}$;*
2. *$D^+f(y) \leq D^-f(y)$ $(y \in \mathrm{Int}(K))$;*
3. *$D^-f\ (<) \leq\ 0$;*
4. *$\tilde{f}$ is (strictly) decreasing.* ◇

Proof 1–3. As r_f is concave, r_f, and therefore f too, is semi-differentiable at each interior point of its domain. As r_f is (strictly) concave, it holds for all $y \in \mathrm{Int}(K)$ that $D^+r_f(y) \leq D^-r_f(y)\ (<) \leq \frac{r_f(y)-r_f(0)}{y-0}$. So $D^+f(y)y + f(y) \leq D^-f(y)y + f(y)\ (<) \leq f(y)$ and thus $D^+f(y) \leq D^-f(y)\ (<) \leq 0$. Thus the proof is complete if $K \neq [0,m]$.

Now consider the case where $K = [0,m]$ and $D^-f(m) \neq -\infty$. So f is differentiable at m and therefore $Dr_f(m)\ (<) \leq \frac{r_f(m)-r_f(0)}{m-0}$ and $Df(m)\ (<) \leq 0$ follows.

4. By parts 1 and 3. □

Lemma 6 *If r_f is concave and on* Int(K) *decreasing, then* $V := \{y \in K^{\oplus} \mid f(y) > 0\}$ *is a real interval with* $]0, y] \subseteq V$ $(y \in V)$, $D^{-}f < 0$ *on* V *and* f *is strictly decreasing on* V. $\diamond$

Proof By Lemma 5(4), $\tilde{f}$ is decreasing. This implies that $V = \{y \in K^{\oplus} \mid \tilde{f}(y) > 0\} = \{y \in K^{\oplus} \mid f(y) > 0\}$ is a real interval with $]0, y] \subseteq V$ $(y \in V)$. As $r_{\tilde{f}}$ is decreasing on Int(K) and $r_{\tilde{f}}$ is concave, even $r_{\tilde{f}}$ is decreasing. So, for $y \in V$, $0 \geq D^{-}r_{\tilde{f}}(y) = D^{-}\tilde{f}(y)y + \tilde{f}(y)$. As $\tilde{f} > 0$ on V, it follows that $D^{-}f(y) = D^{-}\tilde{f}(y) < 0$ $(y \in V)$. As f is continuous on Int(K), it follows that f is strictly decreasing on V. □

A weaker variant of the next result can be found in Murphy et al. (1982); in particular I note that in this result there is no explicit assumption about the monotonicity properties of f.

Proposition 8

1. *If r_f is (strictly) concave, then for every $k \in K^{\ominus}$ the function $r_{f^{(k)}}$ is (strictly) concave.*
2. *If r_f is concave and $\tilde{f}$ is strictly decreasing, then for every $k \in$* Int(K) *the function $r_{f^{(k)}}$ is strictly concave.* $\diamond$

Proof

1. This is evident for $k = 0$. Now suppose $k \in$ Int(K). Suppose r_f is strictly concave; the proof for the case of concave r_f is analogous. I give the proof by contradiction. So suppose $r_{f^{(k)}}$ is not strictly concave. This implies that there exist $x, y \in K_k$ with $x < y$ and $t \in]0, 1[$ such that for $z = tx + (1-t)y$ the inequality $r_{f^{(k)}}(z) \leq tr_{f^{(k)}}(x) + (1-t)r_{f^{(k)}}(y)$ holds. By Lemma 5(4), the function f is strictly decreasing on $K^{\oplus}$. So $f^{(k)}$ also is strictly decreasing.

   As $z + k \geq 0$, the inequality

   $$r_{f^{(k)}}(z)(z + k) \leq (tr_{f^{(k)}}(x) + (1-t)r_{f^{(k)}}(y))(z + k)$$

   follows. As $x + k, y + k \in K$, also $z + k = t(x + k) + (1-t)(y + k) \in K$. As r_f is strictly concave and $z > 0$, I also have the strict inequality

   $$r_f(z + k)z > (tr_f(x + k) + (1-t)r_f(y + k))z.$$

   These inequalities together with $r_f(z + k)z = r_{f^{(k)}}(z)(z + k)$ imply

   $$(tr_{f^{(k)}}(x) + (1-t)r_{f^{(k)}}(y))(z + k) > (tr_f(x + k) + (1-t)r_f(y + k))z.$$

   Using the definition of r_f and $r_{f^{(k)}}$ this inequality becomes

   $$tx(z+k)f(x+k)+(1-t)y(z+k)f(y+k) > t(x+k)zf(x+k)+(1-t)(y+k)zf(y+k).$$

This implies, by using $y - z = t(y - x)$ and $z - x = (1 - t)(y - x)$, that $kt(1 - t)f(y + k)(y - x) > kt(1 - t)f(x + k)(y - x)$ and, as $k > 0$,

$$f(y + k) > f(x + k).$$

But this is a contradiction with the decreasingness of $f^{(k)}$.

2. This is proved by making obvious changes in the proof of part 1. □

Acknowledgements I would like to thank Ferenc Szidarovzsky and Willem Pijnappel for their helpful comments. Any errors that may be found are my responsibility.

References

Burger E (1959) Einführung in die Theorie der Spiele. Walter de Gruyter & Co., Berlin

Corchón LC (1996) Theories of imperfectly competitive markets. Lecture notes in economics and mathematical systems, vol 442, 2nd edn. Springer, Berlin

Cornes R, Sato T (2016) Existence and uniqueness of Nash equilibrium in aggregative games: an expository treatment. In: von Mouche P, Quartieri F (eds) Contributions to equilibrium theory for Cournot oligopolies and related games: essays in Honour of Koji Okuguchi. Springer, Cham

Ewerhart C (2014) Cournot games with biconcave demand. Games Econ Behav 85:37–47

Finus M, von Mouche PHM, Rundshagen B (2014) On uniqueness of coalitional equilibria. In: Petrosjan LA, Zenkevich NA (eds) Contributions to game theory and management, vol VII. Graduate School of Management St Petersburg University, Saint Petersburg, pp 51–60

Folmer H, von Mouche PHM (2004) On a less known Nash equilibrium uniqueness result. J Math Soc 28:67–80

Folmer H, von Mouche PHM (2015) Nash equilibria of transboundary pollution games. In: Ruth M (ed) Handbook of research methods and applications in environmental studies. Edward-Elgar, Cheltenham, pp 504–524

Frank CR, Quandt RE (1963) On the existence of Cournot equilibrium. Int Econ Rev 4:92–96

Gaudet G, Salant SW (1991) Uniqueness of Cournot equilibrium: new results from old methods. Rev Econ Stud 58(2):399–404

Iimura T, Watanabe T (2016) Pure strategy equilibrium in finite symmetric concave games and an application to symmetric discrete Cournot games. In: von Mouche P, Quartieri F (eds) Contributions to equilibrium theory for Cournot oligopolies and related games: essays in Honour of Koji Okuguchi. Springer, Cham

von Mouche PHM (2011) On games with constant Nash sum. In: Petrosjan LA, Zenkevich NA (eds) Contributions to game theory and management, vol IV. Graduate School of Management St. Petersburg, Saint Petersburg, pp 294–310

von Mouche PHM, Quartieri F (2012) Equilibrium uniqueness results for Cournot oligopolies revisited. In: Petrosjan LA, Zenkevich NA (eds) Contributions to game theory and management, vol V. Graduate School of Management St. Petersburg, Saint Petersburg, pp 209–229

von Mouche PHM, Quartieri F (2013) On the uniqueness of Cournot equilibrium in case of concave integrated price flexibility. J Glob Opt 57(3):707–718

von Mouche PHM, Quartieri F (2015) Cournot equilibrium uniqueness in case of concave industry revenue: a simple proof. Econ Bull 35(2):1299–1305

Murphy FH, Sherali HD, Soyster AL (1982) A mathematical programming approach for determining oligopolistic market equilibrium. Math Programm 24:92–106

Nikaido H, Isoda K (1955) Note on non-cooperative games. Pac J Math 5:807–815

Okuguchi K, Suzumura K (1971) Uniqueness of the Cournot oligopoly equilibrium: a note. Econ Stud Q 22:81–83

Quartieri, F.: Necessary and sufficient conditions for the existence of a unique Cournot equilibrium. Ph.D. thesis, Siena-Università di Siena, Italy (2008)

Szidarovszky F, Yakowitz S (1982) Contributions to Cournot oligopoly theory. J Econ Theory 28:51–70

Vasin A, Vasina P, Ruleva T (2007) On organization of markets of homogeneous goods. J Comput Syst Sci Int 46:93–106

Vasin A, Kartunova P (2016) Cournot oligopoly theory for simple electricity markets. In: von Mouche P, Quartieri F (eds) Contributions to equilibrium theory for Cournot oligopolies and related games: essays in Honour of Koji Okuguchi. Springer, Cham

Zhou L (1994) The set of Nash equilibria of a supermodular game is a complete lattice. Games Econ Behav 7:295–300

Pure Strategy Equilibria in Finite Symmetric Concave Games and an Application to Symmetric Discrete Cournot Games

Takuya Iimura and Takahiro Watanabe

Abstract We consider a finite symmetric game where the set of strategies for each player is a one-dimensional integer interval. We show that a pure strategy equilibrium exists if the payoff function is concave with respect to the own strategy and satisfies a pair of symmetrical conditions near the symmetric strategy profiles. As an application, we consider a symmetric Cournot game in which each firm chooses an integer quantity of product. It is shown, among other things, that if the industry revenue function is concave, the inverse demand function is nonincreasing, and the cost function is convex, then the payoff function of the firm satisfies the conditions and this symmetric game has a pure strategy equilibrium.

1 Introduction

Many problems in economics can be modeled as games where the sets of strategies for players are one-dimensional compact intervals of real numbers (hereafter referred to as games on real intervals). The strategies correspond to various economic variables, such as selling prices in Bertrand games, quantities of product in Cournot games, and locations of firms in spatial competition games. Moreover, for simplicity, many applications postulate identical players in addition to one-dimensional strategy sets, which are described by symmetric games. If the payoff functions are continuous and concave with respect to the own strategy, then such a concave game on real intervals has a pure strategy equilibrium. This is a result that dates back to Nikaido and Isoda (1955). If the game is also symmetric, then a symmetric pure strategy equilibrium exists, as proved in Moulin (1986) for the more general class of quasiconcave games on real intervals. Results on the existence of pure strategy equilibria for symmetric games on real intervals were extended to those with discontinuous payoffs in Dasgupta and Maskin (1986) and Reny (1999).

T. Watanabe (✉)
Department of Business Administration, Tokyo Metropolitan University, Tokyo 192-0397, Japan
e-mail: contact_nabe10@nabenavi.net

P. von Mouche, F. Quartieri (eds.), *Equilibrium Theory for Cournot Oligopolies and Related Games*, Springer Series in Game Theory,
DOI 10.1007/978-3-319-29254-0_7

This class of games is familiar to us and is useful for formulating problems, but strategies for games in the real world mostly take discrete values. Typically, the prices and quantities of product are integers. Hence, it would be desirable to examine the existence of a pure strategy equilibrium in finite symmetric games where the sets of strategies are one-dimensional finite integer intervals (hereafter referred to as games on integer intervals). The results of some studies (e.g. Kukushkin 1994; Dubey et al. 2006; Jensen 2010) can be applied to this class of games.

We have shown previously that finite symmetric games on integer intervals have pure strategy equilibria when the payoff functions are integrally concave (Iimura and Watanabe, 2014). This result gives a condition for the existence of equilibria, but the concept of integral concavity, defined in Favati and Tardella (1990), is rather technical. In addition, it can be difficult to check whether the payoff functions are integrally concave or not.

In this paper, we extend this result and show that a finite symmetric game on integer intervals has pure strategy equilibria if the payoff function for each player is concave with respect to the own strategy, defined as a nonincreasing marginal payoff with respect to the own strategy, *and* if it satisfies a pair of symmetrical conditions near the symmetric strategy profiles. The pair of conditions is satisfied if the payoff function is integrally concave, as shown in Sect. 5, and checking them is easier than checking integral concavity.

We also apply the results to symmetric Cournot games. There are many studies regarding the existence, uniqueness and stability of equilibria in n-firm Cournot games for one homogeneous good, such as Okuguchi (1964, 1973) and Novshek (1985). In particular, McManus (1964) and Roberts and Sonnenschein (1976) treat Cournot games with n identical firms; that is, symmetric Cournot games. These studies are concerned with Cournot games on real intervals, however. We consider a symmetric Cournot game in which each firm chooses an integer quantity of product. We show that there exists a pure strategy equilibrium if the industry revenue function is concave, the inverse demand function is nonincreasing, and the cost function is convex, among other things. We note that these settings were used in Murphy et al. (1982) for Cournot games on real intervals, and have been well-studied since then; to our knowledge, however, these settings have not been applied to discrete Cournot games.

The rest of the paper is organized as follows. In Sect. 2, we give some definitions. In Sect. 3, we prove our existence theorem. In Sect. 4, we apply the existence theorem to a discrete symmetric Cournot game. In Sect. 5, we give some concluding remarks.

2 Definitions

A game is a three-tuple $(N, (S_i)_{i \in N}, (u_i)_{i \in N})$, where $N = \{1, \ldots, n\}$ $(n \geq 2)$ is a set of players, S_i is a set of strategies for $i \in N$, and u_i is a real-valued payoff function for $i \in N$ defined on the set of strategy profiles $S := S_1 \times \cdots \times S_n$. In this paper, we only consider symmetric games where the set of strategies is a one-dimensional integer interval. The set of strategies for each player $i \in N$ is an identical integer interval $S_i = \{0, \ldots, m\}$ ($m \geq 1$), and u_i satisfies

$$u_i(s) = u_{\pi(i)}(\phi_\pi(s)), \quad \forall \pi \in \Pi, \ \forall s \in S, \ \forall i \in N,$$

where Π is the set of all permutations of N, and ϕ_π is a permutation of $s \in S$ associated with $\pi \in \Pi$ such that

$$\phi_\pi(s) = (s_{\pi^{-1}(1)}, \ldots, s_{\pi^{-1}(n)}).$$

A strategy profile $s = (s_1, \ldots, s_n) \in S$ is also denoted by $s = (s_i, s_{-i})$, where s_i is the strategy of player $i \in N$ and s_{-i} is the $(n-1)$-tuple of the strategies of the other players. Letting e^i be the ith unit vector for each $i \in N$, a strategy profile $(s_i + 1, s_{-i})$ is also denoted by $s + e^i$. A strategy profile $s \in S$ is a pure strategy equilibrium if

$$u_i(s) \geq u_i(s_i', s_{-i}), \quad \forall s_i' \in S_i, \ \forall i \in N.$$

We say that a real-valued function f on the integer interval S_i is *concave* if it satisfies

$$f(x) - f(x-1) \geq f(x+1) - f(x) \tag{1}$$

for all $x \in S_i$. For any function defined on a real interval $[0, m]$, the restriction of the function to the integer interval $\{0, \ldots, m\}$ is concave if the function is concave in the usual sense for continuous functions.

In (1), we have to consider $f(-1)$ and $f(m+1)$, which occur outside the domain of f. To simplify the definition, f in (1) is regarded as an extension of f, the domain of which is extended to $\{-1, \ldots, m+1\}$, the codomain of which is extended to $\mathbb{R} \cup \{-\infty\}$, and we let $f(-1) = f(m+1) = -\infty$. For the same reason, we also consider an extension of u_i to the domain $\{-1, \ldots, m+1\}^n$ and the codomain $\mathbb{R} \cup \{-\infty\}$, where we let $u_i(s) = -\infty$ for any $i \in N$ for which $s_j = -1$ or $s_j = m+1$ for some $j \in N$.

The payoff function u_i is *concave* with respect to the own strategy if $u_i(\cdot, s_{-i})$ is concave for all s_{-i}. In other words, u_i is concave with respect to own strategy if

$$u_i(s) - u_i(s - e^i) \geq u_i(s + e^i) - u_i(s)$$

for all $s \in S$. Thus, the payoff function u_i is concave with respect to the own strategy if and only if the marginal payoff is nonincreasing with respect to the own strategy.

A finite game $(N, (S_i)_{i \in N}, (u_i)_{i \in N})$ is said to be concave if the payoff u_i is concave with respect to the own strategy for every $i \in N$. Note that when $m = 1$—that is, when every player has only two strategies—every finite symmetric game is concave.

As is the case with concavity on a real interval, any local maximum of a concave function defined on an integer interval is a global maximum. Lemma 1 shows this fact formally.

Lemma 1 *Suppose that u_i is concave with respect to the own strategy. Then:*

(i) for all $s \in S$, if $u_i(s) \geq u_i(s - e^i)$, then $u_i(s) \geq u_i(s'_i, s_{-i})$ for any $s'_i \in S_i$ such that $s'_i \leq s_i$, and
(ii) for all $s \in S$, if $u_i(s) \geq u_i(s + e^i)$, then $u_i(s) \geq u_i(s'_i, s_{-i})$ for any $s'_i \in S_i$ such that $s'_i \geq s_i$.

Proof We will show (i). For $s'_i \in S_i$ such that $s'_i < s_i$, let $t = s_i - s'_i$. Then, because u_i is concave with respect to the own strategy, for any $k \in \{1, \ldots, t\}$ we have

$$\begin{aligned} u_i(s - (k-1)e^i) - u_i(s - ke^i) &\geq u_i(s - (k-2)e^i) - u_i(s - (k-1)e^i) \\ &\ \ \vdots \\ &\geq u_i(s) - u_i(s - e^i). \end{aligned}$$

Therefore

$$\begin{aligned} u_i(s) - u_i(s'_i, s_{-i}) &= \sum_{k=1}^{t} \left(u_i(s - (k-1)e^i) - u_i(s - ke^i) \right) \\ &\geq t \left(u_i(s) - u_i(s - e^i) \right). \end{aligned}$$

Thus, if $u_i(s) \geq u_i(s - e^i)$, then $u_i(s) \geq u_i(s'_i, s_{-i})$ and (i) is proved. The proof of (ii) is similar. □

3 Existence of a Pure Strategy Equilibrium

Let $V_z = \{s \in S \mid s_i \in \{z, z+1\},\ i = 1, \ldots, n\}$ for each $z \in \{0, \ldots, m-1\}$. Our main result in the following theorem asserts that there exists an equilibrium in V_z for some $z \in \{0, \ldots, m-1\}$ if the payoff function in a finite symmetric game is concave with respect to the own strategy and satisfies a pair of conditions.

Theorem 1 *Let $(N, (S_i)_{i \in N}, (u_i)_{i \in N})$ be a finite symmetric game. If u_i is concave with respect to the own strategy and satisfies the conditions*

$$u_i(s - e^i) > u_i(s) \implies u_i(s - e^j) > u_i(s - e^j + e^i) \quad \forall s, s - e^j + e^i \in V_z, \tag{2}$$

$$u_i(s) < u_i(s + e^i) \implies u_i(s + e^j - e^i) < u_i(s + e^j) \quad \forall s, s + e^j - e^i \in V_z, \tag{3}$$

for any $i,j \in N$ and $z \in \{0,\dots,m-1\}$, then there exists an equilibrium $s \in S$ in V_x for some $x \in \{0,\dots,m-1\}$.

Proof Let $\hat{s}^z = (z,\dots,z)$ for each $z \in S_1 = \{0,\dots,m\}$. If there exists a z such that $u_1(\hat{s}^z) \geq u_1(\hat{s}^z - e^1)$ and $u_1(\hat{s}^z) \geq u_1(\hat{s}^z + e^1)$, then $u_1(\hat{s}^z) \geq u_1(s_1, \hat{s}^z_{-1})$ for all $s_1 \in S_1$ by Lemma 1. By symmetry, $u_i(\hat{s}^z) \geq u_i(s_i, \hat{s}^z_{-i})$ for all $s_i \in S_i$ and all $i \in N$, so $\hat{s}^z$ is a symmetric equilibrium.

If there is no such z, then by concavity, we have either $u_1(\hat{s}^z - e^1) < u_1(\hat{s}^z) < u_1(\hat{s}^z + e^1)$ or $u_1(\hat{s}^z - e^1) > u_1(\hat{s}^z) > u_1(\hat{s}^z + e^1)$ for each z. Because $u_1(\hat{s}^0) > u_1(\hat{s}^0 - e^1) = -\infty$ and $u_1(\hat{s}^m) > u_1(\hat{s}^m + e^1) = -\infty$, there exists $x \in S_1$, $0 \leq x \leq m-1$, such that $u_1(\hat{s}^x) < u_1(\hat{s}^x + e^1)$ and $u_1(\hat{s}^{x+1} - e^1) > u_1(\hat{s}^{x+1})$. The latter implies that $u_n(\hat{s}^{x+1} - e^n) > u_n(\hat{s}^{x+1})$ by symmetry. Let $s^0, \dots, s^n$ be points in V_x such that $s^0 = \hat{s}^x$ and $s^{k+1} = s^k + e^{k+1}$ for $k = 0,\dots,n-1$. Note that s^k is given by

$$s_i^k = \begin{cases} x+1 & \text{if } i \leq k, \\ x & \text{if } i > k. \end{cases}$$

Then $u_1(s^0) < u_1(s^1)$ and $u_n(s^{n-1}) > u_n(s^n)$. Therefore there exists a $k \in \{1,\dots,n-1\}$ such that

$$u_k(s^{k-1}) < u_k(s^k), \tag{4}$$

$$u_{k+1}(s^k) \geq u_{k+1}(s^{k+1}). \tag{5}$$

We will show that

$$u_{k+1}(s^k) \geq u_{k+1}(s_{k+1}, s^k_{-(k+1)}) \ \forall s_{k+1} \in S_{k+1}$$

and

$$u_k(s^k) \geq u_k(s_k, s^k_{-k}) \ \forall s_k \in S_k,$$

that is, players k and $k+1$ cannot improve their payoffs by deviating from s^k. Because $s_i^k = s_k^k = x+1$ for each player $i \in \{1,\dots,k\}$ and $s_i^k = s_{k+1}^k = x$ for each player $i \in \{k+1,\dots,n\}$, this implies by symmetry that s^k is a pure strategy equilibrium.

Now, $u_{k+1}(s^k) \geq u_{k+1}(s^k + e^{k+1})$ by (5). We also have $u_{k+1}(s^k - e^{k+1}) \leq u_{k+1}(s^k)$. To see this, suppose by way of contradiction that $u_{k+1}(s^k - e^{k+1}) > u_{k+1}(s^k)$. Then (2) implies that $u_{k+1}(s^k - e^k) > u_{k+1}(s^k - e^k + e^{k+1})$. We then have $u_{k+1}(s^k - e^k) = u_k(s^{k-1})$ and $u_{k+1}(s^k + e^{k+1} - e^k) = u_k(s^k)$ by symmetry, and so $u_k(s^{k-1}) > u_k(s^k)$. However, this contradicts (4). Hence $u_{k+1}(s^k) \geq u_{k+1}(s^k + e^{k+1})$ and $u_{k+1}(s^k - e^{k+1}) \leq u_{k+1}(s^k)$. Thus, by Lemma 1, $u_{k+1}(s^k) \geq u_{k+1}(s_{k+1}, s^k_{-(k+1)})$ for all $s_{k+1} \in S_{k+1}$.

Consider (4), that $u_k(s^{k-1}) < u_k(s^k)$, again. This means that $u_k(s^k - e^k) < u_k(s^k)$. We also have $u_k(s^k) \geq u_k(s^k + e^k)$. To see this, suppose by way of contradiction that

$u_k(s^k) < u_k(s^k + e^k)$. Then (3) implies that $u_k(s^k + e^{k+1} - e^k) < u_k(s^k + e^{k+1})$. We have $u_k(s^k+e^{k+1}-e^k) = u_{k+1}(s^k)$ and $u_k(s^k+e^{k+1}) = u_{k+1}(s^{k+1})$ by symmetry, and so $u_{k+1}(s^k) < u_{k+1}(s^{k+1})$. However, this contradicts (5). Hence $u_k(s^k - e^k) < u_k(s^k)$ and $u_k(s^k) \geq u_k(s^k+e^k)$. Therefore, by Lemma 1, $u_k(s^k) \geq u_k(s_k, s^k_{-k})$ for all $s_k \in S_k$. Hence $s^k \in V_x$ is a pure strategy equilibrium. □

For $n = 2$, namely two-person symmetric games, (2) and (3) can be expressed more compactly. For $i = 1$, both s and $s - e^2 + e^1$ belong to V_z if and only if $s = (z, z+1)$, and both s and $s + e^2 - e^1$ belong to V_z if and only if $s = (z+1, z)$. Condition (2) is effective only if $z \geq 1$ because $s - e^1 \notin S$ for $z = 0$. Similarly, (3) is effective only if $z \leq m-2$, because $s + e^1 \notin S$ for $z = m-1$. If (2) and (3) hold for $i = 1$, then they also hold for any $i \in N$ by symmetry. Thus we obtain the following corollary.

Corollary 1 *Let* $(\{1,2\}, (S_i)_{i\in N}, (u_i)_{i\in N})$ *be a two-person finite symmetric game. If* u_1 *is concave with respect to the own strategy and satisfies the conditions*

$$\begin{aligned} &u_1(z-1, z+1) > u_1(z, z+1) \\ &\implies u_1(z,z) > u_1(z+1, z) \quad \forall z \in \{1, \ldots, m-1\}, \end{aligned} \tag{6}$$

$$\begin{aligned} &u_1(z+1, z) < u_1(z+2, z) \\ &\implies u_1(z, z+1) < u_1(z+1, z+1)\ \forall z \in \{0, \ldots, m-2\}, \end{aligned} \tag{7}$$

then there exists an equilibrium $s \in S$ *in* V_x *for some* $x \in \{0, \ldots, m-1\}$.

None of (2), (3), and concavity are redundant for ensuring the existence of an equilibrium. We show this using examples of two-person symmetric games with more than two strategies.[1] In Fig. 1, the symmetric game on the left with three strategies satisfies (7) and concavity, but not (6), while the symmetric game on the right satisfies (6) and concavity, but not (7). Neither of these has a pure strategy equilibrium.

In Fig. 2, the symmetric game with four strategies satisfies (6) and (7), but this game is not concave: the payoff function for player 1 is not concave with respect to the own strategy at $s_2 = 3$. This game also has no pure strategy equilibrium.[2]

[1] Every symmetric game with two strategies has a pure strategy equilibrium as shown in Cheng et al. (2004).

[2] It can be shown that any two-person symmetric game with three strategies satisfying (6) and (7) has a pure strategy equilibrium, even if it is not concave.

1\2	0	1	2
0	(−1, −1)	(−1, 1)	(1, 0)
1	(1, −1)	(0, 0)	(0, 1)
2	(0, 1)	(1, 0)	(−1, −1)

1\2	0	1	2
0	(−1, −1)	(1, 0)	(0, 1)
1	(0, 1)	(0, 0)	(1, −1)
2	(1, 0)	(−1, 1)	(−1, −1)

Fig. 1 The symmetric concave game on the left violates (6) and the symmetric concave game on the right violates (7)

1\2	0	1	2	3
0	(1, 1)	(0, 2)	(−1, 1)	(2, 0)
1	(2, 0)	(1, 1)	(0, 2)	(−1, 1)
2	(1, −1)	(2, 0)	(1, 1)	(0, 2)
3	(0, 2)	(1, −1)	(2, 0)	(1, 1)

Fig. 2 A symmetric game that satisfies (6) and (7), but is not concave

4 An Application: Discrete Cournot Games

Let $G = (N, (S_i)_{i \in N}, (u_i)_{i \in N})$ be a symmetric game such that u_i is defined for each $i \in N$ by

$$u_i(s) = f(s_1 + \cdots + s_n)s_i - c(s_i), \quad s \in S,$$

where f is an *inverse demand function* and c is a *cost function* that is identical for all players (firms). We call G a symmetric discrete Cournot game.

Following Murphy et al. (1982), we call the function $f(y)y$ the *industry revenue function*, where $y = \sum_{i \in N} s_i$ for $s \in S$. The function $f(a+x)x$ of x will be called the individual revenue function, where $x = s_i$ and $a = \sum_{j \neq i} s_j$. The following lemma is a discrete analogue of Lemma 1 in Murphy et al. (1982).

Lemma 2 *The concave industry revenue implies the concave individual revenue if the inverse demand is either a nonincreasing function or a convex function. Namely, for any integers $x \geq 1$, $a \geq 0$, and $y = a + x$,*

$$f(y)y - f(y-1)(y-1) \geq f(y+1)(y+1) - f(y)y$$

implies

$$f(a+x)x - f(a+x-1)(x-1) \geq f(a+x+1)(x+1) - f(a+x)x, \qquad (8)$$

if either

$$f(y-1) \geq f(y) \ \forall y \geq 1 \quad (\textit{nonincreasing}\, f)$$

or

$$f(y) - f(y-1) \leq f(y+1) - f(y) \ \forall y \geq 1 \quad (convex\, f).$$

Proof It is immediate if $a = 0$. Assume that $a \geq 1$ and observe that

$$\begin{aligned} & f(a+x)x - f(a+x-1)(x-1) \\ =& \{f(y)y - f(y)a\} - \{f(y-1)(y-1) - f(y-1)a\} \\ =& \{f(y)y - f(y-1)(y-1)\} - \{f(y) - f(y-1)\}a, \end{aligned} \tag{9}$$

and

$$\begin{aligned} & f(a+x+1)(x+1) - f(a+x)x \\ =& \{f(y+1)(y+1) - f(y+1)a\} - \{f(y)y - f(y)a\} \\ =& \{f(y+1)(y+1) - f(y)y\} - \{f(y+1) - f(y)\}a. \end{aligned} \tag{10}$$

Hence, if $f(y)y$ is concave and $f(y)$ is convex, we have (8). For the case where $f(y)$ is nonincreasing, suppose by way of contradiction that (8) is not satisfied, and subtract (10) from (9), to obtain

$$\begin{aligned} 0 >& \{f(a+x)x - f(a+x-1)(x-1)\} - \{f(a+x+1)(x+1) - f(a+x)x\} \\ =& [\{f(y)y - f(y-1)(y-1)\} - \{f(y+1)(y+1) - f(y)y\}] \\ & - [\{f(y) - f(y-1)\} - \{f(y+1) - f(y)\}]a. \end{aligned}$$

Then, because $\{f(y)y - f(y-1)(y-1)\} - \{f(y+1)(y+1) - f(y)y\} \geq 0$, we have

$$\{f(y) - f(y-1)\} - \{f(y+1) - f(y)\} > 0.$$

However, we also have that

$$\begin{aligned} 0 >& \{f(a+x)x - f(a+x-1)(x-1)\} - \{f(a+x+1)(x+1) - f(a+x)x\} \\ =& \{f(y)x - f(y-1)(x-1)\} - \{f(y+1)(x+1) - f(y)x\} \\ =& [\{f(y) - f(y-1)\} - \{f(y+1) - f(y)\}]x + [f(y-1) - f(y+1)] \\ \geq& [\{f(y) - f(y-1)\} - \{f(y+1) - f(y)\}]x, \end{aligned}$$

which, with $x \geq 1$, implies that

$$\{f(y) - f(y-1)\} - \{f(y+1) - f(y)\} < 0,$$

a contradiction. Hence (8) must hold. □

Now, we prove the following theorem.

Theorem 2 *A symmetric discrete Cournot game has a pure strategy equilibrium if the industry revenue function is concave, the cost function is convex, and the inverse demand function is nonincreasing.*

Proof For $s = (s_i, s_{-i}) \in S$ such that $s_i \in \{1, \dots, m-1\}$ (recall that $S_i = \{0, \dots, m\}$ with $m \geq 1$), let $x = s_i$, $a = \sum_{j \neq i} s_j$, and $y = a + x$. By Lemma 2, the individual revenue $f(a + x)x$ is concave with respect to x. Then u_i is concave with respect to the own strategy, because

$$\begin{aligned} u_i(s) - u_i(s - e^i) &= [f(a+x)x - c(x)] - [f(a+x-1)(x-1) - c(x-1)] \\ &= [f(a+x)x - f(a+x-1)(x-1)] - [c(x) - c(x-1)] \\ &\geq [f(a+x+1)(x+1) - f(a+x)x] - [c(x+1) - c(x)] \\ &= [f(a+x+1)(x+1) - c(x+1)] - [f(a+x)x - c(x)] \\ &= u_i(s + e^i) - u_i(s), \end{aligned}$$

where we have used the inequalities

$$\begin{aligned} f(a+x)x - f(a+x-1)(x-1) &\geq f(a+x+1)(x+1) - f(a+x)x, \\ c(x) - c(x-1) &\leq c(x+1) - c(x). \end{aligned} \tag{11}$$

To see that u_i satisfies (2) of Theorem 1, suppose by way of contradiction that $u_i(s - e^i) > u_i(s)$ and $u_i(s - e^j) \leq u_i(s - e^j + e^i)$. Then

$$\begin{aligned} f(a+x)x - c(x) &< f(a + (x-1))(x-1) - c(x-1) \text{ and} \\ f((a-1) + (x+1))(x+1) - c(x+1) &\geq f((a-1) + x)x - c(x). \end{aligned}$$

Thus, by subtracting and rearranging,

$$f(a+x-1) + [c(x+1) - c(x)] < f(a+x) + [c(x) - c(x-1)].$$

This is impossible because f is nonincreasing and because of (11). Hence (2) follows. Condition (3) can be shown similarly. □

It is well known that if the best reply correspondence of each player has either decreasing or increasing selection, then a pure strategy equilibrium always exists (see Kukushkin 1994, Dubey et al. 2006, and Jensen 2010). We note that the assumptions of Theorem 2 do not imply decreasingness nor increasingness of a selection of the best reply correspondence. To see this, consider the symmetric and discrete Cournot duopoly with $N = \{1, 2\}$, $S = \{0, \dots, 100\}$, $f(y) = \frac{8}{y+10}$, and

$c(x) = \frac{x}{10}$.[3] This Cournot duopoly game satisfies the assumptions of Theorem 2: (i) the industry revenue function is concave, (ii) the cost function is convex, and (iii) the inverse demand is nonincreasing. The best reply correspondence of player 1, b_1, has the following values:

$$b_1(30) = \{17\}, b_1(20) = \{19\}, b_1(0) = \{18\}.$$

To check the values of b_1 it suffices to verify that

$$\begin{array}{ll} u_1(17,30) - u_1(16,30) = \frac{1}{3990} > 0, & u_1(18,30) - u_1(17,30) = -\frac{53}{16530} < 0, \\ u_1(19,20) - u_1(18,20) = \frac{1}{490} > 0, & u_1(20,20) - u_1(19,20) = -\frac{1}{490} < 0, \\ u_1(18,0) - u_1(17,0) = \frac{11}{1890} > 0, & u_1(19,0) - u_1(18,0) = -\frac{3}{2030} < 0. \end{array}$$

Then, we find that any selection of b_1 is neither decreasing nor increasing.

By Lemma 2, the concavity of u_i is also obtained under the concave industry revenue, the convex inverse demand, and the convex cost. However, for the conditions (2) and (3) to be satisfied by u_i, we also need a nonincreasing inverse demand function.

Another sufficient set of conditions is nonincreasing concave inverse demand and convex cost without assuming concave industry revenue. In fact, we have

$$\begin{aligned} & u_i(s) - u_i(s - e^i) \\ =& (f(a+x)x - c(x)) - (f(a+x-1)(x-1) - c(x-1)) \\ =& (f(a+x) - f(a+x-1))x + f(a+x-1) - (c(x) - c(x-1)) \\ \geq& (f(a+x+1) - f(a+x))x + f(a+x+1) - (c(x+1) - c(x)) \\ =& (f(a+x+1)(x+1) - c(x+1)) - (f(a+x)x - c(x)) \\ =& u_i(s+e^i) - u_i(s), \end{aligned}$$

by using the corresponding inequalities. Conditions (2) and (3) follow from f being nonincreasing and the convexity of c, as shown in the proof.

A simple algorithm for locating an equilibrium is the following. First, search through the symmetric profiles to find a symmetric equilibrium by checking whether

$$u_1(\hat{s}^z - e^1) \leq u_1(\hat{s}^z) \quad \text{and} \quad u_1(\hat{s}^z) \geq u_1(\hat{s}^z - e^1)$$

(see the proof of Theorem 1). Second, if this fails, find two adjacent symmetric profiles $\hat{s}^x$ and $\hat{s}^{x+1}$ such that

$$u_1(\hat{s}^x) < u_1(\hat{s}^x + e^1) \quad \text{and} \quad u_1(\hat{s}^{x+1} - e^1) > u_1(\hat{s}^{x+1}),$$

[3]The authors owe this example to Federico Quartieri.

and then search through $s^1, s^2, \ldots s^{n-1}$ as defined in the proof of Theorem 1 to find an asymmetric equilibrium. Our result is also applicable to the restriction of a symmetric continuous Cournot game to the integer lattice. For example, suppose that the inverse demand function and the cost function are given by

$$f(y) = 2000 - 0.5y^{1.6} \quad \text{and} \quad c(x) = 10 + 0.01x^{1.2}.$$

This is an example of a nonincreasing concave function f and a convex function c. Then, letting $S_i = \{0, \ldots, 180\}$, our algorithm locates an equilibrium for the three-firm symmetric discrete Cournot game at $(46, 46, 45)$.

5 Concluding Remarks

As mentioned in the introduction, Theorem 1 is an extension of the result that finite symmetric games on integer intervals have pure strategy equilibria when the payoff functions are integrally concave (Iimura and Watanabe, 2014). In the following we show this formally.

Denote by conv(S) the convex hull of S, that is, $\text{conv}(S) = [0, m]^n$. Let $\mathrm{N}(y) := \{z \in S \mid |z_i - y_i| < 1\ \forall i = 1, \ldots, n\}$ for $s \in S$. The *local concave extension* of the payoff function $u_i : S \to \mathbb{R}$ is a piecewise-linear function $\bar{u}_i : \text{conv}(S) \to \mathbb{R}$ defined for each $y \in \text{conv}(S)$ (using the points $z \in \mathrm{N}(y)$) by

$$\bar{u}_i(y) := \max\{\sum_{z\in\mathrm{N}(y)} \alpha_z f(z) \mid \sum_{z\in\mathrm{N}(y)} \alpha_z z = y,\ \sum_{z\in\mathrm{N}(y)} \alpha_z = 1,\ \alpha_z \geq 0 \text{ for all } z \in \mathrm{N}(y)\}.$$

Definition 1 (Favati and Tardella 1990) u_i is *integrally concave* if its local concave extension is concave on conv(S).

If u_i is integrally concave, then u_i satisfies (2) and (3). To see this, suppose by way of contradiction that u_i does not satisfy (2) for some $z \in V_z$. Then $u_i(s - e^i) > u_i(s)$ and $u_i(s - e^j + e^i) \geq u_i(s - e^j)$ for some $j \neq i$. Then, considering the local concave extension $\bar{u}_i$ of u_i, we have

$$\begin{aligned}\frac{u_i(s - e^i) + u_i(s - e^j + e^i)}{2} &> \frac{u_i(s) + u_i(s - e^j)}{2}\\ &= \bar{u}_i\Big(\frac{s + (s - e^j)}{2}\Big)\\ &= \bar{u}_i\Big(\frac{(s - e^i) + (s - e^j + e^i)}{2}\Big),\end{aligned}$$

where the first equality follows from the fact that $\mathrm{N}\left(\frac{s+(s-e^j)}{2}\right) = \{s, s - e^j\}$. On the other hand, the concavity of $\bar{u}_i$ implies that

$$\frac{\bar{u}_i(s - e^i) + \bar{u}_i(s - e^j + e^i)}{2} \leq \bar{u}_i\left(\frac{(s - e^i) + (s - e^j + e^i)}{2}\right).$$

This is a contradiction and hence we conclude that u_i satisfies (2). We can similarly prove that u_i satisfies (3).

As a final remark, we note that Theorem 1 is also applicable to symmetric discrete games whose payoff functions are of the form

$$u_i(s) = f(s_i) + g(\sum_{j \in N} s_j),$$

where f and g are two increasing concave functions. This is the case, in particular, for games of private provision of pure public goods.

Acknowledgements This work was supported by JSPS KAKENHI Grant Number 25380233. The authors thank valuable comments from the referees and editors. Especially, we thank Federico Quartieri for providing a good example. Of course all remaining errors are our responsibility.

References

Cheng SF, Reeves DM, Vorobeychik Y, Wellman MP (2004) Notes on equilibria in symmetric games. In: Proceedings of the 6th international workshop on game theoretic and decision theoretic agents, New York

Dasgupta P, Maskin E (1986) The existence of equilibrium in discontinuous economic games, I: theory. Rev Econ Stud 53(1):1–26

Dubey P, Haimanko O, Zapechelnyuk A (2006) Strategic complements and substitutes, and potential games. Games Econ Behav 54(1):77–94

Favati P, Tardella F (1990) Convexity in nonlinear integer programming. Ric Oper 53:3–44

Iimura T, Watanabe T (2014) Existence of a pure strategy equilibrium in finite symmetric games where payoff functions are integrally concave. Discret Appl Math 166:26–33

Jensen M (2010) Aggregative games and best-reply potentials. Econ Theory 43(1):45–66

Kukushkin NS (1994) A fixed-point theorem for decreasing mappings. Econ Lett 46(1):23–26

McManus M (1964) Equilibrium, numbers and size in Cournot oligopoly. Yorks Bull Soc Econ Res 16(2):68–75

Moulin H (1986) Game theory for the social sciences. NYU Press, New York

Murphy F, Sherali H, Soyster A (1982) A mathematical programming approach for determining oligopolistic market equilibrium. Math Programm 24(1):92–106

Nikaido H, Isoda K (1955) Note on noncooperative convex games. Pac J Math 5(1):807–815

Novshek W (1985) On the existence of Cournot equilibrium. Rev Econ Stud 52(1):85–98

Okuguchi K (1964) The stability of the Cournot oligopoly solution: a further generalization. Rev Econ Stud 31(2):143–146

Okuguchi K (1973) Quasi-competitiveness and Cournot oligopoly. Rev Econ Stud 40(1):145–148

Reny PJ (1999) On the existence of pure and mixed strategy Nash equilibria in discontinuous games. Econometrica 67(5):1029–1056

Roberts J, Sonnenschein H (1976) On the existence of Cournot equilibrium without concave profit functions. J Econ Theory 13(1):112–117

On a Discontinuous Cournot Oligopoly

Ferenc Szidarovszky and Akio Matsumoto

Abstract A modified Cournot oligopoly is introduced and examined in which the firms can treat their wastes up to a certain amount, and if the amount of waste is even higher then an outside facility is used with a given fixed cost and higher unit cost. The resulting payoff functions become discontinuous. The best response functions of firms are nonincreasing and also might be discontinuous. The best response functions can be modified to depend on the total industry output, which also might be discontinuous as well as multiple valued. The existence of at least one equilibrium is proved and numerical examples show the possibility of a unique equilibrium as well as that of multiple equilibria.

1 Introduction

Oligopoly theory is one of the most frequently studied subjects in mathematical economics. It dates back to the pioneering work of Cournot (1838) and since then a large number of researchers devoted their efforts to this interesting area. The monograph of Okuguchi (1976) is considered as one of the most important summary of the earlier results which also includes some of his own fundamental contributions. The multi-product extension of the classical theory with some applications is presented in Okuguchi and Szidarovszky (1999). In recent years nonlinear models occupied the main research focus, a comprehensive summary of the more recent developments is presented in Bischi et al. (2010). Most earlier models assumed differentiable payoff functions, where the analytic treatment was straightforward. In Szidarovszky and Yakowitz (1982) only the continuity of the price and cost functions are assumed with the possibility of infinitely many equilibria where the industry output is unique. The uniqueness of the equilibrium is guaranteed if the price function is differentiable at this point. More recently, Zhao and Szidarovszky (2008) introduced production adjustment costs resulting in non-differentiable payoff functions, where the existence of usually infinitely many equilibria is proved.

F. Szidarovszky (✉)
Department of Applied Mathematics, University of Pécs, Ifjúság u. 6., H-7624, Pécs, Hungary
e-mail: szidarka@gmail.com

P. von Mouche, F. Quartieri (eds.), *Equilibrium Theory for Cournot Oligopolies and Related Games*, Springer Series in Game Theory,
DOI 10.1007/978-3-319-29254-0_8

Similar situation is found in Burr et al. (2014) where the output adjustments are limited from both above and below. The payoff functions are non-differentiable and the best responses discontinuous in this case and there are again infinitely many equilibria except in special cases.

In this paper another variant of the Cournot model is introduced, where the payoff functions are discontinuous, therefore the best responses as functions of the output of the rest of the industry are also discontinuous. Furthermore the best responses as functions of the total industry output are not only discontinuous but might have multiple values.

This paper develops as follows. In Sect. 2, the mathematical model is introduced and the best responses are determined. In Sect. 3, the existence of at least one equilibrium is proved and numerical examples show the possibility of both unique and multiple equilibria. Section 4 concludes the paper and further research directions are mentioned.

2 The Model and Best Responses

Consider an N-firm single-product oligopoly without product differentiation. For mathematical simplicity assume linear price and cost functions,

$$p(s) = A - Bs$$

and

$$C_k(x_k) = c_k x_k$$

where x_k is the output of firm k and $s = \sum_{k=1}^{N} x_k$ is the industry output. Assume that each firm produces some waste proportional to its production level, which can be cleaned or deposited by the firm until a certain amount, and if the waste amount is larger than a certain threshold, then it has to be shipped to be cleaned or deposited by a contractor with higher unit cost than that if the firm itself does the cleaning or depositing. In addition the firm also has to pay a certain fixed cost, which can be interpreted as the setup or transportation cost. So the payoff of firm k can be written as follows:

$$\varphi_k = x_k(A - Bs_k - Bx_k) - c_k x_k - \begin{cases} \alpha_k x_k & \text{if } x_k \leq K_k, \\ a_k + \beta_k x_k & \text{if } x_k > K_k \end{cases} \tag{1}$$

where $s_k = \sum_{\ell \neq k}^{N} x_\ell$ is the output of the rest of the industry for firm k, α_k and β_k are the unit costs of cleaning or depositing the waste per unit production (since amount of waste is proportional to the output level), and a_k is the fixed cost and K_k

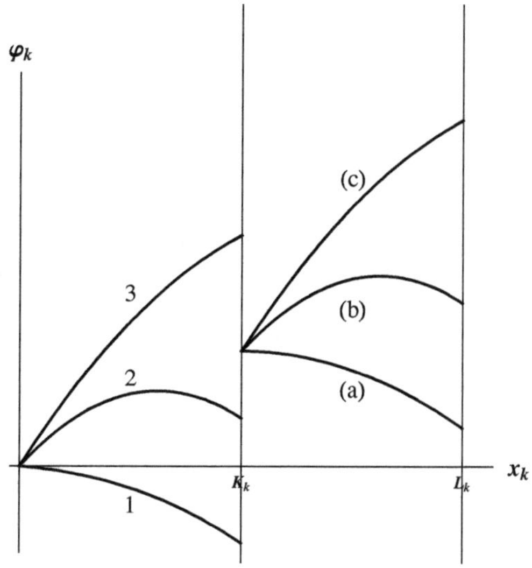

Fig. 1 Possible shapes of the payoff functions

is maximum output level which generates the maximum waste amount that the firm can treat. There are several possibilities for the shape of φ_k, which are summarized in Fig. 1, where L_k is the capacity limit of the firm.

Case 1: $\frac{\partial \varphi_k}{\partial x_k} \leq 0$ at $x_k = 0$.
This is the case when

$$A - Bs_k - c_k - \alpha_k \leq 0$$

that is,

$$s_k \geq \frac{A - c_k - \alpha_k}{B}. \tag{2}$$

In this case the best response of the firm is

$$R_k(s_k) = 0. \tag{3}$$

Case 2: $\frac{\partial \varphi_k}{\partial x_k} > 0$ at $x_k = 0$ and $\frac{\partial \varphi_k}{\partial x_k} \leq 0$ at $x_k = K_k-$.
That is, when (2) is violated with strict inequality and

$$A - Bs_k - 2BK_k - c_k - \alpha_k \leq 0,$$

so

$$\frac{A - c_k - \alpha_k}{B} - 2K_k \leq s_k < \frac{A - c_k - \alpha_k}{B}, \tag{4}$$

and in this case the best response is the stationary point:

$$R_k(s_k) = \frac{A - c_k - \alpha_k}{2B} - \frac{s_k}{2}. \tag{5}$$

Case 3: $\frac{\partial \varphi_k}{\partial x_k} > 0$ at both $x_k = 0$ and $x_k = K_k-$.
So

$$s_k < \frac{A - c_k - \alpha_k}{B} - 2K_k. \tag{6}$$

There are three subcases depending on the signs of the derivative

$$\frac{\partial \varphi_k}{\partial x_k} \text{ at } x_k = K_{k+} \text{ and } x_k = L_k.$$

Case 3(a): $\frac{\partial \varphi_k}{\partial x_k} \leq 0$ at $x_k = K_{k+}$.
This can be written as

$$s_k \geq \frac{A - c_k - \beta_k}{B} - 2K_k, \tag{7}$$

and clearly

$$R_k(s_k) = K_k \tag{8}$$

in this case.
Case 3(b): $\frac{\partial \varphi_k}{\partial x_k} > 0$ at $x_k = K_{k+}$ and $\frac{\partial \varphi_k}{\partial x_k} < 0$ at $x_k = L_k$.
That is,

$$\frac{A - c_k - \beta_k}{B} - 2L_k < s_k < \frac{A - c_k - \beta_k}{B} - 2K_k. \tag{9}$$

The stationary point between K_k and L_k is

$$x_k^* = \frac{A - c_k - \beta_k}{2B} - \frac{s_k}{2} \tag{10}$$

and now the function values $\varphi_k(x_k^*)$ and $\varphi_k(K_k)$ have to be compared. Notice first that

$$\varphi_k(x_k^*) = \left(\frac{A - c_k - \beta_k}{2B} - \frac{s_k}{2}\right)\left(A - Bs_k - \frac{A - c_k - \beta_k - Bs_k}{2} - c_k - \beta_k\right) - a_k$$

$$= \frac{B}{4}\left(\frac{A - c_k - \beta_k}{B} - s_k\right)^2 - a_k.$$

Since

$$\varphi_k(K_k) = K_k\left(A - Bs_k - BK_k - c_k - \alpha_k\right),$$

we have

$$\varphi_k(x_k^*) > \varphi_k(K_k)$$

if and only if

$$\left(\frac{A - c_k - \beta_k}{B} - s_k\right)^2 > \frac{4K_k}{B}\left(A - Bs_k - BK_k - c_k - \alpha_k\right) + \frac{4a_k}{B},$$

which is a quadratic inequality for s_k:

$$0 < s_k^2 + s_k\left(4K_k - 2\frac{A - c_k - \beta_k}{B}\right)$$

$$+\left(-\frac{4AK_k}{B} + 4K_k^2 + \frac{4K_k(c_k + \alpha_k)}{B} + \left(\frac{A - c_k - \beta_k}{B}\right)^2 - \frac{4a_k}{B}\right).$$

The discriminant of the right hand side is

$$D = \frac{16}{B}\left(a_k + K_k(\beta_k - \alpha_k)\right),$$

so its two roots are

$$s_k^{\pm} = \frac{A - c_k - \beta_k}{B} - 2K_k \pm \sqrt{\frac{4}{B}\left(a_k + K_k(\beta_k - \alpha_k)\right)}. \tag{11}$$

The root s_k^+ violates (9), so the only feasible root is s_k^-, if it satisfies the left hand side of (9),

$$\frac{A - c_k - \beta_k}{B} - 2K_k - \sqrt{\frac{4}{B}\left(a_k + K_k(\beta_k - \alpha_k)\right)} > \frac{A - c_k - \beta_k}{B} - 2L_k$$

which can be rewritten as

$$L_k - K_k > \sqrt{\frac{a_k + K_k(\beta_k - \alpha_k)}{B}}. \tag{12}$$

In the case of $s_k < s_k^-$,

$$R_k(s_k) = x_k^*.$$

If $s_k > s_k^-$, then

$$R_k(s_k) = K_k$$

and if $s_k = s_k^-$, then

$$R_k(s_k) = \{K_k, x_k^*\}.$$

If s_k^- is not feasible, then

$$s_k^- \leq \frac{A - c_k - \beta_k}{B} - 2L_k,$$

so in the entire interval (9),

$$R_k(s_k) = K_k.$$

Notice that s_k^- satisfies the right hand side of (9).
Case 3(c): $\frac{\partial \varphi_k}{\partial x_k} \geq 0$ at $x_k = L_k$.
So

$$0 \leq s_k \leq \frac{A - c_k - \beta_k}{B} - 2L_k \tag{13}$$

and in this case both K_k and L_k might be best response, so we have to compare the values of $\varphi_k(K_k)$ and $\varphi_k(L_k)$. Notice first that

$$\varphi_k(K_k) > \varphi_k(L_k)$$

if and only if

$$K_k\,(A - Bs_k - BK_k - c_k - \alpha_k) > L_k\,(A - Bs_k - BL_k - c_k - \beta_k) - a_k$$

which can be written as

$$s_k > s_k^* = \frac{A - c_k - \beta_k}{B} - (L_k + K_k) - \frac{a_k + K_k(\beta_k - \alpha_k)}{B(L_k - K_k)}. \tag{14}$$

It is easy to see that

$$s_k^* \leq \frac{A - c_k - \beta_k}{B} - 2L_k$$

if and only if

$$L_k - K_k \leq \sqrt{\frac{a_k + K_k(\beta_k - \alpha_k)}{B}}. \tag{15}$$

It is easy to see that $s_k^* \leq s_k^-$, and equality holds if and only if

$$s_k^- = s_k^* = \frac{A - c_k - \beta_k}{B} - 2L_k.$$

In comparing (12) and (15), we can conclude the followings:

- If s_k^- is interior in interval (9), then (12) holds with strict inequality, then (15) is violated, so

$$s_k^* > \frac{A - c_k - \beta_k}{B} - 2L_k$$

  implying that $R_k(s_k) = L_k$ in interval (13).
- If

$$s_k^- = \frac{A - c_k - \beta_k}{B} - 2L_k,$$

  then (12) holds with equality, so (15) implies that $s_k^- = s_k^*$ at this point, so

$$R_k(s_k^-) = \{K_k; L_k\}$$

  and

$$R_k(s_k) = L_k \text{ again in all other points of interval (13).}$$

- If

$$s_k^- < \frac{A - c_k - \beta_k}{B} - 2L_k,$$

  then (12) is violated with strict inequality, so

$$s_k^* < \frac{A - c_k - \beta_k}{B} - 2L_k.$$

- If $s_k^* > 0$, then

$$R_k(s_k^*) = \{K_k; L_k\},$$

$$R_k(s_k) = \begin{cases} K_k & \text{for } s_k > s_k^* \\ L_k & \text{for } s_k < s_k^*. \end{cases}$$

- If $s_k^* = 0$, then

$$R_k(s_k) = K_k \text{ for all positive } s_k \text{ values in interval (13),}$$

$$R_k(0) = \{K_k; L_k\}.$$

- If $s_k^* < 0$, then

$$R_k(s_k) = K_k \text{ for all } s_k \text{ from interval (13).}$$

The possible shapes of $R_k(s_k)$ are shown in Fig. 2, where 2(a) represents the case when neither s_k^* nor s_k^- is feasible, part 2(b) shows the case, when only s_k^- is feasible and 2(c) the case when only s_k^* is feasible which includes the extreme case of

$$s_k^* = s_k^- = \frac{A - c_k - \beta_k}{B} - 2L_k.$$

In case (a) the best response function is continuous, however in cases (b) and (c), it is discontinuous by having jump at $s_k = \bar{s}_k$ and $s_k = s_k^*$, respectively.

We can also rewrite the best responses as functions of the total industry output s. In case 1, $\bar{R}_k(s) = 0$ with

$$s \geq \frac{A - c_k - \alpha_k}{B}. \tag{16}$$

In case 2, from (5) we have

$$s = s_k + x_k = \frac{A - c_k - \alpha_k}{2B} + \frac{s_k}{2}$$

therefore the range of s is

$$\frac{A - c_k - \alpha_k}{B} - K_k \leq s < \frac{A - c_k - \alpha_k}{B} \tag{17}$$

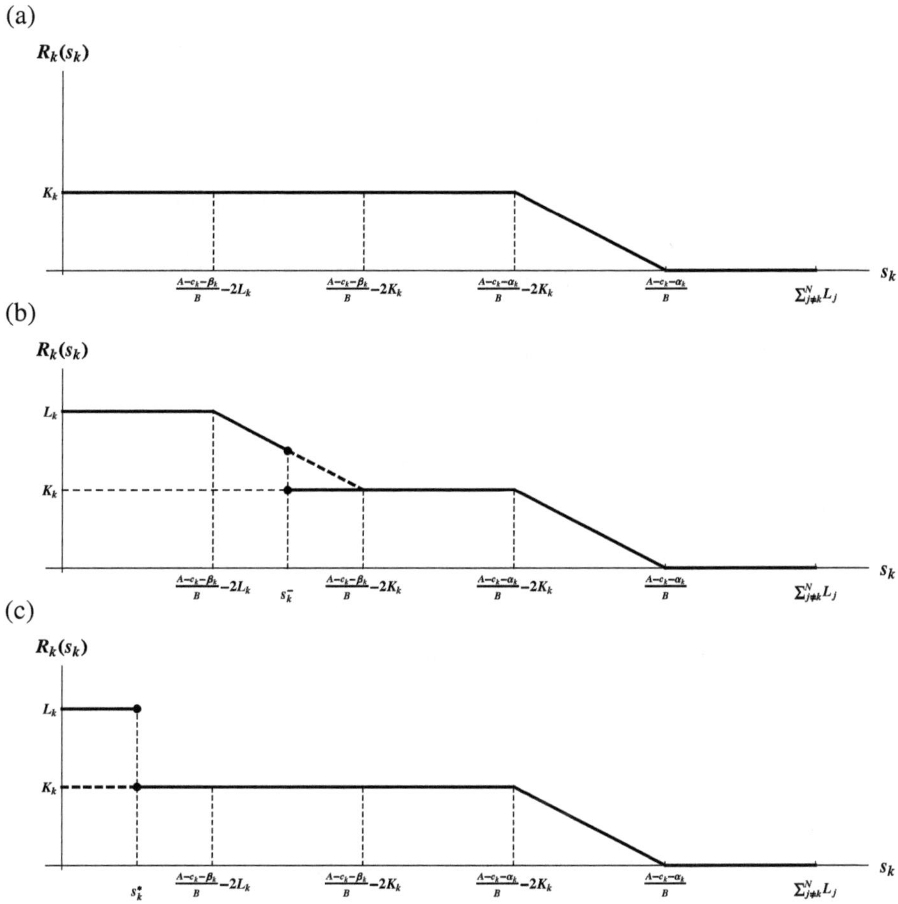

Fig. 2 Possible shapes of best responses

and the best response satisfies equation

$$x_k = \frac{A - c_k - \alpha_k}{2B} - \frac{s - x_k}{2}$$

which implies that

$$\bar{R}_k(s) = \frac{A - c_k - \alpha_k}{B} - s. \tag{18}$$

In case 3(a), $\bar{R}_k(s) = K_k$ and since s_k satisfies (6) and (7),

$$\frac{A - c_k - \beta_k}{B} - K_k \leq s < \frac{A - c_k - \alpha_k}{B} - K_k. \tag{19}$$

In case 3(b) we have two subcases. If s_k^- is not feasible, then $\bar{R}_k(s_k) = K_k$ and from (9),

$$\frac{A - c_k - \beta_k}{B} - 2L_k + K_k < s < \frac{A - c_k - \beta_k}{B} - K_k.$$

If s_k^- is feasible, then in the case of $s_k > s_k^-$, $\bar{R}_k(s) = K_k$ with the domain

$$s_k^- + K_k < s < \frac{A - c_k - \beta_k}{B} - K_k. \tag{20}$$

If $s_k < s_k^-$, then $\bar{R}_k(s) = x_k^*$, so from (10),

$$s = x_k^* + s_k = \frac{A - c_k - \beta_k}{2B} + \frac{s_k}{2}$$

with domain

$$\frac{A - c_k - \beta_k}{B} - L_k < s < \frac{A - c_k - \beta_k}{2B} + \frac{s_k^-}{2} \tag{21}$$

and the best response is obtained from equation

$$s = \frac{A - c_k - \beta_k}{2B} + \frac{s - x_k}{2}$$

so

$$\bar{R}_k(s) = \frac{A - c_k - \beta_k}{B} - s. \tag{22}$$

In case 3(c) we also have two subcases. Assume first that s_k^* is not feasible, then $R_k(s_k) = K_k$ if s_k^- is not feasible as well, or $R_k(s_k) = L_k$ if s_k^- is feasible. In the first case,

$$K_k \leq s \leq \frac{A - c_k - \beta_k}{B} - 2L_k + K_k, \; \bar{R}_k(s) = K_k \tag{23}$$

and in the second case

$$L_k \leq s \leq \frac{A - c_k - \beta_k}{B} - L_k, \; \bar{R}_k(s) = L_k. \tag{24}$$

If s_k^* is feasible, then $R_k(s_k^*) = \{K_k; L_k\}$, $R_k(s_k) = K_k$ for $s_k > s_k^*$ and $R_k(s_k) = L_k$ for $s_k < s_k^*$. The last two cases are as follows:

$$s_k^* + K_k \leq s \leq \frac{A - c_k - \beta_k}{B} - 2L_k + K_k, \; \bar{R}_k(s) = K_k \tag{25}$$

and

$$L_k \le s \le s_k^* + L_k,\ \bar{R}_k(s) = L_k. \tag{26}$$

Figure 3 shows the possible shapes of $\bar{R}_k(s)$, where part (a) gives the case when neither s_k^- nor s_k^* is feasible, part (b) shows the graph when only s_k^- is feasible and part (c) is the case when only s_k^* is feasible or the border line case of

$$s_k^- = s_k^* = \frac{A - c_k - \beta_k}{B} - 2L_k.$$

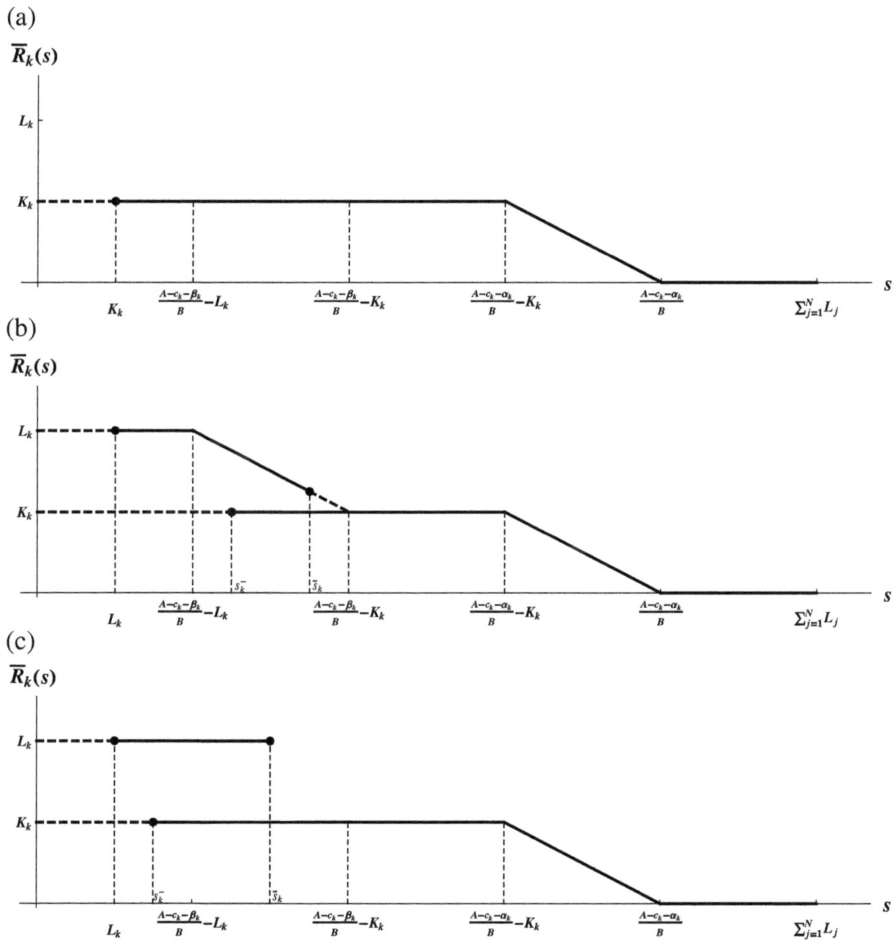

Fig. 3 Possible shapes of $\bar{R}_k(s)$

In particular cases one or more segments for smaller or larger values of s might be missing.

In Fig. 3b

$$\bar{s}_k = s_k^- + K_k \text{ and } \tilde{s}_k = \frac{A - c_k - \beta_k}{2B} + \frac{s_k^-}{2}.$$

Notice that from (9),

$$\tilde{s}_k = \frac{A - c_k - \beta_k}{2B} + \frac{s_k^-}{2} \leq \frac{A - c_k - \beta_k}{2B} + \frac{A - c_k - \beta_k}{2B} - K_k = \frac{A - c_k - \beta_k}{B} - K_k$$

and

$$\tilde{s}_k \geq \frac{A - c_k - \beta_k}{2B} + \frac{A - c_k - \beta_k}{2B} - L_k = \frac{A - c_k - \beta_k}{B} - L_k.$$

In addition,

$$\tilde{s}_k = \frac{A - c_k - \beta_k}{2B} + \frac{s_k^-}{2} > s_k^- + K_k = \bar{s}_k$$

since it can be written as

$$s_k^- < \frac{A - c_k - \beta_k}{B} - 2K_k,$$

which is true by (11).

Notice that in Fig. 3b,c there are jumps between the left end points of the lower segments and the right end points of the upper segments with 45° slopes, since in Fig. 3b, by (10),

$$x_k^*(\tilde{s}_k) - K_k = \left(\frac{A - c_k - \beta_k}{B} - \tilde{s}_k\right) - K_k = \frac{A - c_k - \beta_k}{2B} - \frac{s_k^-}{2} - K_k$$

and

$$\tilde{s}_k - \bar{s}_k = \left(\frac{A - c_k - \beta_k}{2B} + \frac{s_k^-}{2}\right) - \left(s_k^- + K_k\right)$$

are equal. Furthermore in Fig. 3c,

$$\bar{s}_k = s_k^* + K_k \text{ and } \tilde{s}_k = s_k^* + L_k$$

$$\tilde{s}_k - \bar{s}_k = (s_k^* + L_k) - (s_k^* + K_k) = L_k - K_k.$$

3 Equilibrium Analysis

The equilibrium is the solution of equation

$$\sum_{k=1}^{N} \bar{R}_k(s) = s. \tag{27}$$

Consider the left hand side, which can be denoted by $H(s)$. With any value $s > 0$ an $\bar{R}_k(s)$ exists if $s \geq R_k(0)$. So $\left[R_k(0),\ \sum_{\ell=1}^{N} L_\ell\right]$ is the interval for s such that $\bar{R}_k(s)$ is defined. So all $\bar{R}_k(s)$ values exist if s is greater than or equal to the largest value of the left end points of these intervals. Clearly, at the minimal s value $H(s) \geq s$, since for at least one k, that minimal value equals s. At $s = \sum_{\ell=1}^{N} L_\ell$, the value of $H(s)$ is below s, since for all k, $\bar{R}_k(s) \leq L_k$. The only way of having no equilibrium is when the 45 degree line skips through a jump created by at least one function $\bar{R}_k(s)$.

We will next show that $\sum_{k=1}^{N} \bar{R}_k(s)$ cannot have jumps with more than 45° slope. The simple structure of the best response function implies the following simple fact. Let s_A and s_B be two points of the domain of $H(s)$ such that $s_A < s_B$, and let $\bar{R}_B$ be the value of $\bar{R}(s_B)$ or one of the values. Then there is a value $\bar{R}_A$ of $\bar{R}(s_A)$ such that $\bar{R}_A \geq \bar{R}_B$. Clearly the same holds if we add up some or all of the $\bar{R}_k(s)$ functions. Let now $g_1(s)$ be sum of some $\bar{R}_k(s)$ functions and $g_2(s)$ the sum of all others. We will show that the slopes of the jumps in $g_1(s)$ $(g_2(s))$ cannot increase by adding $g_2(s)$ $(g_1(s))$, that is, the slopes of the jumps of $H(s)$ cannot exceed 45°. Consider now a jump of $g_1(s)$ with points, (A, R_A) and (B, R_B) such that $A < B$ and $R_A < R_B$. Let R'_B be a value of $g_2(s)$ at $s = B$, so the corresponding point of $g_1(s) + g_2(s)$ is $(B, R_B + R'_B)$. However there is a point (A, R'_A) on $g_2(s)$ such that $R'_A \geq R'_B$, and the corresponding point on $g_1(s) + g_2(s)$ is $(A, R_A + R'_A)$, so the height of the jump between these points becomes

$$\left(R_B + R'_B\right) - \left(R_A + R'_A\right) \leq R_B - R_A,$$

so the height of this new jump cannot be larger than that of the jump of $g_1(s)$ in $[A, B]$, so its slope cannot be larger either. Therefore adding the best responses $\bar{R}_k(s)$ the slopes of their jumps cannot increase, so the 45° line must cross the curve of $H(s)$ implying the existence of at least one equilibrium.

If all firms can be described by Fig. 3a, then there is a unique equilibrium. The next two examples show cases of a unique equilibrium and multiple equilibria.

Example 1 Assume $N = 2,\ A = 12,\ B = 2,\ a_k = 1,\ c_k = 3,\ \alpha_k = 1,\ \beta_k = 7,\ K_k = 0.5$ and $L_k = 2$ for both firms. In this case $s_1 = x_2,\ s_2 = x_1$, and

$$\sum_{k=1}^{2} L_k = 4,\ \frac{A - c_k - \alpha_k}{B} = 4,\ \frac{A - c_k - \alpha_k}{B} - 2K_k = 3,$$

$$\frac{A - c_k - \beta_k}{B} - 2K_k = 0,\ \frac{A - c_k - \beta_k}{B} - 2L_k = -3.$$

The two best response functions are shown in Fig. 4, from which it is clear that $x_1 = x_2 = 0.5$ is the unique equilibrium.

Example 2 Assume again $N = 2,\ A = 14,\ B = 2,\ a_k = \alpha_k = c_k = 1,\ \alpha_k = 1,\ K_k = 1,\ \beta_k = L_k = 2$. Then

$$\frac{A - c_k - \alpha_k}{B} = 6,\ \frac{A - c_k - \alpha_k}{B} - 2K_k = 4,$$

$$\frac{A - c_k - \beta_k}{B} - 2K_k = 3.5,\ \frac{A - c_k - \beta_k}{B} - 2L_k = 1.5$$

$$s_k^- = s_k^* = 1.5.$$

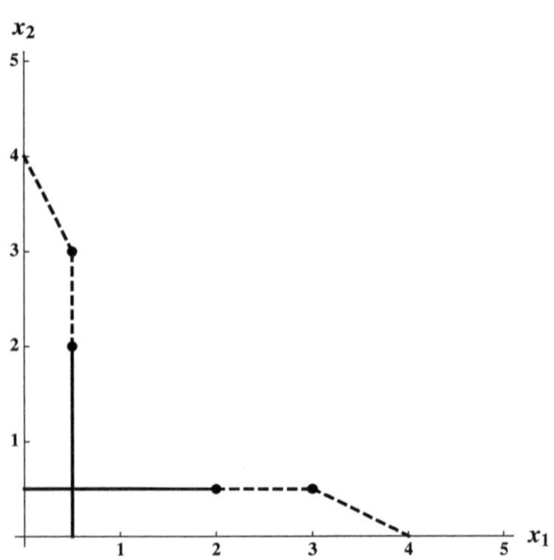

Fig. 4 Equilibrium in Example 1

So

$$R_k(s_k) = \begin{cases} 2 & \text{if } 0 \le s_k < 1.5, \\ \{1;2\} & \text{if } s_k = 1.5, \\ 1 & \text{if } s_k > 1.5. \end{cases}$$

Consider next the duopoly when both firms have the same parameters as given above. In this case $s_1 = x_2$ and $s_2 = x_1$. Their best responses are shown in Fig. 5, where notice that $\sum_{\ell \neq k} L_\ell = 2$ for both firms. It is clear that $x_1 = 1, x_2 = 2$ and $x_1 = 2, x_2 = 1$ are the two equilibria shown by points E_1 and E_2 in the figure.

Example 3 We can redo the previous examples by using the best responses as functions of the total industry output.

(a) In the case of Example 1,

$$\frac{A - c_k - \alpha_k}{B} = 4, \quad \frac{A - c_k - \alpha_k}{B} - K_k = 3.5,$$
$$\frac{A - c_k - \beta_k}{B} - K_k = 0.5, \quad \frac{A - c_k - \beta_k}{B} - L_k < 0.$$

Function $\bar{R}_k(s)$ is shown in Fig. 6. In a duopoly with the identical firms with the above parameters,

$$\sum_{k=1}^{2} \bar{R}_k(s) = 2\bar{R}_k(s),$$

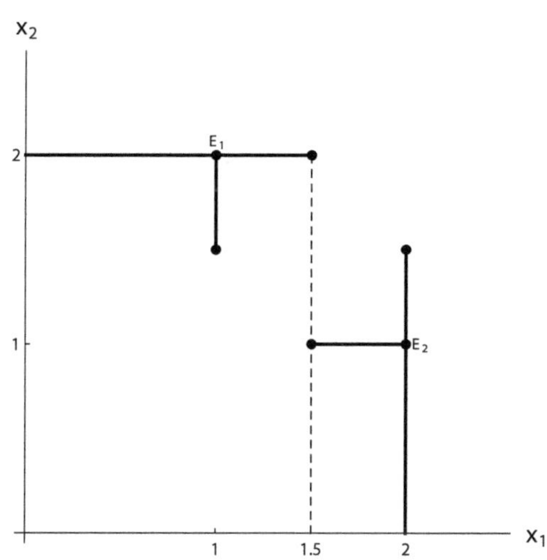

Fig. 5 Equilibria in Example 2

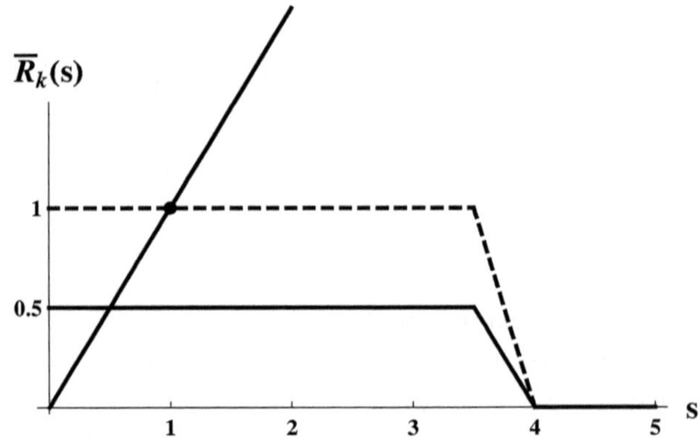

Fig. 6 Equilibrium of Part (a) of Example 3

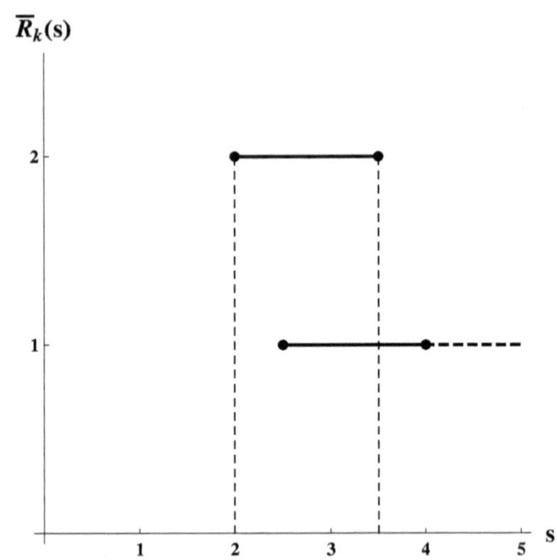

Fig. 7 Best response $\bar{R}_k(s)$ in Part (b) of Example 3

and it is also shown with broken lines. It has a unique intersection with the 45° line at $s = 1$ and since $\bar{R}_k(1) = 0.5$, the unique equilibrium is $x_1 = x_2 = 0.5$.

(b) In the case of Example 2, the best response $\bar{R}_k(s)$ is shown in Fig. 7.

If the same kind of firms from a duopoly, then

$$\bar{R}_1(s) + \bar{R}_2(s) = \begin{cases} 4 & \text{if } 2 \le s < 2.5, \\ \{2,3,4\} & \text{if } 2.5 \le s \le 3.5, \\ 2 & \text{if } s > 3.5 \end{cases}$$

Since $\bar{R}_k(s)$ is either 1 or 2 in the second case, the possible combinations for $\bar{R}_1(s) + \bar{R}_2(s)$ are $1+1,\ 1+2,\ 2+1$ and $2+2$. $H(s) = \bar{R}_1(s) + \bar{R}_2(s)$ is shown in Fig. 8,

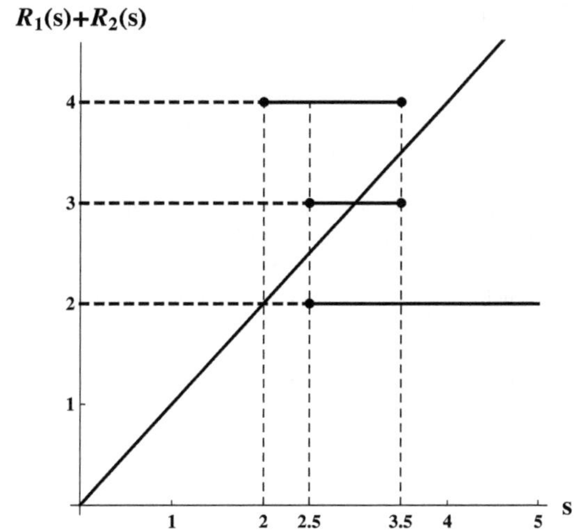

Fig. 8 Equilibrium of Part (b) of Example 3

from which it is clear that $s = 3$ is the unique solution. Notice that at $s = 3$, both $\bar{R}_k(3) = 1$ and $\bar{R}_k(3) = 2$ are feasible, so both $x_1 = 1,\ x_2 = 2$ and $x_1 = 2,\ x_2 = 1$ are equilibria.

4 Conclusion

A modified N-firm single-product oligopoly without product differentiation was examined in which the payoff functions of the firms were discontinuous. The best responses of the firms were determined as functions of the output of the rest of the industry and then they were modified as functions of the total industry output. Based on these modified best responses the existence of at least one equilibrium was proved, and numerical examples showed the possibility of a unique as well as of multiple equilibria. The more general cases with nonlinear price and cost functions will be the subjects of our next research project.

Acknowledgements This paper has been prepared while the first author was a visiting professor at the Department of Economics of Chuo University, Tokyo, Japan. The authors highly appreciate the financial supports from the MEXT-Supported Program for the Strategic Research Foundation at Private Universities 2013–2017, the Japan Society for the Promotion of Science (Grant-in-Aid for Scientific Research (C) 24530202, 25380238 and 26380316) and Chuo University (Joint Research Grant). The usual disclaimers apply.

References

Bischi G-I, Chiarella C, Kopel M, Szidarovszky F (2010) Nonlinear oligopolies: stability and bifurcations. Springer, Heidelberg/Dordrecht/London/New York

Burr C, Gardini L, Szidarovszky F (2014) Discrete time dynamic oligopolies with adjustment constraints. J Dyn Games 2(1):65–87

Cournot A (1838) Recherches sur les Principes Mathématiques de la Théorie des Richesses. Hachette, Paris (English translation (1960): Researches into the mathematical principles of the theory of wealth. Kelley, New York)

Okuguchi K (1976) Expectations and stability in oligopoly models. Springer, Berlin

Okuguchi K, Szidarovszky F (1999) The theory of oligopoly with multi-product firms, 2nd edn. Springer, Berlin

Szidarovszky F, Yakowitz S (1982) Contributions to Cournot oligopoly theory. J Econ Theory 28(1):51–70

Zhao J, Szidarovszky F (2008) N-firm oligopolies with production adjustment costs: best responses and equilibrium. J Econ Behav Organ 68:87–99

Interpreting Markups in Spanish Manufacturing: The Exponential Model

Luis C. Corchón and Lourdes Moreno

Abstract In this paper, we present a model of oligopolistic competition where demand functions are exponential. We show that in such a model there is a unique Cournot equilibrium. This model predicts a monotonic relationship between the markup of a firm and its output. We present evidence, based on a sample of Spanish manufacturing firms, that this relationship is inverse.

1 Introduction

Intuition suggests that the markup of a firm is influenced, among other things, by its output. In this paper, we present a model of oligopolistic competition that predicts such a relationship and we present some empirical evidence that supports such a relationship.

Our model has standard features like demand derived from utility maximization of a representative consumer, constant marginal costs and quantity-setter firms. The new feature is that the representative consumer has an exponential utility function. The exponential form has been used extensively in macroeconomics (see, e.g., Lucas, 1985), welfare economics (see, e.g., Pollak 1971; Kolm 1976a,b) and insurance (Pratt, 1964), but as far as we know, there is only one paper, by Behrens and Murata (2007), that deals with this functional form in imperfectly competitive markets. This paper assumes a continuum of identical price-making firms and thus our models are different because we assume a finite number of possibly different quantity-setter firms. We prove the existence and the uniqueness of a Cournot equilibrium. The meat of the exponential form is that it implies that, in equilibrium, there is a monotonic relationship between the output of a firm and its markup. When the product is homogeneous, this relationship is linear. And when a certain

L.C. Corchón (✉)
Universidad Carlos III, Madrid, Spain
e-mail: lcorchon@eco.uc3m.es

P. von Mouche, F. Quartieri (eds.), *Equilibrium Theory for Cournot Oligopolies and Related Games*, Springer Series in Game Theory,
DOI 10.1007/978-3-319-29254-0_9

parameter is close to zero, our equation predicts an almost constant markup among firms, which is what happens in models of constant elasticity of demand.[1]

We test our model with data from a panel of Spanish manufacturing firms for the period 1990–2005 obtained from the Encuesta Sobre Estrategias Empresariales (Survey on Business Strategies). We estimate individual markups for 947 firms that belong to 20 manufacturing sectors, using the methodology of Roeger (1995). We test the first order condition of the model by a nonlinear least-squares estimation where the independent variable is the previous estimated markups for 851 firms with positive markups. We discard firms with negative profits because this fact is incompatible with profit maximization and the possibility of inaction.

Our results confirm the existence of product differentiation and show a negative relationship between the markup and the output of firms. This may seem counterintuitive because one would expect large firms to have strong market power and, thus, charge high markups. But under product differentiation, firms face different demand elasticities. And small firms may cover market niches in which these elasticities are small.

A criticism of our procedure is that we perform a test of a functional form by running regressions involving markups and outputs. Both variables are determined jointly by profit maximization, so our estimation procedure might be plagued with endogeneity problems.[2] But our exponential model does not admit a closed form solution even under assumptions like product homogeneity and identical firms. This means that all we can do is to test the first order condition of profit maximization, which implies a monotonic relationship between markups and the corresponding outputs. Thus, our procedure can be seen as a test of the necessary conditions of equilibrium.

The rest of the paper goes as follows. In Sect. 2, we present the exponential model. In Sect. 3, we present our empirical findings. Section 4 concludes and outlines some lines of research.

2 The Theoretical Model

A market is composed of n firms that produce a single output, each denoted by $x_1, x_2, \ldots, x_n$. A representative consumer consumes these goods at given prices $p_1, p_2, \ldots, p_n$. In addition, there is an outside good whose consumption is denoted by M and that is the numeraire. The representative consumer has preferences

[1] The first model that predicts variable markups is by Melitz and Ottaviano (2008).

[2] Due to the methodology used to estimate the markups, our sample is a cross section and we cannot use lagged variables as instruments. It is also difficult to find exogenous instruments for the output that are not correlated with markups.

representable by a utility function of the following form

$$U = -\frac{A}{B}e^{-Bm} + M, \quad A > 0, B \neq 0. \tag{1}$$

where $m \equiv \sum_{i=1}^{n} x_i^{\rho}$, $\rho \in (0, 1]$. Thus, utility is quasi-linear in M (as is standard in the literature) but exponential in m, a number that aggregates the consumption of all the goods supplied in the market under consideration. ρ measures the degree of product differentiation since the marginal rate of substitution between good i and j is $(x_j/x_i)^{1-\rho}$. When $\rho = 1$, the marginal rate of substitution is 1 so the product is homogeneous. For the time being, we assume that $B > 0$. The budget constraint is $\sum_{i=1}^{n} p_i x_i + M = I$ where I is the exogenous income. Substituting the value of M in (1) and dropping I, which is constant, we obtain

$$U = -\frac{A}{B}e^{-Bm} - \sum_{i=1}^{n} p_i x_i. \tag{2}$$

The right hand side of (2) can be shown to be concave in outputs. Thus, first order conditions of maximization yield inverse demand functions, namely

$$p_i = A\rho e^{-Bm} x_i^{\rho-1}, \quad i = 1, \ldots, n. \tag{3}$$

Firms have constant marginal costs denoted by c_i. The markup for firm i, denoted by μ_i, is defined as

$$\mu_i = \frac{p_i - c_i}{p_i} \tag{4}$$

Profits for a typical firm, say i, are

$$\pi_i = A\rho e^{-Bm} x_i^{\rho} - c_i x_i. \tag{5}$$

First order conditions of profit maximization when $x_i > 0$ are

$$\frac{\partial \pi_i}{\partial x_i} = A\rho^2 e^{-Bm} x_i^{\rho-1} - AB\rho^2 x_i^{\rho} e^{-Bm} x_i^{\rho-1} - c_i = 0, \tag{6}$$

which can be written as

$$A\rho^2 e^{-Bm} x_i^{\rho-1}(1 - Bx_i^{\rho}) - c_i = 0. \tag{7}$$

The second order condition of profit maximization holds because the left hand side of (7) is decreasing in x_i.

We will refer to the previous model as the Exponential Model and now we investigate its basic properties. We look for a Nash equilibrium in quantities where each firm maximizes profits, taking the output of the other firms as given.

Proposition 1 *The exponential model with $B > 0$ has a unique Nash equilibrium. In this equilibrium, all firms produce a positive output.*

Proof Existence follows from the concavity and continuity of the profit function plus the compactness and convexity of the strategy sets (see, e.g., Friedman, 1977, pp. 152–3).

Interiority follows from the fact that when $x_i \to 0$, if $\rho < 1$, $\frac{\partial \pi_i}{\partial x_i} \to \infty$ and if $\rho = 1$, the left hand side of (7) is positive. Thus, all firms produce a positive output in equilibrium.

Finally, let us prove uniqueness. By defining a new variable $y_i \equiv x_i^\rho$, the game becomes aggregative, i.e., payoffs and first order conditions for each firm, say i, can be written in terms of y_i and $\sum_{j=1}^n y_j$; indeed Eqs. (5) and (6) become

$$\pi_i = A\rho e^{-B\sum_{j=1}^n y_j} y_i - c_i y_i^{\frac{1}{\rho}}. \tag{8}$$

$$\frac{\partial \pi_i}{\partial x_i} = A\rho^2 y_i^{\frac{\rho-1}{\rho}} e^{-B\sum_{j=1}^n y_j}(1 - By_i) - c_i. \tag{9}$$

We see that $\frac{\partial \pi_i}{\partial x_i}$ is decreasing in $\sum y_j$ given y_i and decreasing in y_i given $\sum y_j$. Thus, assumptions A1 and A2 in Corchón (1994) are satisfied. The compactness requirement also holds because y_i can be taken such that $1 \geq By_i$. Thus, as observed in Corchón op. cit. p. 156, the model has a unique equilibrium.[3]

However, we will see in the next section that in our estimation, $B < 0$. This creates two problems that, fortunately, are solvable.

The first problem is that inverse demand is not always decreasing in x_i;

$$\frac{\partial p_i}{\partial x_i} = A\rho e^{-Bm} x_i^{\rho-2}(\rho - 1 - B\rho x_i^\rho). \tag{10}$$

Thus,

$$\frac{\partial p_i}{\partial x_i} < 0 \Leftrightarrow -Bx_i^\rho < \frac{1-\rho}{\rho}. \tag{11}$$

The easiest way to fix this problem is to redefine m as follows:

$$\bar{m} \equiv \sum_{i=1}^n \min(x_i^\rho, \frac{1-\rho}{-B\rho}). \tag{12}$$

[3]These properties also imply well-behaved comparative statics, see Corchón (1994).

Note that $\bar{m}$ is concave in outputs as before. Now firms have the following upper bound to output:

$$x_i \leq (\frac{1-\rho}{-B\rho})^{\frac{1}{\rho}}. \tag{13}$$

Thus, the strategy space for firms is compact (the bound of the case $B > 0$, namely, $1 \geq Bx_i^\rho$, is always satisfied here). This, together with the continuity of the profit function, implies that a solution to profit maximization exists. In the next section, we will see that for the values of the estimated parameters ρ and B, (13) holds for all outputs in our sample. Finally note that (13) implies that markups are non-negative because $\mu_i = 1 - \rho + B\rho x_i^\rho \geq 1 - \rho - (1 - \rho) = 0$.

The second problem is that the second order condition of profit maximization does not necessarily hold because $e^{-B\bar{m}}$ is increasing in x_i. Indeed,

$$\frac{\partial^2 \pi_i}{\partial x_i^2} = A\rho^2 e^{-B\bar{m}} x_i^{\rho-2}(-Bx_i^\rho(3\rho - 1 - \rho B x_i^\rho) + \rho - 1). \tag{14}$$

Thus, we have that

$$\frac{\partial^2 \pi_i}{\partial x_i^2} < 0 \Leftrightarrow -Bx_i^\rho(3\rho - 1 - \rho B x_i^\rho) < 1 - \rho. \tag{15}$$

To ease notation, let $w_i \equiv -Bx_i^\rho$. Thus, (15) now reads as

$$w_i(3\rho - 1 + \rho w_i) < 1 - \rho. \tag{16}$$

We see that this inequality holds when $w_i \simeq 0$ but not for large values of w_i. The maximum value of w_i for which this inequality holds, denoted by $\bar{w}_i$, is

$$\bar{w}_i = \frac{1 - 3\rho + \sqrt{-2\rho + 5\rho^2 + 1}}{2\rho}. \tag{17}$$

It is easy to see that $\bar{w}_i < \frac{1-\rho}{\rho}$ where the latter was the bound on $\bar{w}_i$ in (13). Thus, we have to introduce a tighter bound on x_i, namely,

$$x_i \leq (\frac{\bar{w}_i}{-B})^{\frac{1}{\rho}}. \tag{18}$$

If x_i satisfies this bound, the second order condition of profit maximization holds. Thus, we have:

Proposition 2 *Consider the exponential model with $B < 0$ and where production is bounded by (18). Then, there is a Nash equilibrium in which all firms produce a positive output. If $\rho \geq 0.5$, the equilibrium is unique.*

Proof Since $x_i \in [0, (-\bar{w}_i/B)^{\frac{1}{\rho}}]$, the strategy space of any firm is convex and compact. The profit function is continuous and by (18) concave in x_i. Thus, using the argument in the proof of Proposition 1, an equilibrium exists and it is interior. Consider that in Eq. (6) m and x_i are independent. Totally differentiating this equation, we get

$$\frac{\partial \pi_i}{\partial x_i} dx_i + \frac{\partial \pi_i}{\partial m} dm = 0.$$

We see that $\partial \pi_i / \partial m > 0$. We also see that

$$\frac{\partial \pi_i}{\partial x_i} > 0 \leftrightarrow -Bx_i^{\rho}(2\rho - 1) > 1 - \rho.$$

Suppose $\rho \geq 0.5$. Thus, if $\bar{w}_i < (1-\rho)/(2\rho - 1)$, the relationship between m and x_i is decreasing. And, using graphical methods, we can show that (18) implies the latter inequality. But in this case, there cannot be two equilibria because in one of them, m would be larger than in the other, but all outputs corresponding to the first equilibrium would be smaller than in the second equilibrium.

For small values of ρ, equilibrium may be not unique. For instance let $A = -B = c_i = 1$ for all i and $n = 10$. We look for a symmetric equilibrium output denoted by y. Then making Eq. (9) equal to zero we obtain:

$$\rho^2 y^{\frac{\rho - 1}{\rho}} e^{10y}(1 + y) = 1$$

which is pictured in Fig. 1 below. We see that for $\rho = 0.04$ there are two values of y fulfilling the first order condition. In order to show that these two values are Nash equilibria we have to check that the bound (18) holds. Indeed, $\bar{w}_i = 23.042$ and thus $(23.042)^{\frac{1}{0.04}} = 1.156\,1 \times 10^{34}$. We see that the values of y in Fig. 1 are less than this number.

One possible check of condition (18) is by noting that

$$\mu_i = 1 - \rho - \rho z \geq 1 - \rho - \frac{1}{2}\left(1 - 3\rho + \sqrt{-2\rho + 5\rho^2 + 1}\right). \tag{19}$$

For $\rho = 0.770$, which is the value of ρ estimated in the next section, μ_i must be larger than 0.106. This holds for 635 firms in a sample of 851 firms. If firms are producing near full capacity, the previous argument shows that the second order condition holds globally.

What to do with the 216 firms in the sample whose markups are smaller than 0.106? From the first order condition we obtain

$$e^{-B\bar{m}} x_i^{\rho - 1}(1 - Bx_i^{\rho}) = \frac{c_i}{A\rho^2}. \tag{20}$$

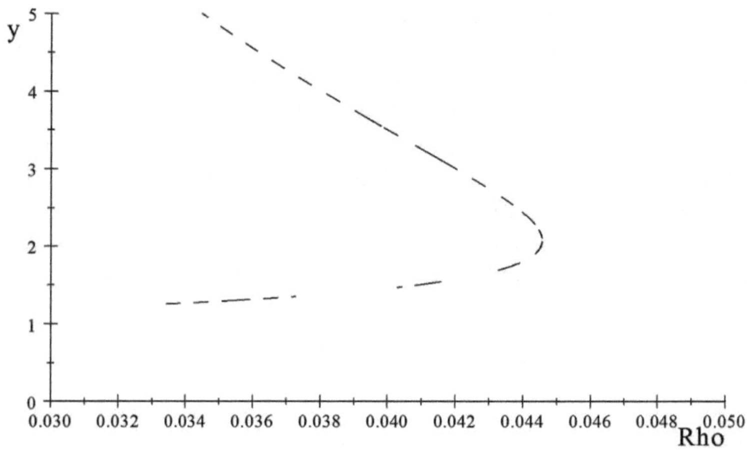

Fig. 1 Multiplicity of equilibria

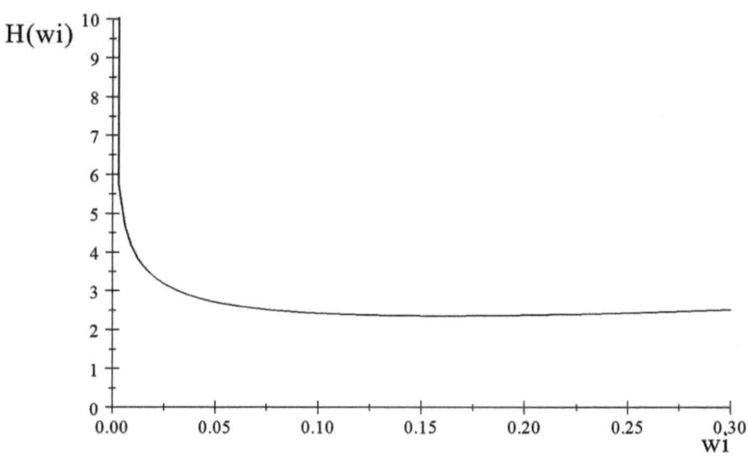

Fig. 2 Graph of $H(w)$

Setting $m_i = \bar{m} - x_i^{\rho}$, the first order condition can be written as

$$e^{w_i} w_i^{\frac{\rho-1}{\rho}} (1 + w_i) = \frac{-B^{\frac{\rho-1}{\rho}}}{A\rho^2} \times \frac{c_i}{e^{-Bm_i}}. \tag{21}$$

Let the left hand side of (21) be denoted by $H(w_i)$. For $\rho = 0.770$, we picture $H(w_i)$ in Fig. 2. Note that at the upper bound of w_i, $H(0.298) = 2.511$.

We now prove that, for the values of the parameters in the sample, there is a unique solution to (21) and that this solution is a global maximum. This proof consists of several steps.

1. *$H(w_i)$ is first decreasing and then increasing.*
   Since $\rho = 0.770$, $H(w_i) = e^{w_i} w_i^{-0.298}(1 + w_i)$, see Fig. 2.[4] Using numerical methods, it can be shown that $H(w_i)$ is decreasing up to $w_i \simeq 0.201$ and increasing up to the upper bound where $H(0.298) = 2.511$. This implies that there are, at most, two solutions in w_i in (21). And that if the right hand side of (21) takes a value larger than 2.511, there is a unique solution in w_i in (21).
2. *Setting the values of the main parameters in (21).*
   Using the estimated value of B (see next section) and the calculated value of m from Eq. (12), A can be calibrated from (3) by adding all equations.

$$A = \frac{\sum_{i=1}^{n} p_i}{0.770 \times e^{-Bm} \times \sum_{i=1}^{n} x_i^{\rho-1}}.$$

   We find that $A = 0.002$. When we consider only the 216 firms with markups smaller than 0.106, we obtain the same estimated A.
3. *Evaluating (21).*
   Using the calibrated value of A, we calculate the right hand side of (21) for the largest value of m_i (11,368.1) and the smallest value of c_i (0.496) in the sample. The obtained value, 3.528, is larger than $H(0.3)$. Thus, the first order condition has a unique solution.
4. *Evaluating profits at the extremes.*
   Now we work out profits at the upper bound for output, namely $(\frac{1-\rho}{-B\rho})^{\frac{1}{\rho}}$

$$A\rho e^{-B\bar{m}} x_i^{\rho} - c_i x_i = Ae^{-B\bar{m}}(\frac{1-\rho}{-B}) - c_i(\frac{1-\rho}{-B\rho})^{\frac{1}{\rho}}$$

   Thus, for profits to be positive $c_i \leq 0.337$ which we don't have in the sample because the smallest value of c_i is 0.496. Thus, profits at the upper bound are negative. They are zero when $w_i = 0$.

From steps 1–4 above, it follows that a solution to profit maximization exists and this solution is not at the extremes; it must be the unique solution to (21). Thus, this solution is indeed the unique global maximum. We obtain the following result.

Proposition 3 *For the estimated values of the parameters, there is a Nash equilibrium in quantities to the exponential model.*

Proof The best reply correspondence exists, it is upper-hemi continuous (by Berge's maximum theorem) and since it is single valued (by 1–4 above), it is continuous. Thus, best replies continuously map a compact interval in itself and by the Brouwer fixed point theorem, a fixed point exists. Clearly, this fixed point is an equilibrium.

[4]We have approximated $(1 - \rho)/\rho = 0.298$ by 0.3.

Proposition 3 is a check of the consistency of our estimations. We estimated under the implicit assumption that an equilibrium exists. Proposition 3 shows that for the estimated values, an equilibrium exists.

We note that the findings leading to Proposition 3 hold not only for the values of the parameters given there but for all the values in the sample except for 15 firms whose marginal cost is smaller than 0.337.

We end this section by noting that first order conditions do not yield a closed form solution. Thus, in order to test the model, we take a different route: from the definition of a markup in (4) and (3), (6) can be written as

$$\mu_i = 1 - \rho + B\rho x_i^{\rho}, \tag{22}$$

which is the equation which will be estimated in the next section. Note that when the product is homogeneous (i.e., $\rho = 1$), this relationship is linear. And when $B \simeq 0$, markups are (almost) constant, like in models in which the elasticity of demand is constant.

3 The Empirical Implementation

We estimate markups using the methodology of Roeger (1995). This approach allows us to obtain markups by estimating one parameter and controlling for potential endogeneity of the productivity shocks.[5] The availability of a long time period for each firm allows them to estimate not only an average markup of the industry but also individual markups.

The estimations are carried out using an unbalanced panel of Spanish manufacturing firms for the period 1990–2005, obtained from the Encuesta Sobre Estrategias Empresariales.[6] The sampling scheme of this survey is conducted for each two-digit level manufacturing NACE class. Companies which employ between 10 and 200 employees are chosen by random sampling and the rate of participation is around 4 %. For firms which employ more than 200 employees, the rate of participation is about 60 %.

The set of variables required to apply Roeger's methodology are sales, nominal labour cost, nominal value of intermediate consumption -where the intermediate consumptions are raw materials, energy and services purchased-, and the nominal

[5]Some papers have used this approach to analyze the effect of trade liberalization on average markups using panel data (see, for example Konings and Vandenbussche 2005; Konings et al. 2005). Moreno and Rodríguez (2011) used this methodology to study the impact of offshoring on margins.

[6]This survey was carried out by the Fundación SEPI.

value of capital services. These variables are defined as follows:

Capital stock of equipment goods: Net stock of capital for equipment goods in real terms. This is calculated by using the perpetual inventory formula:

$$K_t = (1-d)K_{t-1}(P_t/P_{t-1}) + I_t,$$

where P is the price index for equipment, d is the depreciation rate, and I is the investment in equipment.

Effective hours of work: Normal hours plus overtime hours minus lost hours.

Intermediate consumption: Raw material purchases, energy costs and other external services.

Output (x_i): Nominal output—Sales plus stock variation—is deflated using Paasche-type firm individual indices, constructed starting from the price changes in output reported by firms.

Price (p_i): The surveyed firms give annual information about markets served (up to five), identifying their relative importance (in percentage) in total sales of the firm. This information allows us to calculate a Paasche-type price index for all markets and for each market, using the proportions with respect to total sales as weighting.

As in Moreno and Rodríguez (2011), we estimate individual markups for firms with at least nine observations. In this sense, we have one estimated markup for each firm. Specifically, we have information for 947 firms which belong to 20 manufacturing sectors. Figure 3 presents the distribution of these markups for all firms. As can be seen, the dispersion is large and the distribution is slightly skewed,

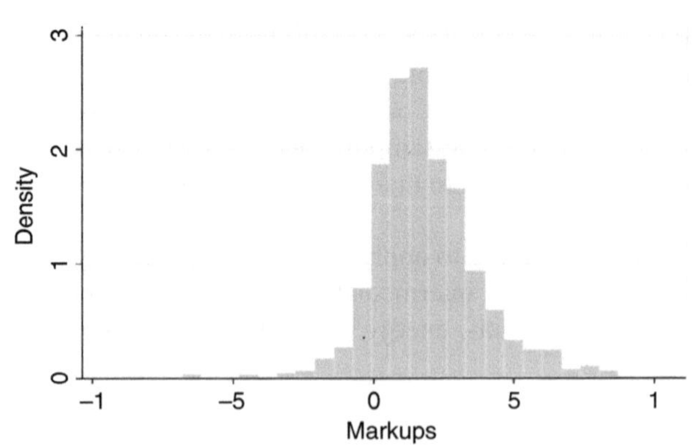

Fig. 3 Markups distribution

Table 1 Average Individual Positive Markups by sectors

	Mean	S.D	Firms	% Exporters	% Large firms
Meat related products	0.123	0.134	26	69	20
Food and tobacco	0.183	0.130	90	59	26
Beverages	0.258	0.165	16	69	44
Textiles and clothing	0.225	0.162	83	78	24
Leather, fur and footwear	0.255	0.165	18	89	0
Timber	0.212	0.161	22	64	14
Paper	0.226	0.191	25	96	32
Printing and publishing	0.247	0.195	47	65	19
Chemicals	0.214	0.162	66	91	30
Plastic and rubber products	0.177	0.127	40	90	23
Non-mineral products	0.286	0.191	64	70	33
Basic metal products	0.184	0.138	36	90	56
Fabricated metal products	0.232	0.164	73	77	15
Ind. and agric. equipment	0.253	0.178	64	86	27
Office mach. and similar	0.328	0.237	8	100	43
Electric materials and accessories	0.253	0.165	51	86	40
Vehicles & motors	0.190	0.134	50	95	58
Other transport equipment	0.220	0.187	17	94	35
Furniture	0.230	0.157	34	71	9
Miscellaneous	0.244	0.182	21	95	19
All firms	0.224	0.162	851	79	28

with a large proportion of firms on the right tail. The average markup for these firms is 0.190.[7]

Almost 10 % of the sample presents negative markups. When these firms are not considered, the average markup increases to 0.224. Table 1 below presents the average of the estimated individual markups for 20 manufacturing sectors when only positive markups are considered. The average of estimated Lerner indexes ranges from 0.123 to 0.328. Most firms in our sample operate in international markets: almost 80 % of them are exporters. About 30 % are large firms with more than 200 employees.

We also test that inequality (13) holds for all firms in the sample.

We estimate (22) by a nonlinear least-squares estimation. The independent variable is the previous estimated markup. We present our results in Table 2 for the case of positive markups (851 firms).

As can be seen, the estimated ρ takes a value equal to 0.770 and it is significant at 1 %. We reject the null hypothesis that is, equal to 1. That is we confirm the existence of product differentiation. The B coefficient (significant at 5 %) presents a negative

[7]It is quite similar to one obtained by the standard approach. In complementary regressions (available upon request), using panel data, the average margin is 0.170.

Table 2 Nonlinear least-squares estimation (Eq. 22)

	Coefficient	*p-value*
ρ	0.770	0.000
B	−0.0006	0.050
R^2	0.656	
Number of observations	851	
Test: $\rho = 1$	p-value: 0.000	

sign indicating a negative relationship between the markups and the firmsoutput in our sample. The value of R^2 is relatively high (taking into account that our regression is a cross section).

4 Conclusions

In this paper, we have presented a new functional form for the inverse demand, namely, the exponential one. We have shown that this form produces encouraging results for a sample of Spanish manufacturing data. Our model relies on assumptions like quantity competition, quasi-linear utility and stable preferences that must be subject to further inquiry before definitive conclusions are drawn. We also do not consider factors that are likely to play an important role in the determination of markups such as entry and sunk costs (Sutton 1991), collusion (Bresnahan 1989), international trade (Melitz and Ottaviano 2008; Feenstra and Kee 2008; De Loecker and Warzynski 2009; Moreno and Rodríguez 2010), non-linear pricing (Miravete and Röller 2004) and location. Our model is admittedly simple but might serve as a useful starting point for future developments. See Corchón and Moreno (2010) for an extension of the exponential model and the corresponding empirical evidence.

Summing up, the exponential model presented in this paper is workable and provides us with a new tool for analyzing markets characterized by imperfect competition.

Acknowledgements We thank Cesar Alonso, Carmen Beviá, Luis Cabral, Rosario Crino, Manuel Domínguez, Natalia Fabra, Philippe Gaignepain, Miguel Gonzalez-Maestre, Paco Marhuenda, Pierre von Mouche, Philipp Schmidt-Dengler, José María Sastre, Galina Zudenkova, an anonymous referee and participants in seminars at U. Carlos III, U. Complutense, Madrid, IEB U. of Barcelona, the 2011 EARIE meeting and the Jornadas de Economia Industrial 2011 for very helpful comments, Diego Rodríguez for programming support and María Hernández for research assistance. The first author acknowledges financial support from ECO2011-25330. The second author acknowledges financial support from ECO2010-19847.

References

Behrens K, Murata Y (2007) General equilibrium models of monopolistic competition: a new approach. J Econ Theory 136:776–787

Bresnahan TF (1989) Empirical studies of industries with market power. In: Schmalensee R, Willig RD (eds) Handbook of industrial organization. North-Holland, New York

Corchón L (1994) Comparative statics for aggregative games: the strong concavity case. Math Soc Sci 28:151–165

Corchón L, Moreno L (2010) Interpreting markups in Spanish manufacturing: the exponential model. Available at SSRN: http://papers.ssrn.com/sol3/papers.cfm?abstract_id=1814011

De Loecker J, Warzynski F (2009) Markups and firm-level export status. NBER working papers, 15198

Feenstra R, Kee HL (2008) Export variety and country productivity: estimating the monopolistic competition model with endogenous productivity. J Int Econ 74:500–518

Friedman JW (1977) Oligopoly and the theory of games. Amsterdam: North Holland

Kolm SC (1976a) Unequal inequalities I. J Econ Theory 12:416–442

Kolm SC (1976b) Unequal inequalities II. J Econ Theory 13:82–111

Konings J, Vandenbussche H (2005) Antidumping protection and markups of domestic firms. J Int Econ 65:151–165

Konings J, Van Cayseele P, Warzybski F (2005) The effects of privatization and competitive pressure on firms' price-cost margins: micro evidence for emerging economies. Rev Econ Stat 87(1):124–134

Lucas R (1985) Models of business cycles. Yrjo Jahnsson lectures. Oxford, Basil Blackwell

Melitz M, Ottaviano G (2008) Market size, trade, and productivity. Rev Econ Stud 75:295–316 (2008)

Miravete E, Röller L-H (2004) Estimating price-cost markups under nonlinear pricing competition. J Eur Econ Assoc 2(2/3):526–535

Moreno L, Rodríguez D (2010) Export activity, persistence and markups. Appl Econ 42:475–488

Moreno L, Rodríguez D (2011) Markups, bargaining power and offshoring: an empirical assessment. World Econ 34(9):1593–1627

Pollak R (1971) Additive utility functions and linear engel curves. Rev Econ Stud 38(116):401–414

Pratt JW (1964) Risk aversion in the small and the large. Econometrica 32(1–2):122–136

Roeger W (1995) Can imperfect competition explain the difference between primal and dual productivity measures? Estimates for U.S. Manufacturing J Pol Econ 103(2):316–330

Sutton J (1991) Sunk costs and market structure. MIT, London

References

[illegible]

Privatization Neutrality Theorem and Discriminatory Subsidy Policy

Kojun Hamada

Abstract This study revisits the privatization neutrality theorem that claims that social welfare is exactly the same before and after privatization when the government gives the optimal subsidy to both public and private firms in a mixed oligopoly. Unlike the existing literature that has assumed that a uniform subsidy is given to public and private firms, we demonstrate that if the discriminatory subsidy rates are adopted even when there is firm asymmetry between public and private firms, the privatization neutrality theorem continues to hold. First, we show that even if the cost of the public firm differs from those of private firms, the privatization neutrality theorem holds by appropriately subsidizing both public and private firms at the different levels. Second, even if the public firm acts as a Stackelberg leader before and after privatization, the government can attain privatization neutrality by adopting the discriminatory subsidy and, as a result, can achieve social welfare maximization. Our result suggests that even when there exists firm asymmetry between public and private firms, it is not important for privatization authorities to determine whether to privatize the public firm.

1 Introduction

Since a seminal article by De Fraja and Delbono (1989), a growing number of studies have sought the answer on the question about whether privatization increases welfare.[1] As a well-known theoretical consequence in a mixed oligopoly, the privatization neutrality theorem has been examined by many scholars who are interested in whether privatization leads to an increase in social welfare. The

[1]Okuguchi (2012) formulated a general model of Cournot mixed oligopoly with partial privatization of a public firm and clarified the existence and stability of the equilibrium and the optimality of partial privatization.

K. Hamada
Faculty of Economics, Niigata University, Niigata, Japan
e-mail: khamada@econ.niigata-u.ac.jp

P. von Mouche, F. Quartieri (eds.), *Equilibrium Theory for Cournot Oligopolies and Related Games*, Springer Series in Game Theory,
DOI 10.1007/978-3-319-29254-0_10

privatization neutrality theorem insists that when the government gives the optimal subsidy to both public and private firms in a mixed oligopoly, social welfare is exactly the same before and after privatization. However, almost all of the existing studies involving this theorem have examined only the situation in which both public and private firms are identical in technology and produce simultaneously. Furthermore, some articles insist that unless public and private firms are identical, privatization is not neutral from the viewpoint of social welfare maximization. In this study, we revisit the privatization neutrality theorem when public and private firms are not identical and differ in production cost and/or the timing of production. Challenging the conventional wisdom that firm heterogeneity collapses privatization neutrality, this study attempts to demonstrate that even if both public and private firms are not identical, the privatization neutrality theorem still holds.

As a seminal article, White (1996) first showed that if subsidies are used before and after privatization, welfare is unchanged by privatization. Since White (1996), a considerable number of articles have investigated the so-called privatization neutrality theorem in various extended situations. Poyago-Theotoky (2001) showed that even when the public firm behaves as a Stackelberg leader, the privatization neutrality theorem holds. Myles (2002) generalized the neutrality result by extending it from the linear demand and quadratic cost functions specified in Poyago-Theotoky (2001) to general functions of demand and cost. Tomaru and Saito (2010) investigated the endogenous timing in a mixed oligopoly and showed that in the resulting Stackelberg duopolies, where the public firm behaves as the leader and private firms as the followers, privatization does not affect welfare. In addition, a number of articles have demonstrated that the privatization neutrality theorem continues to hold in different situations. Tomaru (2006) adopted the partial privatization approach, Kato and Tomaru (2007) introduced nonprofit-maximizing private firms, and Hashimzade et al. (2007) extended the analysis to product differentiation. All of these articles verified the robustness of the privatization neutrality theorem.[2]

However, there are several economic situations in which the privatization neutrality theorem is not satisfied. By means of a tedious calculation, it is well known that when the public firm has a different cost structure from private firms, this theorem does not hold. Likewise, when the public firm behaves as a Stackelberg leader before and after privatization, the different move between public and private firms brings about the difference in social welfare before and after privatization. Fjell and Heywood (2004) considered such a sequential-move situation where the public firm keeps acting as a leader irrespective of privatization and they showed that the privatization neutrality theorem collapses. Several articles have mentioned that firm asymmetry between public and private firms, such as different cost and/or different timing of production, hinders privatization neutrality. Cato and Matsumura (2013) considered a free entry market in which private firms freely enter in a

[2]As a comprehensive survey of privatization neutrality theorem, see Tomaru (2014, Ch. 4) for details.

market and privatization neutrality is not satisfied. Zikos (2007) and Gil-Moltó et al. (2011) introduced R&D competition and showed that the existence of other choice variables than production level undermines privatization neutrality. The privatization neutrality theorem no longer holds with the distortion of taxation that is necessary to fund the subsidy (Matsumura and Tomaru, 2013), the existence of foreign private firms (Matsumura and Tomaru, 2012), and the divergence of objectives between the government and the public firm (Kato, 2008).

Recently, in contrast to the conclusion by Fjell and Heywood (2004) that privatization is not welfare neutral, Matsumura and Tomaru (2013) found that privatization neutrality holds if an output floor is introduced. Their result suggests that neutrality can be recovered by using an output floor regulation, and the economic situation in which the privatization neutrality theorem is robust is enlarged; however, such a regulation seems somewhat artificial. Furthermore, since White (1996) examined the optimal subsidy that is given to both public and private firms in a uniform specific manner in the first place, almost all of the existing literature has assumed that a uniform subsidy is given to public and private firms. If the public firm has a different cost structure and/or a different timing of production, and the government authority that implements the subsidy policy can observe the difference between firms, it is quite natural for such an authority to give the discriminatory subsidies to each firm to maximize social welfare. In this study, by taking the discriminatory subsidy between public and private firms into consideration, we attempt to demonstrate that if different subsidy rates are adopted, even when there is firm asymmetry between public and private firms, the privatization neutrality theorem still holds. The result implies that the privatization neutrality theorem can be recovered even in the situation with firm asymmetry between firms and the maximized social welfare can always be achieved by adopting the discriminatory subsidy policy without relying on any artificial regulation such as an output floor.

In this study, we show the following results. First, we show that even if the cost of the public firm differs from those of private firms, the privatization neutrality theorem holds by appropriately subsiding both public and private firms to maximize social welfare. Second, we show that even if the public firm acts as a Stackelberg leader before and after privatization and there are cost differences between firms, the government can attain privatization neutrality and achieve social welfare maximization by appropriately setting the discriminatory subsidy levels to firms. Our result suggests that, when the government can subsidize heterogeneous firms in a mixed oligopoly in an appropriate manner, the decision making about whether to privatize the public firm is not at all important for privatization authorities in almost all situations where the firm asymmetry between public and private firms is caused for any reason.

The remainder of this chapter is organized as follows. Section 2 describes the model of mixed oligopoly in which the public firm and private firms engage in quantity competition in a homogeneous-good market. Section 3 demonstrates that, even if there are cost differences between public and private firms, the privatization neutrality theorem always holds by adopting the discriminatory specific subsidy. Section 4 considers Stackelberg competition as another firm asymmetry between public and private firms and we show that, even if there is a different timing of production between public and private firms, welfare neutrality is satisfied. Section 5 concludes the study with some remarks.

2 The Model

We basically follow the setting of White (1996). Consider the oligopolistic market of a homogeneous good in which the public firm and n private firms engage in quantity competition. We denote the index of firms by firm $i = \{0, 1, \cdots, n\}$. The public firm is indexed by firm 0 and aims to maximize social welfare. All n private firms have identical technology and each private firm is indexed by firm $i = \{1, 2, \cdots, n\}$. Each private firm aims to maximize its profit. q_i denotes the output of firm i and, because all private firms are identical, the output is also identical, i.e., $q \equiv q_i$, $i = \{1, \cdots, n\}$. The inverse demand function is assumed to be linear, i.e., $p = p(Q) = a - Q; a > 0$, where $Q \equiv q_0 + \sum_{i=1}^{n} q_i = q_0 + nq$ is the total output and p is the price.

The cost function of the firm is denoted by $C(q_i) = c_i + \frac{k_i}{2}q_i^2, c_i \geq 0, k_i > 0$. For brevity, we assume no fixed cost, i.e., $c_i = 0$ and assume that all private firms have the identical cost function, i.e., $k_i \equiv k$. To focus on firm heterogeneity between the public firm and private firms, k_0 is not equal to k.[3] s_i denotes the specific subsidy per unit given to firm i. Thus, s_0 denotes the subsidy to the public firm and $s_i \equiv s$; $i \in \{1, 2, \cdots, n\}$ denotes the subsidy to private firm i, which is equal for private firms due to the homogeneity of private firms. It should be noted that White (1996) and the subsequent studies examined only the uniform specific subsidy, i.e., $s_0 = s$. We investigate the discriminatory specific subsidy that differs between public and private firms to clarify the effect of the discriminatory subsidy. We assume that subsidies do not have any welfare cost that might be caused by the taxation inefficiency. The profit of firm i is $\pi_i = p(Q)q_i - \frac{k_i}{2}q_i^2 + s_i q_i = (a - Q + s_i)q_i - \frac{k_i}{2}q_i^2$. Because all private firms are identical, the profits are the same, i.e., $\pi \equiv \pi_i$, $i = \{1, \cdots, n\}$. Consumer surplus and producer surplus are $CS \equiv \int_0^Q p(x)dx - p(Q)Q = \frac{1}{2}Q^2$ and $PS \equiv \sum_{i=0}^{n} \pi_i = p(Q)Q - \frac{1}{2}\sum_{i=0}^{n} k_i q_i^2 + s_0 q_0 + snq$, respectively. Social welfare is defined by the sum of the consumer surplus and the producer surplus, i.e., $W \equiv CS + PS - s_0 q_0 - s\sum_{i=1}^{n} q_i = aQ - \frac{1}{2}Q^2 - \frac{1}{2}\sum_{i=0}^{n} k_i q_i^2$.

The timing of the mixed oligopoly consists of a two- or three-stage game as follows. In the first stage, the government determines the optimal discriminatory specific subsidy levels to both public and private firms (s_0, s) to maximize social welfare. If the subsequent quantity competition between firms takes place in a Cournot (Stackelberg) manner, then the total number of stages is two (three), respectively. After the first stage, all firms observe the optimal subsidy levels (s_0, s). When Cournot competition takes place, in the second stage, the public firm and

[3] We assume that privatization does not change the cost function of the public firm in order to avoid the complexity caused by cost efficiency.

the private firms engage in simultaneous-move quantity competition. When the competition takes place in a Stackelberg manner, in which the public firm behaves as the leader, in the second stage, the public firm determines its own output level and, in the third stage, after observing the public firm's output, the private firms determine their own output levels.

In the above setting, we consider the situation in which although not only all firms but also the government or the regulatory authority have complete information, the government cannot directly control the production of firms by using the incentive compatible mechanism. What the government can do at best is to determine the appropriate subsidy level in order to induce firms to properly produce. The equilibrium concept follows the subgame perfect Nash equilibrium (SPNE). We solve the equilibrium by backward induction.

In the subsequent sections, we clarify the following results in turn. In Sect. 3, we demonstrate that, even if the public firm has a different cost from those of the private firms, the privatization neutrality theorem still holds with the optimal discriminatory specific subsidy. In Sect. 4, we demonstrate that, even if the public firm behaves as a Stackelberg leader before and after privatization, the privatization neutrality theorem holds with the optimal discriminatory specific subsidy.

3 The Cost Difference Between the Public Firm and Private Firms

3.1 Before Privatization

In the second stage, the public firm maximizes social welfare with regard to q_0. The first-order condition for the public firm to maximize social welfare is as follows:

$$\frac{\partial W}{\partial q_0} = a - Q - k_0 q_0 = 0 \Leftrightarrow q_0 = r_0(q) = \frac{a - nq}{k_0 + 1}. \tag{1}$$

The reaction function $q_0 = r_0(q)$ does not depend on s_0 because the subsidy level s_0 only affects the distribution between economic agents without changing the social welfare level.

Private firms maximize their own profits. The first-order condition for a private firm is as follows:

$$\frac{\partial \pi_i}{\partial q_i} = (a - Q + s) - q_i - kq_i = 0 \Leftrightarrow q_i = \frac{a + s - q_0 - \sum_{j \neq i,0}^{n} q_j}{k + 2}. \tag{2}$$

Table 1 Cournot-Nash equilibrium in the second stage before privatization

Public firm's output	q_{0B}	$\frac{(k+1)a-ns}{nk_0+(k_0+1)(k+1)}$
Private firm's output	q_B	$\frac{k_0a+(k_0+1)s}{nk_0+(k_0+1)(k+1)}$
Total output	Q_B	$\frac{(nk_0+k+1)a+nk_0s}{nk_0+(k_0+1)(k+1)}$
Price	p_B	$\frac{k_0((k+1)a-ns)}{nk_0+(k_0+1)(k+1)}$
Public firm's profit	π_{0B}	$\frac{[k_0(k+1)a+2(nk_0+(k_0+1)(k+1))s_0-nk_0s]((k+1)a-ns)}{2(nk_0+(k_0+1)(k+1))^2}$
Private firm's profit	π_B	$\frac{(k+2)(k_0a+(k_0+1)s)^2}{2(nk_0+(k_0+1)(k+1))^2}$
Social welfare	W_B	$\frac{1}{2}Q_B^2+\pi_{0B}+n\pi_B-s_0q_{0B}-nsq_B$

By substituting $q_i \equiv q$ into (2) because of the identical output of all private firms, we obtain the following equation:

$$q = r(q_0) \equiv \frac{a+s-q_0}{n+k+1}. \tag{3}$$

The reaction function of the identical private firm, (3), depends on their own subsidy level s. By solving the reaction functions, (1) and (3), we obtain the equilibrium variables in Table 1.[4]

Note that $p_B = k_0q_{0B}$, $p_B + s = (k+1)q_B$, $\pi_{0B} = s_0q_{0B} + \frac{k_0}{2}q_{0B}^2$, and $\pi_B = \frac{k+2}{2}q_B^2$. The size relationship between q_{0B} and q_B depends on the relative size of s. That is, $q_{0B} \gtreqless q_B$ if and only if $(k-k_0+1)a \gtreqless (n+k_0+1)s$. Likewise, the size relationship between π_{0B} and π_B also depends on the relative size of s. The equilibrium result shown in Table 1 claims that the increase in a private firm's subsidy before privatization brings about the increase in q_B, Q_B, π_B, and CS_B and the decrease in q_{0B} and p_B. Whether π_{0B} and W_B increase with the rise in s depends on the situation.

In the first stage, the government sets the discriminatory specific subsidy to the public firm and the private firms (s_0, s) to maximize social welfare. The social welfare depends only on s as follows:

$$W_B(s) = \frac{1}{2}Q_B^2 + \pi_{0B} + n\pi_B - s_0q_{0B} - nsq_B. \tag{4}$$

Because $\pi_{0B} = s_0q_{0B} + \frac{k_0}{2}q_{0B}^2$ and q_{0B}, q_B, Q_B, and π_B do not depend on s_0, the social welfare before privatization does not depend on s_0 at all. The optimal subsidy level to the public firm is indeterminate from the viewpoint of social welfare

[4] Throughout the study, for brevity, we focus only on the interior solution and guarantee that the second-order condition is always satisfied. We denote the equilibrium variables before and after privatization by subscript B (before) and A (after), respectively.

maximization. We solve the optimal subsidy to private firms before privatization s_B^*. By simple calculation, we obtain the derivatives of the equilibrium variables with respect to s as follows:

$$\frac{\partial q_{OB}}{\partial s} = -\frac{n}{nk_O + (k_O+1)(k+1)} < 0, \ \frac{\partial q_B}{\partial s} = \frac{k_O+1}{nk_O + (k_O+1)(k+1)} > 0,$$
$$\frac{\partial Q_B}{\partial s} = \frac{nk_O}{nk_O + (k_O+1)(k+1)} > 0,$$
$$\frac{\partial \pi_{OB}}{\partial s} = (s_O + k_O q_{OB})\frac{\partial q_{OB}}{\partial s}, \ \frac{\partial \pi_B}{\partial s} = (k+2)q_B\frac{\partial q_B}{\partial s} > 0. \tag{5}$$

The first-order condition for the government with respect to s is as follows:

$$\frac{\partial W_B}{\partial s} = Q_B\frac{\partial Q_B}{\partial s} + \frac{\partial \pi_{OB}}{\partial s} + n\frac{\partial \pi_B}{\partial s} - s_O\frac{\partial q_{OB}}{\partial s} - nq_B - ns\frac{\partial q_B}{\partial s} = 0$$
$$\Leftrightarrow s = \frac{Q_B\frac{\partial Q_B}{\partial s} + (s_O + k_O q_{OB})\frac{\partial q_{OB}}{\partial s} + n(k+2)q_B\frac{\partial q_B}{\partial s} - s_O\frac{\partial q_{OB}}{\partial s} - nq_B}{n\frac{\partial q_B}{\partial s}}$$
$$= \frac{nk_O Q_B - nk_O q_{OB} + n(k_O+1)(k+2)q_B - n(nk_O + (k_O+1)(k+1))q_B}{n(k_O+1)}$$
$$= \frac{nk_O Q_B - nk_O q_{OB} + n(k_O + 1 - nk_O)q_B}{n(k_O+1)} = \frac{k_O a + (k_O+1)s}{nk_O + (k_O+1)(k+1)} = q_B$$
$$\Leftrightarrow s_B^* = \frac{k_O a}{nk_O + k(k_O+1)}. \tag{6}$$

By (6), we obtain that the optimal subsidy to private firms is equal to the level of their own output.

By substituting the optimal subsidy level to private firms into the equilibrium variables in Table 1, we obtain the SPNE before privatization as shown in Table 2.

Note that $s_B^* = q_B = \frac{k_O}{k}q_{OB}$, $p_B = k_O q_{OB}$, $p_B + s = (k+1)q_B$, $\pi_{OB} = s_O q_{OB} + \frac{k_O}{2}q_{OB}^2$, and $\pi_B = \frac{k+2}{2}q_B^2$. The output of a more cost-efficient firm is larger than that of less efficient ones, i.e., $q_{OB} \gtrless q_B \Leftrightarrow k \gtrless k_O$. Although the size of π_{OB} depends on s_O, s_O is not determined endogenously.

3.2 *After Privatization*

In the second stage, the privatized public firm aims to maximize its profit with regard to q_O. The first-order condition for the public firm is as follows:

$$\frac{\partial \pi_O}{\partial q_O} = (a - Q + s_O) - q_O - k_O q_O = 0 \Leftrightarrow q_O = r_O(q) = \frac{a + s_O - nq}{k_O + 2}. \tag{7}$$

Table 2 SPNE before privatization

Private firm's subsidy	s_B^*	$\frac{k_0 a}{nk_0+k(k_0+1)}$
Public firm's output	q_{0B}	$\frac{ka}{nk_0+k(k_0+1)}$
Private firm's output	q_B	$\frac{k_0 a}{nk_0+k(k_0+1)}$
Total output	Q_B	$\frac{(nk_0+k)a}{nk_0+k(k_0+1)}$
Price	p_B	$\frac{k_0 ka}{nk_0+k(k_0+1)}$
Public firm's profit	π_{0B}	$\frac{k[k_0 ka+2(nk_0+k(k_0+1))s_0]a}{2(nk_0+k(k_0+1))^2}$
Private firm's profit	π_B	$\frac{k_0^2(k+2)a^2}{2(nk_0+k(k_0+1))^2}$
Social welfare	W_B	$\frac{(nk_0+k)a^2}{2(nk_0+k(k_0+1))}$

Table 3 Cournot-Nash equilibrium in the second stage after privatization

Public firm's output	q_{0A}	$\frac{(k+1)a+(n+k+1)s_0-ns}{n(k_0+1)+(k_0+2)(k+1)}$
Private firm's output	q_A	$\frac{(k_0+1)a-s_0+(k_0+2)s}{n(k_0+1)+(k_0+2)(k+1)}$
Total output	Q_A	$\frac{(n(k_0+1)+k+1)a+(k+1)s_0+n(k_0+1)s}{n(k_0+1)+(k_0+2)(k+1)}$
Price	p_A	$\frac{(k_0+1)(k+1)a-(k+1)s_0-n(k_0+1)s}{n(k_0+1)+(k_0+2)(k+1)}$
Public firm's profit	π_{0A}	$\frac{(k_0+2)((k+1)a+(n+k+1)s_0-ns)^2}{2(n(k_0+1)+(k_0+2)(k+1))^2}$
Private firm's profit	π_A	$\frac{(k+2)((k_0+1)a-s_0+(k_0+2)s)^2}{2(n(k_0+1)+(k_0+2)(k+1))^2}$
Social welfare	W_A	$\frac{1}{2}Q_A^2+\pi_{0A}+n\pi_A-s_0 q_{0A}-nsq_A$

Different from the case before privatization, the reaction function of the privatized public firm, (7), depends on s_0 because the subsidy stimulates the increase in output and leads to the increase in profit.

As the behavior of private firms maximizing their own profit is the same before and after privatization, the first-order condition for a private firm is also the same as that before privatization, i.e., (2). The reaction function of the identical private firms is given by (3), which depends on their own subsidy level s. By solving the reaction functions, (7) and (3), we obtain the equilibrium variables in Table 3.

Note that $p_A+s_0=(k_0+1)q_{0A}$, $p_A+s=(k+1)q_A$, $\pi_{0A}=\frac{k_0+2}{2}q_{0A}^2$, $\pi_A=\frac{k+2}{2}q_A^2$, and $CS_A=\frac{1}{2}Q_A^2$. The size relationship between q_0 and q depends on the relative size of (s_0,s), i.e., $q_0 \gtreqless q \Leftrightarrow (k-k_0)a+(n+k+2)s_0-(n+k_0+2)s \gtreqless 0$. Likewise, the size relationship between π_{0A} and π_A also depends on the relative size of (s_0,s). The equilibrium result in Table 3 claims that the increase in the public firm's subsidy s_0 after privatization brings about the increase in q_{0A}, Q_A, π_{0A}, and CS_A and the decrease in q_A, p_A, and π_A. Whether W_A increases with the s_0 depends on the situation. On the other hand, the increase in a private firm's subsidy s after privatization brings about the increase in q_A, Q_A, π_A, and CS_A and the decrease in q_{0A}, p_A, and π_{0A}. Whether W_A increases with the s also depends on the situation.

In the first stage, the government sets the discriminatory specific subsidy to the public firm and the private firms to maximize social welfare. From (3), the social welfare depends on (s_0, s) as follows:

$$W_A(s_0, s) = \frac{1}{2}Q_A^2 + \pi_{0A} + n\pi_A - s_0 q_{0A} - nsq_A. \tag{8}$$

Unlike the case before privatization, the social welfare after privatization depends on s_0. We solve the optimal subsidy levels to both public and private firms after privatization (s_{0A}^*, s_A^*). By simple calculation, we obtain the partial derivatives of the equilibrium variables with respect to s_0 and s respectively as follows:

$$\begin{aligned}
&\frac{\partial q_{0A}}{\partial s_0} = \frac{n+k+1}{n(k_0+1)+(k_0+2)(k+1)} > 0, \quad \frac{\partial q_A}{\partial s_0} = -\frac{1}{n(k_0+1)+(k_0+2)(k+1)} < 0,\\
&\frac{\partial Q_A}{\partial s_0} = \frac{k+1}{n(k_0+1)+(k_0+2)(k+1)} > 0,\\
&\frac{\partial \pi_{0A}}{\partial s_0} = (k_0+2)q_{0A}\frac{\partial q_{0A}}{\partial s_0} > 0, \quad \frac{\partial \pi_A}{\partial s_0} = (k+2)q_A\frac{\partial q_A}{\partial s_0} < 0.
\end{aligned}\tag{9}$$

$$\begin{aligned}
&\frac{\partial q_{0A}}{\partial s} = -\frac{n}{n(k_0+1)+(k_0+2)(k+1)} < 0, \quad \frac{\partial q_A}{\partial s} = \frac{k_0+2}{n(k_0+1)+(k_0+2)(k+1)} > 0,\\
&\frac{\partial Q_A}{\partial s} = \frac{n(k_0+1)}{n(k_0+1)+(k_0+2)(k+1)} > 0,\\
&\frac{\partial \pi_{0A}}{\partial s} = (k_0+2)q_{0A}\frac{\partial q_{0A}}{\partial s} < 0, \quad \frac{\partial \pi_A}{\partial s} = (k+2)q_A\frac{\partial q_A}{\partial s} > 0.
\end{aligned}\tag{10}$$

The first-order conditions for the government with respect to (s_0, s) are as follows:

$$\begin{aligned}
\frac{\partial W_A}{\partial s_0} &= Q_A\frac{\partial Q_A}{\partial s_0} + \frac{\partial \pi_{0A}}{\partial s_0} + n\frac{\partial \pi_A}{\partial s_0} - q_{0A} - s_0\frac{\partial q_{0A}}{\partial s_0} - ns\frac{\partial q_A}{\partial s_0} = 0\\
\Leftrightarrow s_0 &= \frac{Q_A\frac{\partial Q_A}{\partial s_0} + (k_0+2)q_{0A}\frac{\partial q_{0A}}{\partial s_0} + n(k+2)q_A\frac{\partial q_A}{\partial s_0} - q_{0A} - ns\frac{\partial q_A}{\partial s_0}}{\frac{\partial q_{0A}}{\partial s_0}}\\
&= \frac{(n+k+1)q_{0A} - nq_A + ns}{n+k+1}\\
\Leftrightarrow &(n+k+1)s_0 - ns = (n+k+1)q_{0A} - nq_A.
\end{aligned}\tag{11}$$

$$\begin{aligned}
\frac{\partial W_A}{\partial s} &= Q_A\frac{\partial Q_A}{\partial s} + \frac{\partial \pi_{0A}}{\partial s} + n\frac{\partial \pi_A}{\partial s} - s_0\frac{\partial q_{0A}}{\partial s} - nq_A - ns\frac{\partial q_A}{\partial s} = 0\\
\Leftrightarrow s &= \frac{Q_A\frac{\partial Q_A}{\partial s} + (k_0+2)q_{0A}\frac{\partial q_{0A}}{\partial s} + n(k+2)q_A\frac{\partial q_A}{\partial s} - s_0\frac{\partial q_{0A}}{\partial s} - nq_A}{n\frac{\partial q_A}{\partial s}}
\end{aligned}$$

$$= \frac{-q_{0A} + (k_0 + 2)q_A + s_0}{k_0 + 2}$$

$$\Leftrightarrow s_0 - (k_0 + 2)s = q_{0A} - (k_0 + 2)q_A. \tag{12}$$

By solving the simultaneous Eqs. (11) and (12) with respect to (s_0, s), we can arrange the equations as follows:

$$s_0 = q_{0A} = \frac{(k+1)a + (n+k+1)s_0 - ns}{n(k_0+1) + (k_0+2)(k+1)}$$

$$\Leftrightarrow (nk_0 + (k_0+1)(k+1))s_0 + ns = (k+1)a. \tag{13}$$

$$s = q_A = \frac{(k_0+1)a - s_0 + (k_0+2)s}{n(k_0+1) + (k_0+2)(k+1)} \Leftrightarrow$$

$$s_0 + (n(k_0+1) + k(k_0+2))s = (k_0+1)a. \tag{14}$$

Solving the simultaneous Eqs. (13) and (14) with respect to (s_0, s), we obtain the optimal subsidy levels as follows:

$$(s_{0A}^*, s_A^*) = \left(\frac{ka}{nk_0 + k(k_0+1)}, \frac{k_0 a}{nk_0 + k(k_0+1)}\right). \tag{15}$$

By we obtain that the optimal subsidy level to each firm is equal to the firm's output.

By substituting the optimal subsidy levels into the equilibrium variables in Table 3, we obtain the SPNE after privatization as shown in Table 4.

Table 4 SPNE after privatization

Public firm's subsidy	s_{0A}^*	$\frac{ka}{nk_0+k(k_0+1)}$
Private firm's subsidy	s_A^*	$\frac{k_0 a}{nk_0+k(k_0+1)}$
Public firm's output	q_{0A}	$\frac{ka}{nk_0+k(k_0+1)}$
Private firm's output	q_A	$\frac{k_0 a}{nk_0+k(k_0+1)}$
Total output	Q_A	$\frac{(nk_0+k)a}{nk_0+k(k_0+1)}$
Price	p_A	$\frac{k_0 ka}{nk_0+k(k_0+1)}$
Public firm's profit	π_{0A}	$\frac{k^2(k_0+2)a^2}{2(nk_0+k(1+k_0))^2}$
Private firm's profit	π_A	$\frac{k_0^2(k+2)a^2}{2(nk_0+k(1+k_0))^2}$
Social welfare	W_A	$\frac{(nk_0+k)a^2}{2(nk_0+k(k_0+1))}$

Note that $s_{0A}^* = q_{0A}$, $s_A^* = q_A$, $p_A + s_{0A}^* = (k_0 + 1)q_{0A}$, $p_A + s_A^* = (k + 1)q_A$, $\pi_{0A} = \frac{k_0+2}{2}q_{0A}^2$, and $\pi_A = \frac{k+2}{2}q_A^2$. The output of the more cost-efficient firm is larger than that of the less efficient one, i.e., $q_{0A} \gtreqless q_A \Leftrightarrow k \gtreqless k_0$. In the case before privatization, the public firm's subsidy s_{0B} and also its profit π_{0B} are not determined endogenously. In contrast, after privatization, the optimal public firm's subsidy and the resulting profit are uniquely determined. As $q_{0A} \gtreqless q_A \Leftrightarrow k \gtreqless k_0$, the following relationship also follows: $\pi_{0A} \gtreqless \pi_A \Leftrightarrow k \gtreqless k_0$.

3.3 *Privatization Neutrality When There Are Cost Differences Between Firms*

Comparing the SPNE before and after privatization as shown in Tables 2 and 4, we can immediately demonstrate the privatization neutrality theorem when the public firm and the private firms have different costs.

Proposition 1 *Suppose that the government gives the optimal discriminatory subsidy to both the public and the private firms with different costs. The private firm's subsidy, the public firm's output, the private firm's output, and the social welfare are the same before and after privatization. While the public firm's subsidy is not endogenously determined before privatization, it is uniquely determined after privatization. That is,* $s_B^* = s_A^* = \frac{k_0 a}{nk_0+k(k_0+1)}$, $q_{0B} = q_{0A} = \frac{ka}{nk_0+k(k_0+1)}$, $q_B = q_A = \frac{k_0 a}{nk_0+k(k_0+1)}$, $W_B = W_A = \frac{(nk_0+k)a^2}{2(nk_0+k(k_0+1))}$. s_{0B}^* *is indeterminate, whereas* $s_{0A}^* = \frac{ka}{nk_0+k(k_0+1)}$.

Proof The proof is immediately obtained by comparing the equilibrium variables in Tables 2 and 4.

Proposition 1 implies that if the government optimally sets the distortionary subsidies to the public and the private firms with cost difference, the privatization neutrality theorem still holds. This result suggests that the economic situation in which privatization neutrality holds is enlarged. Almost all the existing literature on privatization neutrality, including White (1996), Poyago-Theotoky (2001), and Myles (2002), limits the discussion to only the situation in which both the public firm and the private firms have identical technology, perhaps to avoid the complexity of the analysis. Several articles implicitly suggest that when there is a cost difference between the public firm and the private firms, the privatization neutrality theorem collapses; this suggestion is correct because if both the public firm and the private firms are given a uniform specific subsidy, $s_0 = s$, the social welfare after privatization is less than that before privatization. However, this result is attributed only to the technical constraint that the government cannot provide different subsidy levels between firms. It would be more natural that the government sets the discriminatory subsidy when there is firm asymmetry between firms and the government itself can realize such a firm asymmetry in advance. Our result demonstrates for the first time that even if there is firm asymmetry such as a cost difference, privatization neutrality always holds by adopting the discriminatory subsidy.

4 Stackelberg Competition When the Public Firm Acts as the Leader

In this section, we focus on the Stackelberg competition when the public firm acts as a leader. Fjell and Heywood (2004) showed that when the public firm behaves as a Stackelberg leader before and after privatization, the privatization neutrality theorem breaks down and social welfare decreases by privatization. However, they assumed that the government adopts the uniform specific subsidy to both firms, which differ in the timing of production. Contrary to the result of Fjell and Heywood (2004), we demonstrate that even if the public firm behaves as a leader irrespective of privatization, the privatization neutrality theorem is satisfied by adopting the discriminatory specific subsidy between the leader and the followers.

We examine the SPNE result before and after privatization in the model described in Sect. 2. As in the previous section, we allow a cost difference between the public firm and the private firms.[5] The timing of the game consists of three stages because the stages of the output choice are separated into two under Stackelberg competition. In the first stage, the government determines the optimal discriminatory specific subsidy levels to both the public and private firms to maximize social welfare. All firms observe the optimal subsidy levels after the first stage. In the second stage, the public firm, behaving as a Stackelberg leader, determines its production output q_0. In the third stage, after observing the public firm's output, all identical private firms determine their output q_i independently and noncooperatively.

4.1 Before Privatization

In the third stage, each private firm as a Stackelberg follower maximizes its profit with regard to q_i. The first-order condition for private firm i is as follows:

$$\frac{\partial \pi_i}{\partial q_i} = (a - Q + s) - q_i - kq_i = 0 \Leftrightarrow q_i = r_i(q_0, \{q_j\}_{j \neq i}) \equiv \frac{a + s - q_0 - \sum_{j \neq i,0}^{n} q_j}{k + 2}. \tag{16}$$

[5] Therefore, in this section, we show that even if there are double firm differences—in not only the timing of production but also the cost—only the adoption of the discriminatory specific subsidy is sufficient to satisfy the privatization neutrality theorem.

By substituting $q \equiv q_i$ into (16) because all firms are identical, we obtain the following reaction function:

$$q = r(q_0) \equiv \frac{a + s - q_0}{n + k + 1}. \tag{17}$$

The reaction function of an identical private firm (17) depends on their own subsidy level s and its derivative is $r'(q_0) = -\frac{1}{n+k+1} < 0$.

In the second stage, taking into consideration that the identical output of private firms follows the reaction function (17), the public firm maximizes the social welfare with regard to q_0. The first-order condition for the public firm is as follows:

$$\frac{\partial W}{\partial q_0} + n\frac{\partial W}{\partial q}r'(q_0) = a - Q - k_0 q_0 + n(a - Q - kq)r'(q_0) = 0$$
$$\Leftrightarrow q_0 = \frac{(k+1)a - nq}{nk_0 + (k_0+1)(k+1)}$$
$$\Leftrightarrow q_0 = \frac{(nk + (k+1)^2)a - ns}{(n+k+1)(nk_0 + (k_0+1)(k+1)) - n}. \tag{18}$$

Similarly to the case in Sect. 3, the public firm's output before privatization does not depend on s_0 because the subsidy level s_0 affects only the distribution between economic agents without changing the social welfare level.

By substituting (18) into (17), we obtain the private firm's output in the Stackelberg equilibrium. From simple calculation, we summarize the equilibrium variables in Table 5.

Note that $p_B^S + s = (k+1)q_B^S$, $\pi_{0B}^S = (p_B^S + s_0 - \frac{k_0}{2}q_{0B}^S)q_{0B}^S$, and $\pi_B^S = \frac{k+2}{2}(q_B^S)^2$.

The size relationship between q_{0B}^S and q_B^S depends on the relative size of s, i.e., $q_{0B}^S \gtreqless q_B^S \Leftrightarrow [(k-k_0)(n+k+1) + (k+1)]a \gtreqless (k_0+1)(n+k+1)s$. Likewise, the size relationship between π_{0B}^S and π_B^S also depends on the relative size of s. The equilibrium result shown in Table 5 claims that the increase in a private firm's

Table 5 Stackelberg equilibrium in the second stage before privatization

Public firm's output	q_{0B}^S	$\frac{(nk+(k+1)^2)a-ns}{(n+k+1)(nk_0+(k_0+1)(k+1))-n}$
Private firm's output	q_B^S	$\frac{k_0(n+k+1)a+(nk_0+(k_0+1)(k+1))s}{(n+k+1)(nk_0+(k_0+1)(k+1))-n}$
Total output	Q_B^S	$\frac{[n(k_0(n+k+1)+k)+(k+1)^2]a+n(k_0(n+k+1)+k)s}{(n+k+1)(nk_0+(k_0+1)(k+1))-n}$
Price	p_B^S	$\frac{k_0(k+1)(n+k+1)a-n(k_0(n+k+1)+k)s}{(n+k+1)(nk_0+(k_0+1)(k+1))-n}$
Public firm's profit	π_{0B}^S	$(\frac{k_0[n(k+2)+(k+1)^2]a+[nk_0(2k+3)+2(k_0+1)(k+1)^2]s}{2[(n+k+1)(nk_0+(k_0+1)(k+1))-n]} + s_0)q_{0B}^S$
Private firm's profit	π_B^S	$\frac{(k+2)(k_0(n+k+1)a+(nk_0+(k_0+1)(k+1))s)^2}{2[(n+k+1)(nk_0+(k_0+1)(k+1))-n]^2}$
Social welfare	W_B^S	$\frac{1}{2}(Q_B^S)^2 + \pi_{0B}^S + n\pi_B^S - s_0 q_{0B}^S - nsq_B^S$

subsidy before privatization brings about the increase in q_B^S, Q_B^S, π_B^S, and CS_B^S and the decrease in q_{0B}^S and p_B^S. Whether π_{0B}^S and W_B^S increase with the increase in s depends on the situation.

In the first stage, the government sets the discriminatory specific subsidy to the public firm and private firms to maximize social welfare. From (5), the social welfare depends on (s_0, s) as follows:

$$W_B^S(s) = \frac{1}{2}(Q_B^S)^2 + \pi_{0B}^S + n\pi_B^S - s_0 q_{0B}^S - nsq_B^S. \tag{19}$$

Because $\pi_{0B}^S = (p_B^S + s_0 - \frac{k_0}{2}q_{0B}^S)q_{0B}^S$ and q_{0B}^S, q_B^S, Q_B^S, and π_B^S do not depend on s_0, then the social welfare before privatization does not depend on s_0 either. The optimal public firm's subsidy is indeterminate from the viewpoint of social welfare maximization. By simple calculation, we obtain the derivatives of the equilibrium variables with respect to s as follows:

$$\begin{aligned}
&\frac{\partial q_{0B}^S}{\partial s} = -\frac{n}{(n+k+1)(nk_0+(k_0+1)(k+1))-n} < 0,\ \frac{\partial q_B^S}{\partial s} = \frac{nk_0+(k_0+1)(k+1)}{(n+k+1)(nk_0+(k_0+1)(k+1))-n} > 0,\\
&\frac{\partial Q_B^S}{\partial s} = \frac{n(k_0(n+k+1)+k)}{(n+k+1)(nk_0+(k_0+1)(k+1))-n} > 0,\ \frac{\partial p_B^S}{\partial s} = -\frac{n(k_0(n+k+1)+k)}{(n+k+1)(nk_0+(k_0+1)(k+1))-n} < 0,\\
&\frac{\partial \pi_{0B}^S}{\partial s} = (p_B^S + s_0 - k_0 q_{0B}^S)\frac{\partial q_{0B}^S}{\partial s} - \frac{\partial Q_B^S}{\partial s}q_{0B}^S,\ \frac{\partial \pi_B^S}{\partial s} = (k+2)q_B^S\frac{\partial q_B^S}{\partial s} > 0.
\end{aligned} \tag{20}$$

We solve the optimal private firm's subsidy before privatization s_B^*. The first-order condition for the government with respect to s is as follows:

$$\begin{aligned}
\frac{\partial W_B^S}{\partial s} &= Q_B^S\frac{\partial Q_B^S}{\partial s} + \frac{\partial \pi_{0B}^S}{\partial s} + n\frac{\partial \pi_B^S}{\partial s} - s_0\frac{\partial q_{0B}^S}{\partial s} - nq_B^S - ns\frac{\partial q_B^S}{\partial s} = 0\\
\Leftrightarrow s &= \frac{Q_B^S\frac{\partial Q_B^S}{\partial s} + (p_B^S + s_0 - k_0 q_{0B}^S)\frac{\partial q_{0B}^S}{\partial s} - \frac{\partial Q_B^S}{\partial s}q_{0B}^S + n(k+2)q_B^S\frac{\partial q_B^S}{\partial s} - s_0\frac{\partial q_{0B}^S}{\partial s} - nq_B^S}{n\frac{\partial q_B^S}{\partial s}}\\
&= \frac{(nk_0 + (k_0+1)(k+1))q_B^S - p_B^S + k_0 q_{0B}^S}{nk_0 + (k_0+1)(k+1)} = \frac{k_0(n+k+1)q_B^S + k_0 q_{0B}^S + s}{nk_0 + (k_0+1)(k+1)}\\
\Leftrightarrow s &= \frac{k_0((n+k+1)q_B^S + q_{0B}^S)}{k + k_0(n+k+1)} \Leftrightarrow s = \frac{k_0(a+s)}{k + k_0(n+k+1)}\\
\Leftrightarrow s_B^{S*} &= \frac{k_0 a}{nk_0 + k(k_0+1)}.
\end{aligned} \tag{21}$$

By substituting the optimal private firm's subsidy into the equilibrium variables in Table 5, we obtain the SPNE before privatization as shown in Table 6. The optimal private firm's subsidy is equal to its output, i.e., $s_B^* = q_{0B}$.

Note that $s_B^{S*} = q_B^S = \frac{k_0}{k}q_{0B}^S$, $p_B^S = k_0 q_{0B}^S$, $p_B^S + s_B^{S*} = (k+1)q_B^S$, $\pi_{0B}^S = s_0 q_{0B}^S + \frac{k_0}{2}(q_{0B}^S)^2$, and $\pi_B^S = \frac{k+2}{2}(q_B^S)^2$. The output of a more cost-efficient firm is

Table 6 SPNE before privatization

Private firm's subsidy	s_B^{S*}	$\frac{k_0 a}{nk_0+k(k_0+1)}$
Public firm's output	q_{0B}^S	$\frac{ka}{nk_0+k(k_0+1)}$
Private firm's output	q_B^S	$\frac{k_0 a}{nk_0+k(k_0+1)}$
Total output	Q_B^S	$\frac{(nk_0+k)a}{nk_0+k(k_0+1)}$
Price	p_B^S	$\frac{k_0 ka}{nk_0+k(k_0+1)}$
Public firm's profit	π_{0B}^S	$\frac{k[k_0 ka+2(nk_0+k(k_0+1))s_0]a}{2(nk_0+k(k_0+1))^2}$
Private firm's profit	π_B^S	$\frac{k_0^2(k+2)a^2}{2(nk_0+k(k_0+1))^2}$
Social welfare	W_B^S	$\frac{(nk_0+k)a^2}{2(nk_0+k(k_0+1))}$

larger than that of a less efficient one, i.e., $q_{0B}^S \gtrless q_B^S \Leftrightarrow k \gtrless k_0$. Although the size of π_{0B}^S depends on s_0, s_0 is not endogenously determined.

4.2 *After Privatization*

In the third stage, each private firm as a follower maximizes its profit. Because the first-order condition is the same as that before privatization, the reaction function satisfies (17). In the second stage, taking the reaction function (17) into consideration, the privatized public firm aims to maximize its profit. The first-order condition for the public firm is as follows:

$$\frac{\partial \pi_0}{\partial q_0} + n\frac{\partial \pi_0}{\partial q} r'(q_0) = (a - Q + s_0) - q_0 - k_0 q_0 - nq_0 r'(q_0) = 0$$

$$\Leftrightarrow q_0 = \frac{(n+k+1)(a+s_0) - n(n+k+1)q}{n(k_0+1)+(k_0+2)(k+1)}$$

$$\Leftrightarrow q_0 = \frac{(k+1)a + (n+k+1)s_0 - ns}{nk_0 + (k_0+2)(k+1)}. \tag{22}$$

The public firm's output after privatization q_0 depends on both s_0 and s.

By substituting (22) into (17), we obtain the private firm's output in the Stackelberg equilibrium. We summarize the equilibrium variables in Table 7.

It should be noted that $p_A^S + s_0 = \frac{nk_0+(k_0+1)(k+1)}{n+k+1} q_{0A}^S$, $p_A^S + s = (k+1)q_A^S$, $\pi_{0A}^S = \frac{nk_0+(k_0+2)(k+1)}{2(n+k+1)}(q_{0A}^S)^2$, and $\pi_A^S = \frac{k+2}{2}(q_A^S)^2$. The size relationship between q_0^S and q^S depends on the relative size of (s_0, s), i.e., $q_0^S \gtrless q^S \Leftrightarrow (k-k_0)a + (n+k+2)s_0 - (n+k_0+2)s \gtrless 0$. Likewise, the size relationship between π_{0A}^S and π_A^S depends on the relative size of (s_0, s).

Table 7 Stackelberg equilibrium in the second stage after privatization

Public firm's output	q_{0A}^S	$\frac{(k+1)a+(n+k+1)s_0-ns}{nk_0+(k_0+2)(k+1)}$
Private firm's output	q_A^S	$\frac{(nk_0+(k_0+1)(k+1))a-(n+k+1)s_0+(n(k_0+1)+(k_0+2)(k+1))s}{(n+k+1)(nk_0+(k_0+2)(k+1))}$
Total output	Q_A^S	$\frac{[(k+1)(n(k_0+2)+k+1)+n^2k_0]a+(k+1)(n+k+1)s_0+n(nk_0+(k_0+1)(k+1))s}{(n+k+1)(nk_0+(k_0+2)(k+1))}$
Price	p_A^S	$\frac{(k+1)(nk_0+(k_0+1)(k+1))a-(k+1)(n+k+1)s_0-n(nk_0+(k_0+1)(k+1))s}{(n+k+1)(nk_0+(k_0+2)(k+1))}$
Public firm's profit	π_{0A}^S	$\frac{((k+1)a+(n+k+1)s_0-ns)^2}{2(n+k+1)(nk_0+(k_0+2)(k+1))}$
Private firm's profit	π_A^S	$\frac{(k+2)[(nk_0+(k_0+1)(k+1))a-(n+k+1)s_0+(n(k_0+1)+(k_0+2)(k+1))s]^2}{2(n+k+1)^2(nk_0+(k_0+2)(k+1))^2}$
Social welfare	W_A^S	$\frac{1}{2}(Q_A^S)^2+\pi_{0A}^S+n\pi_A^S-s_0q_{0A}^S-nsq_A^S$

The equilibrium result in Table 7 claims that the increase in the public firm's subsidy s_0 after privatization brings about the increase in q_{0A}^S, Q_A^S, π_{0A}^S, and CS_A^S and the decrease in q_A^S, p_A^S, and π_A^S. Whether W_A^S increases with the s_0 depends on the situation. On the other hand, the increase in the private firm's subsidy s after privatization brings about the increase in q_A^S, Q_A^S, π_A^S, and CS_A^S and the decrease in q_{0A}^S, p_A^S, and π_{0A}^S. Whether W_A^S increases with the s also depends on the situation.

In the first stage, the government sets the discriminatory specific subsidy to the public firm and the private firms to maximize social welfare. From (7), the social welfare depends on (s_0, s) as follows:

$$W_A^S(s_0, s) = \frac{1}{2}(Q_A^S)^2 + \pi_{0A}^S + n\pi_A^S - s_0q_{0A}^S - nsq_A^S. \tag{23}$$

Unlike the case before privatization, social welfare after privatization depends on s_0. By simple calculation, we obtain the partial derivatives of the equilibrium variables with respect to s_0 and s respectively as follows:

$$\begin{aligned}
&\frac{\partial q_{0A}^S}{\partial s_0} = \frac{n+k+1}{nk_0+(k_0+2)(k+1)} > 0,\ \frac{\partial q_A^S}{\partial s_0} = -\frac{1}{nk_0+(k_0+2)(k+1)} < 0,\ \frac{\partial Q_A^S}{\partial s_0} = \frac{k+1}{nk_0+(k_0+2)(k+1)} > 0,\\
&\frac{\partial \pi_{0A}^S}{\partial s_0} = \frac{nk_0+(k_0+2)(k+1)}{n+k+1} q_{0A}^S \frac{\partial q_{0A}^S}{\partial s_0} > 0,\ \frac{\partial \pi_A^S}{\partial s_0} = (k+2)q_A^S \frac{\partial q_A^S}{\partial s_0} < 0.
\end{aligned} \tag{24}$$

$$\begin{aligned}
&\frac{\partial q_{0A}^S}{\partial s} = -\frac{n}{nk_0+(k_0+2)(k+1)} < 0,\ \frac{\partial q_A^S}{\partial s} = \frac{(n(k_0+1)+(k_0+2)(k+1))}{(n+k+1)(nk_0+(k_0+2)(k+1))} > 0,\\
&\frac{\partial Q_A^S}{\partial s} = \frac{n(nk_0+(k_0+1)(k+1))}{(n+k+1)(nk_0+(k_0+2)(k+1))} > 0,\\
&\frac{\partial \pi_{0A}^S}{\partial s} = \frac{nk_0+(k_0+2)(k+1)}{n+k+1} q_{0A}^S \frac{\partial q_{0A}^S}{\partial s} < 0,\ \frac{\partial \pi_A^S}{\partial s} = (k+2)q_A^S \frac{\partial q_A^S}{\partial s} > 0.
\end{aligned} \tag{25}$$

We solve the optimal subsidy levels to both the public firm and the private firms after privatization (s_{0A}^*, s_A^*). The first-order conditions for the government with respect to

(s_0, s) are as follows:

$$\frac{\partial W_A^S}{\partial s_0} = Q_A^S \frac{\partial Q_A^S}{\partial s_0} + \frac{\partial \pi_{0A}^S}{\partial s_0} + n\frac{\partial \pi_A^S}{\partial s_0} - q_{0A}^S - s_0\frac{\partial q_{0A}^S}{\partial s_0} - ns\frac{\partial q_A^S}{\partial s_0} = 0$$

$$\Leftrightarrow s_0 = \frac{Q_A^S \frac{\partial Q_A^S}{\partial s_0} + \frac{nk_0+(k_0+2)(k+1)}{n+k+1} q_{0A}^S \frac{\partial q_{0A}^S}{\partial s_0} + n(k+2)q_A^S \frac{\partial q_A^S}{\partial s_0} - q_{0A}^S - ns\frac{\partial q_A^S}{\partial s_0}}{\frac{\partial q_{0A}^S}{\partial s_0}}$$

$$= \frac{(k+1)q_{0A}^S - nq_A^S + ns}{n+k+1}$$

$$\Leftrightarrow (n+k+1)s_0 - ns = (k+1)q_{0A}^S - nq_A^S. \tag{26}$$

$$\frac{\partial W_A^S}{\partial s} = Q_A^S \frac{\partial Q_A^S}{\partial s} + \frac{\partial \pi_{0A}^S}{\partial s} + n\frac{\partial \pi_A^S}{\partial s} - s_0\frac{\partial q_{0A}^S}{\partial s} - nq_A^S - ns\frac{\partial q_A^S}{\partial s} = 0$$

$$\Leftrightarrow s = \frac{Q_A^S \frac{\partial Q_A^S}{\partial s} + \frac{nk_0+(k_0+2)(k+1)}{n+k+1} q_{0A}^S \frac{\partial q_{0A}^S}{\partial s} + n(k+2)q_A^S \frac{\partial q_A^S}{\partial s} - s_0\frac{\partial q_{0A}^S}{\partial s} - nq_A^S}{n\frac{\partial q_A^S}{\partial s}}$$

$$= \frac{-(k+1)q_{0A}^S + (n(k_0+1) + (k_0+2)(k+1))q_A^S + (n+k+1)s_0}{n(k_0+1) + (k_0+2)(k+1)}$$

$$\Leftrightarrow (n+k+1)s_0 - (n(k_0+1) + (k_0+2)(k+1))s$$
$$= (k+1)q_{0A}^S - (n(k_0+1) + (k_0+2)(k+1))q_A^S. \tag{27}$$

By solving the simultaneous Eqs. (26) and (27) with respect to (s_0, s), we can arrange the equations as follows:

$$s_0 = \frac{k+1}{n+k+1} q_{0A}^S = \frac{(k+1)((k+1)a + (n+k+1)s_0 - ns)}{(n+k+1)(nk_0 + (k_0+2)(k+1))}$$

$$\Leftrightarrow (n+k+1)[nk_0 + (k_0+1)(k+1)]s_0 + n(k+1)s = (k+1)^2 a. \tag{28}$$

$$s = q_A^S = \frac{(nk_0 + (k_0+1)(k+1))a - (n+k+1)s_0 + (n(k_0+1) + (k_0+2)(k+1))s}{(n+k+1)(nk_0 + (k_0+2)(k+1))} \Leftrightarrow$$

$$(n+k+1)s_0 + [(n+k)(nk_0 + (k_0+2)(k+1)) - n]s = (nk_0 + (k_0+1)(k+1))a. \tag{29}$$

Solving the simultaneous Eqs. (28) and (29) with respect to (s_0, s), we obtain the optimal subsidy levels as follows:

$$(s_{0A}^{S*}, s_A^{S*}) = \left(\frac{k(k+1)a}{(n+k+1)(nk_0 + k(k_0+1))}, \frac{k_0 a}{nk_0 + k(k_0+1)}\right). \tag{30}$$

Table 8 SPNE after privatization

Public firm's subsidy	s_{0A}^{S*}	$\frac{k(k+1)a}{(n+k+1)(nk_0+k(k_0+1))}$
Private firm's subsidy	s_A^{S*}	$\frac{k_0 a}{nk_0+k(k_0+1)}$
Public firm's output	q_{0A}^S	$\frac{ka}{nk_0+k(k_0+1)}$
Private firm's output	q_A^S	$\frac{k_0 a}{nk_0+k(k_0+1)}$
Total output	Q_A^S	$\frac{(nk_0+k)a}{nk_0+k(k_0+1)}$
Price	p_A^S	$\frac{k_0 ka}{nk_0+k(k_0+1)}$
Public firm's profit	π_{0A}^S	$\frac{k^2(nk_0+(k_0+2)(k+1))a^2}{2(n+k+1)(nk_0+k(k_0+1))^2}$
Private firm's profit	π_A^S	$\frac{k_0^2(k+2)a^2}{2(nk_0+k(1+k_0))^2}$
Social welfare	W_A^S	$\frac{(nk_0+k)a^2}{2(nk_0+k(k_0+1))}$

By (29), we obtain that the optimal private firm's subsidy is equal to its output, whereas the optimal public firm's subsidy is less than its output.

By substituting the optimal subsidy levels into the equilibrium variables in Table 7, we obtain the SPNE after privatization, as shown in Table 8.

It should be noted that $s_{0A}^{S*} = \frac{k+1}{n+k+1}q_{0A}^S$, $s_A^{S*} = q_A^S$, $p_A^S + s_A^{S*} = (k+1)q_A^S$, $\pi_{0A}^S = \frac{nk_0+(k_0+2)(k+1)}{2(n+k+1)}(q_{0A}^S)^2$, $\pi_A^S = \frac{k+2}{2}(q_A^S)^2$, and $W_A^S = \frac{1}{2}[(Q_A^S)^2 + k_0(q_{0A}^S)^2 + nk(q_A^S)^2]$. The output of a more cost-efficient firm is larger than that of a less efficient one, i.e., $q_{0A}^S \gtreqless q_A^S \Leftrightarrow k \gtreqless k_0$. In the case before privatization, the public firm's subsidy s_{0B}^S and also its profit π_{0B}^S are not determined endogenously. In contrast, after privatization, the optimal public firm's subsidy s_{0A}^{S*} and the resulting profit π_{0A}^S are uniquely determined. Because $q_{0A} \gtreqless q_A \Leftrightarrow k \gtreqless k_0$, the following relationship also follows: $\pi_{0A} \gtreqless \pi_A \Leftrightarrow k \gtreqless k_0$. Which is larger, s_{0A}^{S*} or s_A^{S*}, depends on the relative size of k, k_0, and n, i.e., $s_{0A}^{S*} \gtreqless s_A^{S*} \Leftrightarrow (k-k_0)(k+1) > nk_0$. If $k = k_0$, $s_{0A}^{S*} < s_A^{S*}$. That is, when both the public firm and the private firms have the identical cost, the subsidy to a Stackelberg leader is less than that to followers. Likewise, which is larger, π_{0A}^S or π_A^S, depends on the relative size of k, k_0, and n, i.e., $\pi_{0A}^S \gtreqless \pi_A^S \Leftrightarrow nk_0(k^2 - k_0(k+2)) + (k+1)(k-k_0)(k_0k + 2(k+k_0)) > 0$. Interestingly, if $k_0 = k$, $\pi_{0A}^S < \pi_A^S$. That is, when all firms have identical cost, the public firm as a leader decreases the subsidy level and the public firm's profit is less than that for the private firm.

4.3 Privatization Neutrality When the Public Firm Acts as a Stackelberg Leader

Comparing the SPNE before and after privatization, as shown in Tables 6 and 8, we can immediately demonstrate the privatization neutrality theorem when the leader public firm and follower private firms with different costs engage in the Stackelberg competition.

Proposition 2 *Suppose that the government gives the optimal discriminatory subsidy to both public and private firms with different costs when the public firm behaves as a Stackelberg leader. The private firm's subsidy, the public firm's output, the private firm's output, and the social welfare are the same before and after privatization. While the public firm's subsidy is not endogenously determined before privatization, it is uniquely determined after privatization. That is,* $s_B^{S*} = s_A^{S*} = \frac{k_0 a}{nk_0+k(k_0+1)}$, $q_{0B}^S = q_{0A}^S = \frac{ka}{nk_0+k(k_0+1)}$, $q_B^S = q_A^S = \frac{k_0 a}{nk_0+k(k_0+1)}$, $W_B^S = W_A^S = \frac{(nk_0+k)a^2}{2(nk_0+k(k_0+1))}$. s_{0B}^{S*} *is indeterminate, whereas* $s_{0A}^{S*} = \frac{k(k+1)a}{(n+k+1)(nk_0+k(k_0+1))}$.

Proof The proof is immediately obtained by comparing the equilibrium variables in Tables 6 and 8.

Proposition 2 implies that if the government sets the discriminatory subsidy to both the public firm and private firms, the privatization neutrality theorem still holds even in the situation in which there are multiple firm differences in costs and production timing between public and private firms. It should be noted that, even if firm heterogeneity exists between public and private firms, the discriminatory specific subsidy appropriately adjusts the differences between firms and realizes the optimal level of social welfare. As a result, it suggests that by adopting the discriminatory subsidy, there is no need for the government to be concerned about whether to privatize the public firm. Our result demonstrates for the first time that, irrespective of the difference in the move of output choice between the public firm and private firms, the privatization neutrality theorem always holds by adopting the discriminatory subsidy.

Finally, we compare the optimal subsidy level of the privatized public firm in Cournot competition with that in Stackelberg competition when the public firm is the leader. We obtain the following proposition.

Proposition 3 *The privatized public firm's subsidy in Stackelberg equilibrium is less than that in Cournot equilibrium. That is,*

$$s_{0A}^{*} = \frac{ka}{nk_0 + k(k_0+1)} > s_{0A}^{S*} = \frac{k(k+1)a}{(n+k+1)(nk_0+k(k_0+1))}.$$

Proof The proof is immediately obtained by comparing the equilibrium subsidy in Tables 4 and 8.

The privatized public firm loses the first-mover advantage by the reduction of the discriminatory subsidy. As a result, the social welfare is the same in both Cournot and Stackelberg equilibria because the discriminatory subsidy can achieve the maximized social welfare in all cases. This result suggests that, in all cases with firm heterogeneity, the generalized privatization neutrality theorem is satisfied.

5 Concluding Remarks

This study revisited the privatization neutrality theorem that when the government gives the optimal subsidy to both public and private firms in a mixed oligopoly, social welfare is exactly the same before and after privatization. If the public and private firms have different technology before privatization, the uniformity of subsidy for both firms is not justifiable. We demonstrated that if different subsidy rates are adopted, even when there is firm asymmetry between public and private firms, the privatization neutrality theorem continues to hold. Even if the cost of the public firm differs from those of private firms, the privatization neutrality theorem holds by appropriately subsidizing both public and private firms in a discriminatory manner. Furthermore, even if the public firm acts as a Stackelberg leader before and after privatization, the government can attain privatization neutrality and achieve social welfare maximization by appropriately setting the subsidy level for each firm. Our result suggests that, even when the government can subsidize heterogeneous firms in a mixed oligopoly, the application of the discriminatory subsidy policy enables the government to attain the maximized social welfare and brings about privatization neutrality in wider circumstances.

We conclude with a brief discussion of possible extensions. This study focuses only on firm asymmetry between the public firm and private firms as a hindrance to privatization neutrality. However, as briefly described in the Introduction, there are several other hindrances in which the privatization neutrality theorem does not hold. We expect that when adopting the discriminatory subsidies to the public firm and private firms in such cases, the privatization neutrality theorem might still be satisfied. For example, the existence of foreign private firms undermines the neutrality of privatization. In our preliminary results, however, we obtained that in the international mixed oligopoly in which foreign private firms enter in the market, the neutrality theorem does not hold even when discriminatory subsidies are adopted. Furthermore, a very common argument in favor of privatization is the effect of privatization on costs of public firms. If there is a cost efficiency on the privatized public firm, welfare neutrality might not be satisfied and privatization might increase welfare. In addition, if subsidies paid by means of taxation cause taxation inefficiency, would the neutrality still hold? Further consideration of the privatization neutrality theorem under discriminatory subsidies is a future challenge.

Acknowledgements I would like to thank Toshihiro Ichida, Kunizaki Minoru, Tohru Naito, Mitsuyoshi Yanagihara, and an anonymous reviewer as well as other seminar participants of the 40th Nagoya International Economic Study Group (NIESG) at Matsuyama University and the seminar of mixed oligopoly theory at Nagoya University, for valuable comments and suggestions. This study was supported in part by Japan Society for the Promotion of Science (JSPS) Grant-in-Aid for Scientific Research (C) No. 25380286. The author is solely responsible for any errors.

References

Cato S, Matsumura T (2013) Long-run effects of tax policies in a mixed market. FinanzArchiv Public Financ Anal 69(2):215–240
De Fraja G, Delbono F (1989) Alternative strategies of a public enterprise in oligopoly. Oxf Econ Pap 41(2):302–11
Fjell K, Heywood JS (2004) Mixed oligopoly, subsidization and the order of firm's moves: the relevance of privatization. Econ Lett 83(3):411–416
Gil-Moltó MJ, Poyago-Theotoky J, Zikos V (2011) R&D subsidies, spillovers, and privatization in mixed markets. South Econ J 78(1):233–255
Hashimzade N, Khodavaisi H, Myles GD (2007) An irrelevance result with differentiated goods. Econ Bull 8(2):1–7
Kato H (2008) Privatization and Government preference. Econ Bull 12(40):1–7
Kato K, Tomaru Y (2007) Mixed oligopoly, privatization, subsidization, and the order of firms' moves: several types of objectives. Econ Lett 96(2):287–292
Matsumura T, Tomaru Y (2013) Privatization neutrality theorem revisited. Econ Lett 118(2):324–326
Matsumura T, Tomaru Y (2012) Market Structure and privatization policy under international competition. Jpn Econ Rev 63(2):244–258
Matsumura T, Tomaru Y (2013) Mixed duopoly, privatization, and subsidization with excess burden of taxation. Can J Econ 46(2):526–554
Myles GD (2002) Mixed oligopoly, subsidization and the order of firms' moves: an irrelevance result for the general case. Econ Bull 12(1):1–6
Okuguchi K (2012) General analysis of Cournot mixed oligopoly with partial privatization. Eurasian Econ Rev 2(1):48–62
Poyago-Theotoky J (2001) Mixed oligopoly, subsidization and the order of firms' moves: an irrelevance result. Econ Bull 12(3):1–5
Tomaru Y (2006) Mixed oligopoly, partial privatization and subsidization. Econ Bull 12(5):1–6
Tomaru Y (2014) Economic analysis on competition between public and private firms and privatization. Keiso Shobo Publishing, Tokyo (Japan)
Tomaru Y, Saito M (2010) Mixed duopoly, privatization and subsidization in an endogenous timing framework. Manch Sch 78(1):41–59
White MD (1996) Mixed oligopoly, privatization and subsidization. Econ Lett 53(2):189–195
Zikos V (2007) A reappraisal of the irrelevance result in mixed duopoly: a note on R&D competition. Econ Bull 12(8):1–6

Cournot Oligopoly Theory for Simple Electricity Markets

Alexander Vasin and Polina Kartunova

Abstract The present paper surveys and generalizes the previous results on Cournot competition for the classic model and two-node markets in context of electricity markets development. We recall results on existence and computation of Cournot-Nash equilibrium and its relation to the uniform price auction. Then for a two-node market we consider different types of equilibria with binding and unbinding transmission capacity constraints. For each type we examine local equilibria which meet first-order conditions, their computation employing the concept of Cournot supply function, existence and co-existence of different types' equilibria. We search for the optimal transmission capacity that maximizes the social welfare, recall the previous results for a perfectly competitive two-node market and study the problem for the Cournot oligopoly.

1 Introduction

Starting with the seminal paper by Cournot (1838), and up till now, the Cournot oligopoly model develops and finds more and more applications. In particular, Okuguchi (2007, and previous papers) examines its generalization for markets with product differentiation. Another recent direction considers the model with account of markets for production factors (Okuguchi 2010; Bertoletti and von Mouche 2014). The present paper discusses development and application of the Cournot oligopoly theory for electricity markets. Oligopolistic structure is typical for them. In particular, for European electricity markets the share of the largest company in the total production typically exceeds 0.25 (see Newbery 2009). All electricity markets are organized as auctions of some type. The typical architecture is a uniform price auction or its generalization for markets with a network structure. Our paper Vasin and Vasina (2005) justifies application of the Cournot oligopoly theory for analysis of electricity markets organized as a uniform price auction under general assumptions. We study Nash equilibria of the auction and show that only

A. Vasin (✉)
Lomonosov Moscow State University, Moscow, Russia
e-mail: vasin@cs.msu.su

P. von Mouche, F. Quartieri (eds.), *Equilibrium Theory for Cournot Oligopolies and Related Games*, Springer Series in Game Theory,
DOI 10.1007/978-3-319-29254-0_11

the equilibria corresponding to the Cournot-Nash outcome are stable with respect to adaptive dynamics in such auction. (Kukushkin (2016) provides some general results on convergence of such dynamics to the equilibrium.)

Many electricity markets are still in the development process. Computation of the Cournot-Nash outcome and evaluation of its deviation from the Walrasian equilibrium permits to compare different variants of the market structure with respect to the social welfare optimization. For this purpose Vasin and Vasina (2005) provide an efficient method employing the Cournot supply schedules' technique. (Cornes and Sato (2016) consider another approach to the problem.) A typical generating company owns several generators with limited capacities and (approximately) constant or linear marginal costs. For this case our method permits to easily compute the Cournot-Nash outcome. Another important problem relates to a network structure of many electricity markets. Consumers and producers are located at different nodes connected by the lines with limited transmission capacities. The literature on such markets (Hogan 1998; Davidson et al. 2009; Vasin et al. 2007) considers primarily models with a fixed network structure and studies Walrasian and Cournot equilibria of the markets. The classic Cournot model is applicable only for the markets where the transmission capacity constraints and losses are not substantial. Below we recall the results for a generalization of the Cournot model for the simplest network market: a two-node market with a transmission line connecting the submarkets. Vasin et al. (2007) describes possible types of Cournot-Nash equilibria for a two-node market, provides a method for their computation and establishes necessary and sufficient conditions for their existence. Vasin and Shamanaev (2010) examine the equilibria' set structure.

One more research direction relates to optimization of the transmission system. The recent paper by Daylova and Vasin (2014) studies a two-node market with a variable transmission capacity. They find out characteristics of the Cournot-Nash equilibrium depending on the capacity. They also set a problem of the social welfare maximization and determine the optimal transmission capacity under conditions of perfect competition, taking into account producers' profits, consumers' surpluses, losses under transmission and costs of the transmission system expansion.

The present study aims to survey, specify and generalize the previous results on Cournot competition. Section 2 reviews results for the classic Cournot model: propositions on existence and uniqueness of NE, deviation of the Cournot outcome from the Walrasian equilibrium, correspondence of this outcome to the equilibria of other market models, in particular, of the uniform price auction. In Sect. 3 we survey results on Cournot-Nash equilibria for the two-node market: we consider different types of equilibria with binding and unbinding transmission capacity constraints; for each type we examine local equilibria which meet first-order conditions, their computation employing the concept of Cournot supply function, existence and co-existence with equilibria of different types.

We provide conditions for each type of the local equilibrium to be a true Nash equilibrium. Section 4 discusses the social welfare optimization problem and the search for the optimal transmission capacity. We recall the results by Daylova and Vasin (2014) for a perfectly competitive two-node market and study the problem for the Cournot oligopoly.

Below we assume that the transmission system works as if there is a perfect competition among intermediaries who can buy electricity at one node and sell it at the other. For many real markets, the system operators regulate the flows according to this assumption (see Hogan 1998). Hogan (2007) studies the different case and provides a literature review on transmission grid expansion for electricity markets with transmission lines controlled by private transmission companies. The results below relate also to other markets of homogeneous goods, in particular, to natural gas and oil markets.

2 Previous Results on the Standard Cournot Oligopoly Model

2.1 Existence and Computation of Cournot-Nash Equilibria

Many papers (see Novshek 1985; Kukushkin 1994; Amir 1996; Amir and Lambson 2000 and others) study existence and computation problems for the Cournot-Nash equilibria. Vasin et al. (2007) prove existence of the unique Nash equilibrium under non-decreasing demand elasticity and marginal costs for an oligopoly with heterogeneous producers. They develop a constructive method employing the Cournot supply functions to compute Cournot-Nash equilibria.

The formal model and the results are as follows. Consider a market for a homogeneous good with a finite set A of producers. Each producer a is characterized by his cost function $C^a(v)$ with non-decreasing marginal costs for $v \in [0, V^a]$ where V^a is his production capacity. Consumers' behavior is characterized by a demand function $D(p)$, which is continuously differentiable, decreasing in p, tends to 0 as p tends to infinity, and is known to all agents.

Consider a model of Cournot competition in this market. Then a strategy for each producer a is his production volume $v^a \in [0, V^a]$. Producers set these values simultaneously. Let $\overrightarrow{v} = (v^a, a \in A)$ denote a strategy combination. The market price $p(\overrightarrow{v})$ equalizes the demand with the actual supply: $p(\overrightarrow{v}) = D^{-1}(\sum_{a \in A} v^a)$. The payoff function of producer a determines his profit $f^a(\overrightarrow{v}) = v^a p(\overrightarrow{v}) - C^a(v^a)$ Thus, the interaction in Cournot model corresponds to the normal form game $\Gamma_C = \langle A, [0, V^a], f^a(\overrightarrow{v}), \overrightarrow{v} \in \otimes_{a \in A}[0, V^a], a \in A \rangle$ where $[0, V^a]$ is the set of strategies for $a \in A$. Combination $(v^{a*}, a \in A)$ of production volumes is a *Cournot equilibrium* (CE) if it is a NE in the game Γ_C.

Let $(v^{a*}, a \in A)$ denote Nash equilibrium production volumes and let $p^* = D^{-1}(\sum_{a \in A} v^{a*})$ be the corresponding price. A necessary and sufficient condition for this collection to be a Nash equilibrium is that for any a,

$$p^* \in \operatorname{Argmax}_{p\in\left[D^{-1}\left(\sum\limits_{b\neq a} v^{b*}+V^a\right), D^{-1}\left(\sum\limits_{b\neq a} v^{b*}\right)\right]}$$
$$\left\{\left(D(p) - D(p^*) + v^{a*}\right) p - C^a \left(D(p) - D(p^*) + v^{a*}\right)\right\}.$$

Then the F.O.C. for a Nash equilibrium is

$$v^{a^*} \in (p^* - C^{a'}(v^{a^*}))|D'(p^*)| \textit{ for any } a \textit{ s.t. } C^{a'}(0) < p^*, \tag{1}$$

$$v^{a^*} = 0 \textit{ if } C^{a'}(0) \geqslant p^*, \tag{2}$$

where $C^{a'}(v) = [C^{a'}_{-}(v), C^{a'}_{+}(v)]$ in the break points of the marginal cost function. In particular, $C^{a'}_{+}(V^a) = \infty$. Combination $(p^*, v^{a^*}, a \in A)$ is called *a local Cournot equilibrium* if it meets the necessary conditions (1) and (2).

Cournot supply function $S^a_C(p)$ of a producer a is defined for $p > 0$ as a solution of the system (1) and (2). This function determines the optimal production volume of producer a if p is a Cournot equilibrium price. The function is uniquely defined under the given conditions. In particular, consider the case with piece-wise constant marginal costs function: $C^a(0) = 0$, $C^{a'}(v) = c^a_i$ for $\sum\limits_{k\leqslant i-1} V^a_k < v < \sum\limits_{k\leqslant i} V^a_k$, $c^a_{i-1} \leqslant c^a_i, i = 1, \ldots, m, V^a_0 = 0, \sum\limits_{k\leqslant m} V^a_k = V^a$, and affine demand function $D(p) = \max(0, \overline{D} - dp)$. Then

$$S^a_C(p) = \begin{cases} 0, p < c^a_1, \\ (p - c^a_1)d \text{ if } (p - c^a_1)d < V^a_1, \\ V^a_1 \text{ if } (p - c^a_2)d < V^a_1 < (p - c^a_1)d, \\ (p - c^a_2)d \text{ if } V^a_1 \leqslant (p - c^a_2)d < V^a_1 + V^a_2, \\ V^a_1 + V^a_2 \text{ if } (p - c^a_2)d < V^a_1 + V^a_2 < (p - c^a_3)d, \\ \ldots\ldots\ldots \\ V^a \text{ if } (p - c^a_m)d > V^a. \end{cases}$$

Condition (1) implies this system: if $v^{a^*} \in (0, V^a_1)$ then $C^{a'}(v^{a^*}) = c^a_1$; if $v^{a^*} = V^a_1$ then

$$v^{a^*}_1 \in [(p - c^a_2)d, (p - c^a_1)d],$$

and so on. Figure 1 shows a typical form of this function. The Cournot price p^* is determined by the equation $\sum\limits_a S^a_C(p^*) = D(p^*)$. A simple algorithm finds the sum of functions $S^a_C(p)$ and its intersection with D(p).

Fig. 1 Cournot supply function

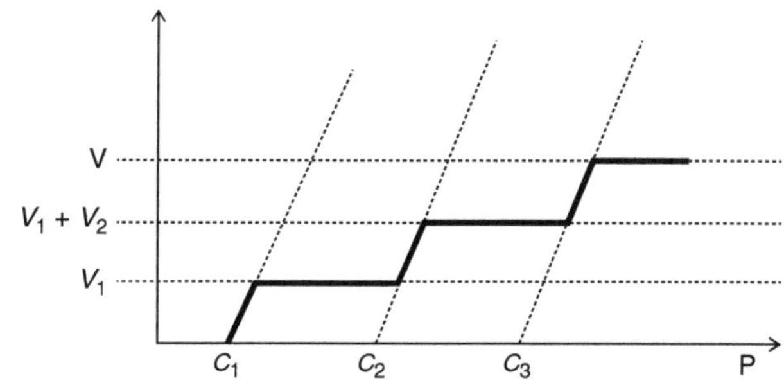

For the considered case, it is obvious that the price is unique: the demand function is decreasing while the functions S^a_C are non-decreasing. In the general case, the following proposition provides a sufficient condition for a unique Cournot price.

Proposition 1 *Let the demand function $D(p)$ and the demand elasticity $e(p) \stackrel{def}{=} p\frac{|D'(p)|}{D(p)}$ meet one of the following conditions:*

(a) $D(p) > 0$ and $e(p)$ does not decrease in p for $p \in (\tilde{p}, M)$, $D(p) = 0$ for $p \geqslant M$
(b) $D(p) > 0$ and $e(p)$ does not decrease in p for $p \geqslant \tilde{p}$, $\lim_{p\to\infty} e(p) > 1/n$, where $\tilde{p}$ is the Walrasian price and n is the total number of producers in the market. Then there exists a unique Nash equilibrium in the game Γ_C.

The idea of the proof is that p^* meets the F.O.C. for Nash equilibrium iff it is a solution of the equation $F(p) \stackrel{def}{=} \sum_a S^a_C(p)/D(p) = 1$. Under the given conditions, the function $F(p)$ is continuous and monotonous in the interval $(0, M)$. Indeed since D(p) decreases, the ratio $S^a_C(p)/D(p)$ increases at any point where $S^a_C(p)$ does not decrease. Otherwise consider the relation $S^a_C(p)/D(p) = e(p)(1 - \frac{C^{a'}_p(S^a_c(p))}{p})$ that follows from (1). Since the demand elasticity and the marginal cost do not decrease, the right-hand side is decreasing in p whenever $S^a_C(p)$ is decreasing.

2.2 Estimate of Deviation from Walrasian Equilibrium

Vasin et al. (2007) show that the relative deviation of the Cournot price from the Walrasian price is less or equal to the share of the largest company in the total production volume, divided by the demand elasticity. This estimate coincides with the Lerner index for the company and is precise if its marginal costs are equal for the Walrasian and the Cournot equilibria. Newbery (2009) considers these results in context of the data for European electricity markets. The share of the largest company typically exceeds 0.25, while the demand elasticity is less than 0.2. Thus, the data obviously contradicts to the Cournot model. Newbery calls it as Lerner Paradox and discusses different explanations.

Consider the formal results. Combination $(v^{a*}, a \in A)$ of production volumes is a *Walrasian equilibrium* (WE) and $\tilde{p}$ is a *Walrasian price* of the market if, for any a,

$$\tilde{v}^a \in S^a(\tilde{p}) \overset{def}{=} Arg \max_{v^a}(v^a \tilde{p} - C^a(v^a)), \tag{3}$$

$\sum_a \tilde{v}^a = D(\tilde{p})$. The *Walrasian supply function* $S^a(p)$ determines the (generally non-unique) optimal production volume of firm a under a given price p. Formally, it is a non-decreasing closed upper semi-continuous point-set mapping with convex values. A trivial result is that the unique Walrasian price exists under the specified assumptions on the demand function.

The next proposition evaluates the deviation of the Cournot outcome from the Walrasian equilibrium proceeding from the demand elasticity and the maximal share of one firm in the total production at the Cournot equilibrium.

Proposition 2 *For every firm a, its Lerner index at the Cournot equilibrium meets equation*

$$L_c^a \overset{def}{=} \frac{p^* - C^{a'}(v^{a^*})}{p^*} = \frac{sh^a(p^*)}{e(p^*)}, \tag{4}$$

where $sh^a(p^*) = \dfrac{v^{a^*}}{D(p^*)}$ *is the share of producer a in the total production volume,* $e(p)$ *is the demand elasticity. Moreover,*

$$\frac{p^* - \tilde{p}}{p^*} \leqslant \max_a \frac{sh^a(p^*)}{e(p^*)}. \tag{5}$$

Proof Relation (4) is equivalent to (1) whenever $C^{a'}$ is continuous at v^{a^*}. For at least one producer b, $C^{b'}(v^{b^*}) \leqslant \tilde{p}$. (Otherwise $v^{a^*} \geqslant S^a(\tilde{p})$ and $p^* < \tilde{p} \leqslant C^{a'}(v^{a^*})$ for every $a \in A$. But, proceeding from (1), $p^* - C^{a'}(v^{a^*}) \geqslant 0$ for every a.) Hence,

$$\frac{p^* - \tilde{p}}{p^*} \leqslant \frac{p^* - C^{b'}(v^{a^*})}{p^*} = \frac{sh^b(p^*)}{e(p^*)} \leqslant \max_a \frac{sh^a(p^*)}{e(p^*)}. \tag{6}$$

Note: condition (5) holds as the equality for a symmetric oligopoly with a fixed marginal cost $c = p$, and also for a large firm with a fixed marginal cost interacting with the competitive environment characterized by a smaller marginal cost and a limited total capacity.

2.3 *Relation of Cournot Oligopoly Model to Some Other Auctions*

Vasin et al. (2007) study a uniform price auction, where a strategy of each producer is a non-decreasing step function that determines the actual supply of electric energy depending on the price. Such auctions are typical for wholesale electricity markets, in particular, for day-ahead markets where the basic production and consumption components are determined. The real auctions differ in the rules for acceptable bids. Russian DAM accepts bids with at most three steps, a different bid for every hour of the next day, while the market of England and Welsh permitted up to 48 steps, but a unique bid for the whole day.

The mentioned study shows that, for any Nash equilibrium of the corresponding game, the market price lies between the competitive equilibrium price and the Cournot price, and vice versa, each price in this range corresponds to a Nash equilibrium. However, only the Nash equilibrium corresponding to the Cournot outcome is stable with respect to the dynamics of adaptive strategies.

Note that Moreno and Ubeda (2006) obtain similar results for the two-step model where at the first step producers set capacities, and at the second step they compete by setting reserve prices. Kreps and Scheinkman (1983) show that the SPE outcome of the two-stage model "first quantities, then prices" also corresponds to the Cournot equilibrium.[1]

The formal model of the uniform price auction and the results are as follows. Every producer $a \in A$ sends to the auctioneer his bid function $R^a(p)$ that determines the amount of energy this producer is ready to sell at price $p, p \geqslant 0$. Acceptable bid $R^a(p)$ is a non-decreasing step function with a limited number of steps. So this is not a usual function but a point-set mapping: at any jump point its value is a stretch, and it obtains the same properties as a Walrasian supply function.

A profile of bids determines the total supply $R(p) = \sum_a R^a(p)$ and the cut-off price $\tilde{c}(R^a, a \in A)$ that meets condition $D(\tilde{c}) \in R(\tilde{c})$. Proceeding from the properties of the demand function, the cut-off price is uniquely determined for any non-zero supply, as well as the Walrasian price in the market model. In order to define payoff functions, consider two cases. Let $R^+(p) \stackrel{def}{=} \max R(p), R^-(p) \stackrel{def}{=} \min R(p)$. If $R^+(\tilde{c}) = D(\tilde{c})$ then each producer sells the volume $R^{a+}(\tilde{c})$ at the cut-off price. Otherwise, first each producer sells $R^{a-}(\tilde{c})$, and then the residual demand $D(\tilde{c}) - R^-(\tilde{c})$ is distributed among producers with $R^{a+}(\tilde{c}) > R^{a-}(\tilde{c})$ according to some rationing rule (typically the proportional rule).

[1]One implication from the mentioned results is that for all these mechanisms the expected deviation of the market price from the Walrasian price is the same as for the Cournot model, and they are not responsible for solution of the Lerner paradox.

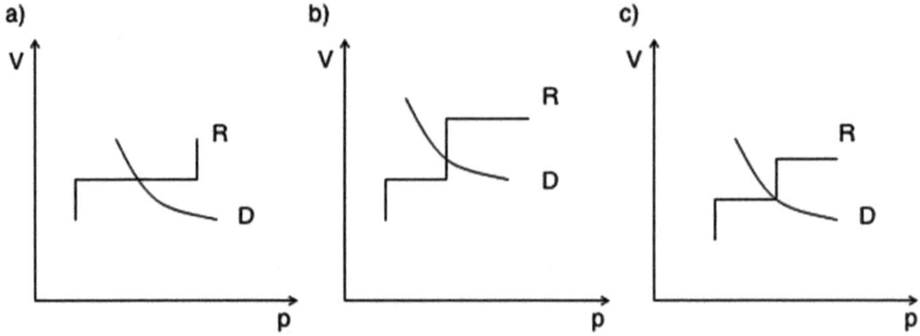

Fig. 2 Three types of Nash equilibria in the uniform price auction

Under a given rationing rule, the profit of producer $b \in A$ is determined as follows:

$$f^b(R^a(.), a \in A) = \tilde{c}(R^a, a \in A)v^b(R^a, a \in A) - C^b(v^b(R^a, a \in A)),$$

where $v^b(R^a, a \in A) \in [R^{b-}(\tilde{c}), R^{b+}(\tilde{c})]$ is the final demand for his production. Thus, we have defined the normal form game Γ_S that corresponds to the auction.

Note that there might be three possible types of Nash equilibria for Γ_S: (a) those Nash equilibria for which $R(\tilde{c}) = D(\tilde{c})$ (Nash equilibria without rationing), (b) those for which $D(\tilde{c}) \in (R^-(\tilde{c}), R^+(\tilde{c}))$ (Nash equilibria with rationing), (c) those for which $D(\tilde{c}) = R^-(\tilde{c}) < R^+(\tilde{c})$ (Nash equilibria with a barrier, see Fig. 2).

Proposition 3 *Let the market meet conditions of Proposition 1. Then:*

(*a*) *For every Nash equilibrium without rationing, the production volumes correspond to the local Cournot equilibrium. Vice versa, if* $(v^a, a \in A)$ *is a Cournot equilibrium, then the corresponding Nash equilibrium exists in* Γ_S.

(*b*) *If* $(R^a, a \in A)$ *is a Nash equilibrium such that* $D(\tilde{c}) \in (R^-(\tilde{c}), R^+(\tilde{c}))$, *then there exists at most one producer* $b \in A$ *such that* $R^{b-}(\tilde{c}) < S^{b-}(\tilde{c})$ *(so* $v^a \in S^a(\tilde{c})$ *for any* $a \neq b$*); the cut-off price lies in the interval* $[\tilde{p}, p^*]$.

(*c*) *For any Nash equilibrium of the type (c), the cut-off price lies in the interval* $[\tilde{p}, p^*]$. *Vise versa, for any* $p \in [\tilde{p}, p^*]$ *there exists a Nash equilibrium* $(R^a, a \in A)$ *such that* $\tilde{c}(R^a, a \in A) = p$.

The first part of (a) is true since all bids $R^a(p)$ are constant in some interval around $\tilde{c}$ for the a-type equilibrium. So $v^a = R^a(\tilde{c})$, $a \in A$, and $\tilde{c}$ meets F.O.C. similar to (1) for the Cournot equilibrium $(v^{a^*}, a \in A, p^*)$. For such equilibrium, the profile of bids $(R^a(p) \equiv v^{a^*}, a \in A)$ is a NE in Γ_S with the same outcome. Typically there exist other NE with this outcome. In particular, Vasin et al. (2007) find out conditions for bids

$$R^a(p) = \begin{cases} S^a(p) \ until \ S^a(p) < v^{a^*}, \\ v^{a^*} otherwise, \end{cases}$$

(secure under the demand random falls), $a \in A$, to form such NE. The construction of the c-type equilibrium corresponding to price $\overline{p} \in [\tilde{p}, p^*]$ is as follow. We find $v^a \in [v_C^a(\overline{p}), S^a(\overline{p})], a \in A$, such that $\sum_a v^a = D(\overline{p})$ (they exist because $S_C^a(p) \leqslant S^a(p)$ for any p) and set

$$R^a(p) = \begin{cases} v^a, p < \overline{p}, \\ v^a + J^a, p > \overline{p}, \end{cases}$$

$a \in A$, where the jumps J^a are so large that it is unprofitable for any player to increase the cut-off price by reducing his supply level in the neighbourhood of the price $\tilde{c}$. For a b—type equilibrium, one company (the leader) makes such barrier for every other company (from the competitive environment). However, keeping this barrier is unprofitable. Reduction of $R^{a+}(\tilde{c})$ to v^a for any $a \in A$ does not change the profits of the players if other strategies are fixed. Moreover, as soon as the barrier is sufficiently small, some player finds it profitable to reduce his supply function and thus increase the profits of the other players. Thus, every Nash equilibrium of the types b and c is unstable and cannot occur as an outcome of the auction until some players act for the interests of some external regulator. The expected outcome of the auction corresponds to the Cournot equilibrium, and the estimates from Proposition 2 also hold in this case.

3 Cournot Equilibria for Two-Node Market

Vasin and Vasina (2005, 2006) consider a simple network electricity market: the market with two nodes. Each local market is characterized by the demand function and the finite set of producers with non-decreasing marginal costs. For every producer, his strategy is a reported supply function that determines his supply of electric power depending on the price. The markets are connected by a transmitting line with fixed share of losses and a limited transmission capacity. For any line in the whole sale electricity market, these values are the known technical characteristics related to the electric resistance of the line. Under given strategies of producers, the network administrator first computes the cut-off prices for the separated markets. If the ratio of the prices is sufficiently close to one then transmission is unprofitable with account of the loss. In this case, the outcome is determined by the cut-off prices for the isolated markets. Otherwise the network administrator sets the flow to the market with the higher cut-off price (for instance market 2). This flow reduces the supply and increases the cut-off price at the market 1. Simultaneously it increases the supply and reduces the cut-off price at the market 2. If the transmitted volume does not exceed the transmission capacity, the network administrator determines this volume so that the ratio of the final cut-off prices corresponds to the loss coefficient. Otherwise, the administrator sets the volume to be equal to the transmission capacity. Thus, he acts as if perfectly competitive intermediaries transmit the good from one market to the other. Such strategy maximizes the total welfare if the reported supply functions (the bids) correspond to the actual costs.

For Cournot competition in this market, the study shows that there exist three possible types of Nash equilibrium: (A) an equilibrium with zero flow between the markets and the ratio of the prices close to 1; such equilibrium is determined as if there are two separated markets; (B) an equilibrium with a positive flow and the ratio of the prices corresponding to the loss coefficient; (C) an equilibrium with a positive flow equal to the transmission capacity and the ratio of the prices exceeding the level for B-type equilibrium. Proceeding from the first order condition, local equilibria of each type are defined, and it is shown how to compute them by means of Cournot supply functions. The study finds out, under what conditions a local equilibrium is a real Nash equilibrium. For the market with constant marginal costs and affine demand functions, it determines the set of Nash equilibria depending on the parameters. One interesting finding is that, in the symmetric case with equal parameters of the local markets and a small loss coefficient, the local equilibrium corresponding to the isolated markets is not a Nash equilibrium, but there exist two asymmetric Nash equilibria with a positive flow of the good.

Vasin et al. (2012) show that in general there exists one more D-type of NE in this model, such that the both constraints for the ratio of the prices and for the transmitted volume are binding. Daylova and Vasin (2014) study the structure of the local equilibria set depending on the transmission capacity. They show that, for some interval of this parameter, only the D-type equilibrium may exist.

The formal results are as follows.

Consider two local markets connected by a transmitting line. Every local market $l = 1, 2$ is characterized by the finite set A^l of producers, $|A^l| = n_l$, the cost functions $C^a(v), a \in A^l$, and demand function $D^l(p)$, in the same way as the local market in Sect. 2. Let $k \in (0, L)$ be the loss coefficient that shows the share of the lost electric power under transmission from one market to the other, Q is the maximal transmitted power. Under Cournot competition, each producer sets $v^a \in [0, V^a]$. Simultaneously and independently each firm reports the auctioneer its strategy. For a given strategy combination, nodal cut-off prices $\tilde{c}^l$ and transmitted volume q are determined as follows.

Let $\overline{c}^l(\overline{v}), l = 1, 2$ denote the cut-off prices for isolated markets, such that $D^l(\overline{c}^l) = \sum_{a \in A} v_l^q$, $\lambda = (1-k)^{-1}$. If $\lambda^{-1} \leqslant \overline{c}^2(\overline{v})/\overline{c}^1(\overline{v}) \leqslant \lambda$ then $q = 0$, the final prices are $\tilde{c}^l(\overline{v}) = \overline{c}^l(\overline{v}), l = 1, 2$, that is, the markets stay isolated (*A-type* outcome below). If $\overline{c}^2(\overline{v})/\overline{c}^1(\overline{v}) > \lambda$ then q (the transmitted volume from market 1 to market 2) is equal to the solution $\overline{q}$ of the system

$$D^1(\tilde{c}^1) = \sum_{A^1} v^a - \overline{q}; \tag{7}$$

$$D^2(\tilde{c}^2) = \sum_{A^2} v^a + \overline{q}/\lambda; \tag{8}$$

$\tilde{c}^2 = \lambda\tilde{c}^1$, until $\overline{q} > Q$ (B_{1-2} type outcome). The relation of the prices means the "no arbitrage" condition under the unbinding transmission capacity constraint. Equations (7) and (8) reflect the supply-demand balances at the nodes. If $\overline{q} > Q$ then $q = Q$, $\tilde{c}^1$, $\tilde{c}^2$ are determined from (7) and (8) with $\overline{q} = Q$, and $\tilde{c}^2 > \lambda\tilde{c}^1$ (C_{1-2} type outcome). The capacity constraint is binding in this case. The unique solution of the system exists because the involved functions are monotonous and continuous. The case $\overline{c}^1/\overline{c}^2 > \lambda$ is treated in the symmetric way.

The first-order conditions for the A-type outcome with prices

$$p_1^*, p_2^* \ such \ that \ \lambda^{-1} < p_2^*/p_1^* < \lambda, \tag{9}$$

to be a Cournot equilibrium are quite similar to the conditions (1) and (2) for the local market:

$$v^{a*} \in \left(p_i^* - C^{a'}(v^{a*})\right) \left|D^{i'}(p_i^*)\right|, \ for \ any \ a \in A^i \ s.t. \ C^{a'}(0) < p_i^*, \tag{10}$$

$$v^{a*} = 0 \ if \ C^{a'}(0) \geqslant p_i^*, \tag{11}$$

where $C^{a'}(v) = \left[C_-^{a'}(v), C_+^{a'}(v)\right]$ in the break points of the marginal cost function. Besides that,

$$\sum_{A^i} v^{a*} = D^i(p_i^*), i = 1, 2. \tag{12}$$

For the B_{1-2}-type outcome with

$$q \in (0, Q), \lambda p_1^* = p_2^*, \tag{13}$$

the first-order conditions of the Cournot equilibrium are obtained in a similar way. Note that, for any small change of the strategy v^{a*}, producer $a \in A^1$ stays in the market with the demand function

$$D^1(p_1(\overline{v})) + \lambda \left(D^2(\lambda p_1(\overline{v})) - \sum_{A^2} v^a\right)$$

where the price $p_1(\overline{v})$ meets equation $\sum_{A^1} v^a = D^1(p_1(\overline{v})) + \lambda\left(D^2(\lambda p_1(\overline{v})) - \sum_{A^2} v^a\right)$.

Thus,

$$v^{a*} \in \left(p_1^* - C^{a'}(v^{a*})\right) \left|D^{1'}(p_1^*) + \lambda^2 D^{2'}(\lambda p_1^*)\right| \tag{14}$$

for any

$$a \in A^1 \ s.t. \ C^{a'}(0) < p_1^*, v^{a*} = 0 \ if \ C^{a'}(0) \geqslant p_1^*. \tag{15}$$

Similarly, producers in the market 2 face the demand $D^2(\lambda p_1) + 1/\lambda(D^1(p_1) - \sum\limits_{A^1} v^a)$, and

$$v^{a*} \in \left(\lambda p_1^* - C^{a'}(v^{a*})\right) \left|D^{2'}(\lambda p_1^*) + D^{1'}(p_1^*)/\lambda^2\right| \tag{16}$$

for any

$$a \in A^2 \ s.t. \ C^{a'}(0) < p_2^*, v^{a*} = 0 \ if \ C^{a'}(0) \geqslant p_2^*. \tag{17}$$

For C_{1-2} type equilibrium, the capacity constraint is binding:

$$q = Q, \lambda p_1^* < p_2^*, \tag{18}$$

then the F.O.C.s are

$$v^{a*} \in \left(p_i^* - C^{a'}(v^{a*})\right) \left|D^{i'}(p_i^*)\right|, \ for \ any \ a \in A^i \ s.t. \ C^{a'}(0) < p_i^*, \tag{19}$$

$$v^{a*} = 0 \ if \ C^{a'}(0) \geqslant p_i^*, \tag{20}$$

$$\sum_{A^1} v^{a*} = D^1(p_1^*) + Q, \tag{21}$$

$$\sum_{A^2} v^{a*} = D^2(p_2^*) - Q/\lambda. \tag{22}$$

Vasin et al. (2012) find one more possible type of Cournot equilibrium. At the D_{1-2}-type equilibrium, $v = C$ and $\lambda p_1^* = p_2^*$. The FOCs for producers at the first node are

$$(p_1^* - C_-^{a'}(v^{a*}))|D^{1'}(p_1^*) + \lambda^2 D^{2'}(\lambda p_1^*)| \geqslant v^{a*} \geqslant (p_1^* - C_+^{a'}(v^{a*}))|D^{1'}(p_1^*)|, a \in A_1.$$

Under continuous marginal costs this is equivalent to $S^a_{1C1-2}(p_1^*) \geq v^{a*} \geq S^a_{1C}(p_1^*)$.

The FOCs for the second node are

$$(\lambda p_1^* - C_-^{a'}(v^{a*}))|D^{2'}(\lambda p_1^*)| \geqslant v^{a*} \geqslant (\lambda p_1^* - C_+^{a'}(v^{a*}))|D^{2'}(\lambda p_1^*) + D^{1'}(p_1^*)/\lambda^2|, a \in A_2.$$

Note that such inequality can hold only if $D^{1'}(p_1^*) = 0$ or v^{a*} is a jump point for $C^{a'}(q)$.

Thus, even in the simplest variant of the network market there exist 7 (with account of the flow direction) possible types of local equilibria. The following proposition shows that some of them are incompatible and reduces the set of variants for examination.

Conditions (10), (11) and (12) mean that A-type equilibrium exists iff there exist local equilibria for separated markets with prices p_1^*, p_2^* that meet condition (9). In order to find such equilibrium, one should solve equations $\sum_{a\in A^i} S_{ic}^a(p_i^*) = D^i(p_i^*)$, $i = 1,2$, and check this condition. Here S_{ic}^a is the Cournot supply function for isolated market i. It is determined by (10) and (11).

Similarly, in order to find C_{1-2} type equilibrium, one should solve equations

$$\sum_{a\in A^1} S_{1c}^a(p_1^*) = D^1(p_1^*) + Q, \sum_{a\in A^2} S_{2c}^a(p_2^*) = D^2(p_2^*) - Q/\lambda \tag{23}$$

and check inequality (18) for the solution.

Below we bound our study with the case where the function $D_1^i(p_i)$ is concave and positive in the interval $(0, M_i)$ where $D^i(p_i) > 0$, $i = 1,2$. Then $|D_1^i(p_i)|$ does not decrease and, proceeding from (10), $S_{ic}^a(p_i)$ is a non-decreasing function in this interval. So each equation for A-type and C-type equilibrium prices has at most one solution. Let there exist an A-type equilibrium with prices p_1^{*0}, p_2^{*0}. Then C_{1-2} type equilibrium does not exist. Indeed, let (p_1^*, p_2^*) be a solution of the system (23). For $i = 1,2$ the difference $S_{ic}(p_i) - D_i(p_i)$ monotonously increases, hence $p_1^* > p_1^{*0}, p_2^* < p_2^{*0}$, and $p_2^*/p_1^* < p_2^{*0}/p_1^{*0} < \lambda$. So (18) does not hold. In a similar way it is easy to check that if C_{1-2} type equilibrium exists then A and C_{2-1}-type equilibria do not exist for this market.

Summing up the discussion, we obtain the following result.

Proposition 4 *Under the given assumption on the demand functions, there exists at most one local equilibrium of the types A,* C_{1-2} *or* C_{2-1}*. In particular, if the prices for separated markets meet condition* $\lambda p_1^{*0} < p_2^{*0}$ *then, for any* $Q > 0$*, equilibria of the types A and* C_{2-1} *do not exist. If* $\lambda^{-1} < p_2^{*0}/p_1^{*0} < \lambda$ *then, for any* $Q > 0$*, there exists an A-type equilibrium, while C-type equilibria do not exist.*

Now consider B-type equilibria. For every $a \in A^1$, relations (14) and (15) implicitly define function $S_{1-2}^a(p_1^*)$ that determines the equilibrium volume v^{a*} depending on the equilibrium price p_1^*. Similarly, for $a \in A^2$ such function is defined by (16) and (17). Proceeding from the balance equations (7) and (8), we obtain the following equation for determination of the equilibrium price:

$$D^1(p_1^*) + \lambda D^2(\lambda p_1^*) = \sum_{a\in A_1} S_{1-2c}^a(p_1^*) + \lambda \sum_{a\in A_2} S_{1-2}^a(\lambda p_1^*). \tag{24}$$

Under the given assumptions on $D^i(p_i)$, the left-hand side is concave and hence the right-hand side does not decrease in interval $(0, \min\{M_1, M_2/\lambda\})$. But the both properties break at the end of it. For instance, let $M_1 < M_2, \lambda = 1, C^a(v^a) = c_1 v^a, a \in A_1, C^a(v^a) = c_2 v^a, a \in A_2, c_1 < c_2, D^i(p_i) = k_i(M_i - p_i), i = 1,2.$

Then for $a \in A_i$,

$$S^a_{1-2c}(p_1) = \begin{cases} \max(0, (k_1 + k_2)(p_1 - c_i)), \; p_1 < M_1 \\ k_2(p_1 - c_i), \; M_1 < p_1 < M_2 \end{cases}$$

Figure 3 shows possible variants of the total demand and the Cournot supply relations for the case $c_1 = c_2 = c, k_1 = k_2 = 1/3, M_2 = M_1$ and one producer at each local market. In general a solution p^* of equation (24) corresponds to B_{1-2} type equilibrium only if $0 < \sum\limits_{a \in A_1} S^a_{1-2c}(p^*_1) - D^1(p^*_1) < Q$. A sufficient condition for uniqueness of the equilibrium is that the minimal value of $\sum\limits_{a \in A_1} S^a_{1-2c}(M_1) + \lambda \sum\limits_{a \in A_2} S^a_{1-2c}(\lambda M_1)$ exceeds $D_2(\lambda M_1)$ or it's maximal value is less then $D_2(\lambda M_1)$ (variants 1 and 3 in Fig. 3). For the given example two equilibria with the opposite flows exist under certain values of c_1, c_2, n_1, n_2. Vasin and Shamanaev (2010) show that B-type equilibria may co-exist with C_{1-2} type, as well as with C_{2-1} type.

Consider conditions for some types of the local equilibrium to be a true Nash equilibrium. For an A-type equilibrium (with zero flow between the markets), under a sufficiently large increase of production volume by player $a \in A^1$, the price in the market 1 reduces to the level $p_1 = p^*_2/\lambda$. The further increasing of the volume permits the player to sell his production also at the market 2. Under a sufficiently large volume, the price may fall below $\widehat{\widehat{p}}_1$ such that $\lambda(D^2(\lambda\widehat{\widehat{p}}_1) - \sum\limits_{A^2} v^{b*}) = Q$, and the transmission capacity constraint becomes binding. Figure 4 shows the (residual) demand function for this producer.

Note that, according to Proposition 4, v^{a*} is the optimal strategy of agent a for the demand function $D^1(p_1) - \sum\limits_{A^1 \backslash a} v^{b*}$. His optimal strategy under the demand $D^1(p_1) - \sum\limits_{A^1 \backslash a} v^{b*} + \lambda \left(D^2(\lambda p_1) - \sum\limits_{A^2} v^{b*} \right)$ corresponds to the price $\overline{p}_1$ such that

$$\begin{aligned} & D^1(\overline{p}_1) - \sum_{A^1 \backslash a} v^{b*} + \lambda \left(D^2(\lambda \overline{p}_1) - \sum_{A^2} v^{b*} \right) = \\ & = \overline{v}^a \in \left(\overline{p}_1 - C^{a'}(\overline{v}^a) \right) \left| D^{1'}(\overline{p}_1) + \lambda^2 D^{2'}(\lambda \overline{p}_1) \right| \end{aligned} \tag{25}$$

Let $\widehat{\widehat{p}}_2$ meet condition $\lambda \left(D^1(\lambda \widehat{\widehat{p}}_2) - \sum\limits_{A^1} v^{b*} \right) = Q$.

Proposition 5 *A local equilibrium point of the type A (that meets conditions (9), (10), (11) and (12)) is not a Nash equilibrium if and only if, for some producer* $a \in A^1$, *the maximal payoff under demand* $D^1(p_1) - \sum\limits_{A^1 \backslash a} v^{b*} + \lambda \left(D^2(\lambda p_1) - \sum\limits_{A^2} v^{b*} \right)$ *in*

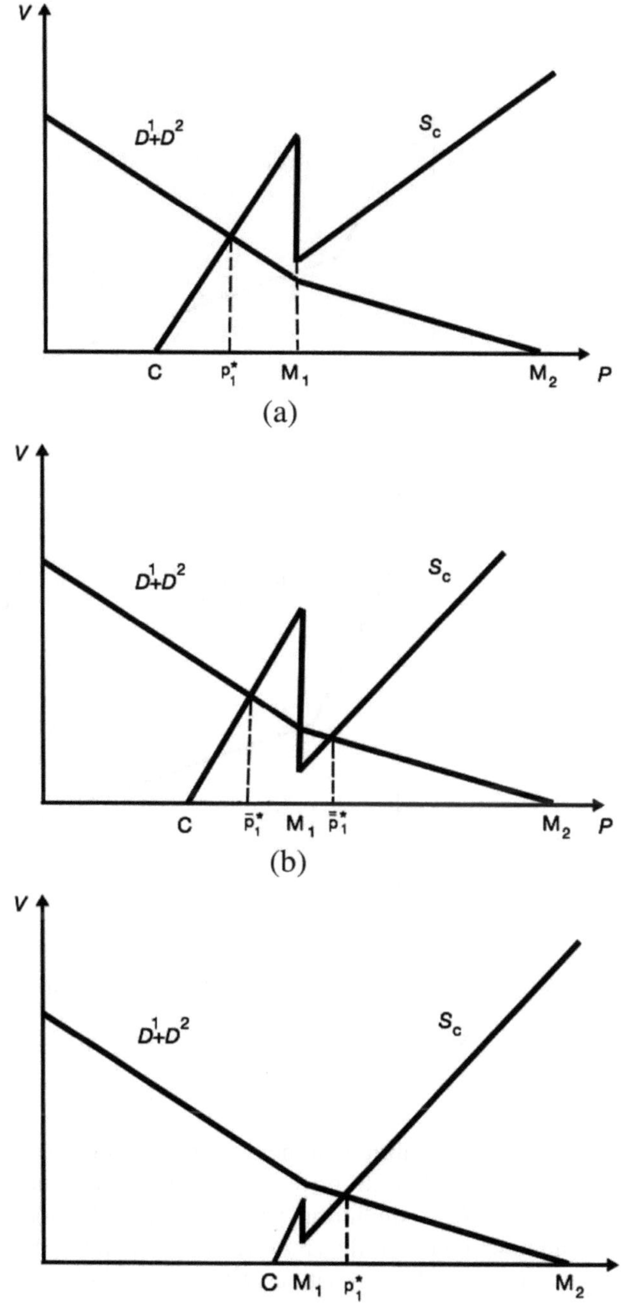

Fig. 3 Total Cournot supply function and equilibrium prices for $c_1 = c_2 = c$: (**a**) $c = M_1/4$; (**b**) $c = 5/8M_1$; (**c**) $c = 4/5M_1$

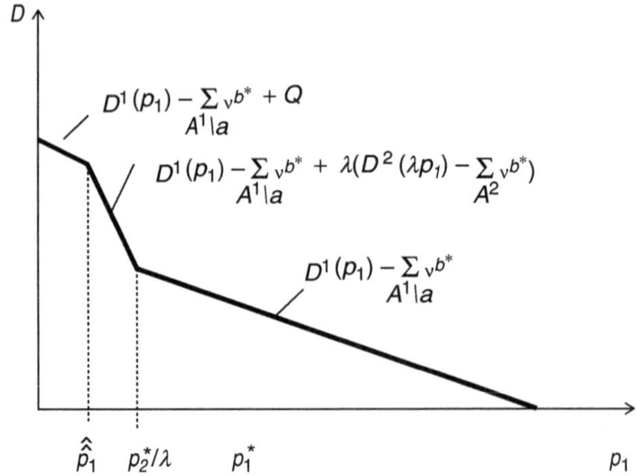

Fig. 4 The residual demand for player $a \in A^1$ and A-type equilibrium

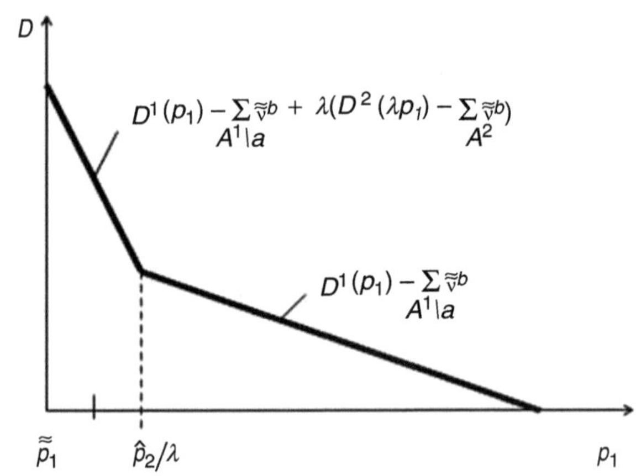

Fig. 5 The residual demand for player $a \in A^1$ and B-type equilibrium

interval $(\widehat{\widehat{p}}_2, p_2^*/\lambda)$ *exceeds* $f^a(\overline{v}^*)$ *or, for some producer* $a \in A^2$, *the maximal payoff under demand* $D^2(p_2) - \sum_{A^2 \backslash a} v^{b*} + \lambda \left(D^1(\lambda p_2) - \sum_{A^1} v^{b*} \right)$ *in interval* $(\widehat{\widehat{p}}_2, p_1^*/\lambda)$ *exceeds* $f^a(\overline{v}^*)$.

Now, consider the local equilibrium of the type (b). In order to avoid confusion, let $\widetilde{\widetilde{p}}_1, \widetilde{\widetilde{v}}$ denote the local equilibrium price and production volumes for this case, while $p_i^*, i = 1, 2$, be the Nash equilibrium prices for isolated markets. In this case the residual demand function for producer $a \in A^1$ is shown in Fig. 5.

The price $\hat{p}_2$ is determined by equation $D^2(\hat{p}_2) = \sum_{A^2} \widetilde{\widetilde{v}}^b$, and $\widetilde{\widetilde{p}}_1 < \hat{p}_2/\lambda$. Reducing his supply, producer $a \in A^1$ can increase the market price to $\hat{p}_2/\lambda$. Under the further reduction, the market splits. The optimal price $\overline{p}_1$ for the demand function $D^1(p_1) - \sum_{A^1\backslash a} \widetilde{\widetilde{v}}^b$ meets relation $D^1(\overline{p}_1) - \sum_{A^1\backslash a} \widetilde{\widetilde{v}}^b = \overline{v}^a \in \left(\overline{p}_1 - C^{a'}(\overline{v}^a)\right) \left|D^{1'}(\overline{p}_1)\right|$.

Proposition 6 *A local equilibrium point of the type (b) (that meets conditions (13), (14), (15), (16) and (17)) is not a Nash equilibrium if and only if for some producer $a \in A^1$ the optimal price $\overline{p}_1$ under demand $D^1(p_1) - \sum_{A^1\backslash a} \widetilde{\widetilde{v}}^b$ is grater than $\hat{p}_2/\lambda$ and the payoff under this price exceeds $f^a(\widetilde{\widetilde{v}})$.*

Note that for $a \in A^2$ it is unprofitable to deviate from the local equilibrium of the type (b) with a positive flow from market 1 to market 2. Under fixed strategies of other players, the demand function looks like in Fig. 6.

By sufficiently large increase of the production volume, agent a can decrease the price and split the market in some cases. However, his payoff will decrease in this case because the demand function for $p < \lambda\hat{p}_1$ lies below the demand curve for the joint market.

For C_{1-2} type local equilibria, the situation is symmetric in some sense. In order to avoid confusion, let $p_i^*(Q), i = 1, 2$, and $v^{a*}(Q), a \in A^1 \cup A^2$, denote equilibrium prices and production volumes in this case. They meet conditions (18), (19), (20), (21) and (22). The demand function for producer $a \in A^2$ looks like in Fig. 7.

Proposition 7 *C_{1-2} type local equilibrium (that meets conditions (18), (19), (20), (21) and 22)) is not a Nash equilibrium if and only if, for some producer $a \in A^2$, the maximal payoff under demand*

$$D^2(\overline{p}_2) - \sum_{A^2\backslash a} v^{b*}(Q) + \lambda^{-1}\left(\sum_{A^1} v^{b*}(Q) - D^1(\lambda^{-1}\overline{p}_2)\right)$$

in interval $(\lambda p_1^*, \lambda p_1^*(Q))$ exceeds $f^a(\overline{v}^*(Q))$.

As to the players from the market 1, under certain conditions some of them can unite the markets by reduction of v^a. However, this is never profitable for the agent.

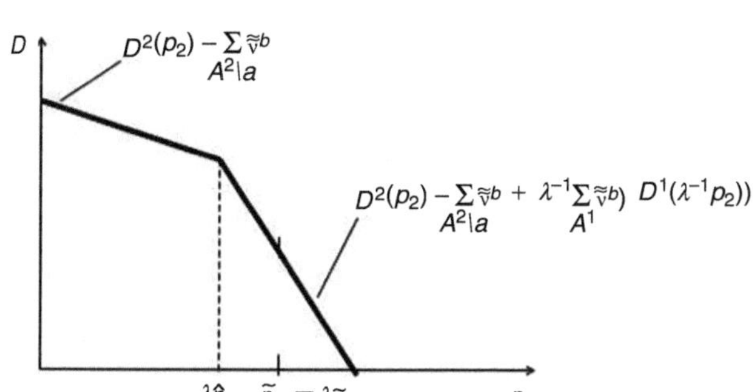

Fig. 6 The residual demand for player $a \in A^2$ and B-type equilibrium

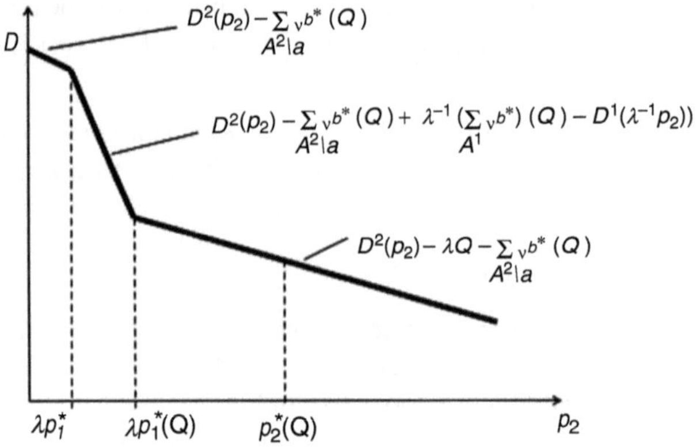

Fig. 7 The residual demand for player $a \in A^2$ and C-type equilibrium

4 Determination of Optimal Transmission Capacity

4.1 Study for Perfectly Competitive Market

This subsection follows Daylova and Vasin (2014) who set the welfare maximization problem and provide a method for computation of the optimal transmission capacity for a perfectly competitive two-node market.

Under assumption of perfect competition, the optimal strategy of each producer is determined by his Walrasian supply function $S^a(p)$ (3). The total supply at the market i is determined as follows: $S_i(p) \stackrel{\text{def}}{=} \sum_{a \in A_i} S^a(p), i = 1, 2.$

Let $\widetilde{p_i}(Q)$, $i = 1, 2$, denote the prices corresponding to Walrasian supply functions under transmission capacity Q. Prices $\widetilde{p_i}(0)$ meet the equations $D_i(\widetilde{p_i}) \in S_i(\widetilde{p_i})$, $i = 1, 2$. We assume that $(1-k)\widetilde{p_2}(0) > \widetilde{p_1}(0)$. Then the equilibrium flow q and the prices satisfy the following conditions:

$$S_1(\widetilde{p_1}) = D_1(\widetilde{p_1}) + q, \tag{26}$$

$$S_2(\widetilde{p_2}) = D_2(\widetilde{p_2}) - (1-k)q, \tag{27}$$

$$\left[\begin{array}{ll} q \leq Q, & \widetilde{p_1} = (1-k)\widetilde{p_2}, \\ q = Q, & \widetilde{p_1} < (1-k)\widetilde{p_2}. \end{array}\right. \tag{28}$$

Consider functions $\widetilde{p_1}^0(q)$ and $\widetilde{p_2}^0(q)$ implicitly determined by (26) and (27) respectively. If $q = Q$, then $\widetilde{p_i}(Q) = \widetilde{p_i}^0(Q)$.

Proposition 8 *Consider the transmission capacity $\widehat{Q}$ such that $\widetilde{p_1}^0(\widehat{Q}) = (1-k)\widetilde{p_2}^0(\widehat{Q})$. If $Q \geq \widehat{Q}$, then at the Walrasian equilibrium,*

$$v = \widehat{Q} \leqslant Q, \quad \widetilde{p_i}(Q) = \widetilde{p_i}^0(\widehat{Q}), \quad i = 1, 2.$$

For $Q < \widehat{Q}$, the equilibrium values meet:

$$q = Q, \quad \widetilde{p_i}(Q) = \widetilde{p_i}^0(Q), \quad i = 1, 2,$$
$$\widetilde{p_1}(Q) < (1-k)\widetilde{p_2}(Q).$$

Without taking into account the construction costs, the social welfare $W(Q)$ includes the benefit of the transmission system $T(Q)$, consumer surpluses $S_i(Q)$ and producers' profits $Pr_i(Q)$, $i = 1, 2$. The benefit of the network system is determined by $T(Q) = \widetilde{p_2}(Q)(1-k)Q - \widetilde{p_1}(Q)Q$ if $Q < \widehat{Q}$. Otherwise it equals zero. Producers at market i get the total profit equal to

$$Pr_i(Q) = \sum_{a \in A_i} (p_i(Q)S^a_{iC}(\tilde{p}_i(Q)) - C^a(S^a_{iC}(\tilde{p}_i(Q)))).$$

Consumer surplus at market i is given by $CS_i(Q) = \int_{\widetilde{p_i}(Q)}^{\infty} D_i(p)dp$.

Function $W(Q)$ is concave and increases in Q if $Q \le \widehat{Q}$. Moreover, $W'(Q) = (1-k)\widetilde{p_2}(Q) - \widetilde{p_1}(Q)$.

Let $B(Q)$ denote the costs of the transmission line construction: $B(Q) = 0$ if $Q = 0$; $B(Q) = b_f + b_v(Q)$ if $Q > 0$, where $b_v(Q)$ is a convex function that determines the variable costs; b_f denotes the fixed costs. Taking into account the construction costs, the total welfare is $TW(Q) = W(Q) - B(Q)$.

Proposition 9 *The optimal transmission capacity Q^*, that maximizes the total welfare, equals zero if $(1-k)\widetilde{p_2}(0) - \widetilde{p_1}(0) \le b'_v(0)$. If this inequality does not hold, the value Q^{*L} corresponding to a local maximum proceeds from equation $(1-k)\widetilde{p_2}(Q^{*L}) - \widetilde{p_1}(Q^{*L}) = b'_v(Q^{*L})$ and satisfies $Q^{*L} < \widehat{Q}$. If $TW(Q^{*L}) > TW(0)$ then $Q^* = Q^{*L}$. Otherwise $Q^* = 0$.*

Who wins and who looses under the connection of the local markets? The transmission system gets the profit equal to $\widetilde{p_2}(Q^*)(1-k)Q^* - \widetilde{p_1}(Q^*)Q^*$, because $Q^{*L} < \widehat{Q}$. At the first market, producers win $\int_{p_1(0)}^{\widetilde{p_1}(Q^*)} S_1(p)dp$, consumers loose $\int_{p_1(0)}^{\widetilde{p_1}(Q^*)} D_1(p)dp$. At the second market consumers win $\int_{\widetilde{p_2}(Q^*)}^{\widetilde{p_2}(0)} D_2(p)dp$ and producers loose $\int_{\widetilde{p_2}(Q^*)}^{\widetilde{p_2}(0)} S_2(p)dp$. The cost of the transmission line construction should be paid by the winners.

4.2 Study for Cournot Competition

Now consider how the social welfare $W(Q)$ depends on the transmission capacity for the Cournot oligopoly considered in Sect. 3. We assume that Cournot prices p_i^{*0} for separated markets meet condition $\lambda p_1^{*0} < p_2^{*0}$. Then C_{1-2} type equilibrium exists for sufficiently small Q. The prices $p_i(Q)$ are uniquely determined by equations $D_1(p_1) + Q = S_{C1}(p_1)$, $D_2(p_2) - Q\lambda_Q^{-1} = S_{C2}(p_2)$. Proceeding from these equations and the Implicit Function Theorem, we obtain

$$p_1'(Q) = \frac{1}{S_{C1}'(p_1) - D_1'(p_1)} > 0, \tag{29}$$

$$p_2'(Q) = \frac{\lambda^{-1}}{D_2'(p_2) - S_{C2}'(p_2)} < 0. \tag{30}$$

So such equilibrium exists for $Q \in (0, \underline{Q}(\lambda))$ where $\underline{Q}(\lambda)$ meets equation $\lambda p_1(\underline{Q}(\lambda)) = p_2(\underline{Q}(\lambda))$. We examine $W(Q)$ in this interval.

Let $\overline{A}_i(p_i) \stackrel{def}{=} \left\{a \in A_i \textit{ such that } C_-^{a'}(S_{Ci}^a(p_i)) = C_+^{a'}(S_{Ci}^a(p_i))\right\}$.

For any $a \in \overline{A}_i(p_i)$, the Cournot supply function meets conditions (10) as a strict equality. Let $P_i = \{p_i \mid \exists a \in A_i \textit{ such that } p_i = C_-^{a'}(S_{Ci}^a(p_i)) < C_+^{a'}(S_{Ci}^a(p_i)) \textit{ or } p_i = C_+^{a'}(S_{Ci}^a(p_i)) > C_-^{a'}(S_{Ci}^a(p_i))\}$, $i = 1, 2$. In general P_1, P_2 are finite sets.

Then the total Cournot supply function derivative is $S_{Ci}'(p_i) = \sum_{a \in \overline{A}_i(p_i)} S_{Ci}^{a'}(p_i)$, where

$$S_{Ci}^{a'}(p) = \frac{|D_i'(p)| - (p - C^{a'}(S_{Ci}^a(p)))D_i''(p)}{1 + C^{a''}(S_{Ci}^a(p))\,|D_i'(p)|}. \tag{31}$$

This expression proceeds from the theorem on implicit function derivative. Note that $S_{Ci}^{a'}(p_i) = 0$ for $a \notin \overline{A}_i(p_i)$ in this case.

Proposition 10 *The derivative $W'(Q)$ exists for any $Q \in (0, \underline{Q})$ except for points Q such that $p_1(Q) \in P_1$ or $p_2(Q) \in P_2$. At any other point*

$$W'(Q) = p_2(Q)\lambda^{-1} - p_1(Q) + p_1'(Q) \sum_{a \in \overline{A}_1(p_1(Q))} \frac{S_{C1}^a(p_1(Q))S_{C1}^{a'}(p_1(Q))}{|D_1'(p_1(Q))|} +$$
$$p_2'(Q) \sum_{a \in \overline{A}_2(p_2(Q))} \frac{S_{C2}^a(p_2(Q))S_{C2}^{a'}(p_2(Q))}{|D_2'(p_2(Q))|},$$

where $p_1'(Q)$, $p_2'(Q)$ and $S_{Ci}^{a'}(p)$ are given by (29), (30) and (31).

Now consider the comparative statics of the social welfare components with respect to transmission capacity Q. Since $p_1(Q)$ increases in Q and $p_2(Q)$ decreases in Q, so the consumer surpluses change in a similar way: the surplus at the market 1 increases while the surplus at the market 2 decreases. Under the concave demand functions, the Cournot supply functions do not decrease in the prices.

Proceeding from

$$Pr_i^{'}(Q) = p_i^{'}(Q) \cdot \left(\sum_{a \in A_i} S_{Ci}^{a}(p_i(Q)) + \sum_{a \in A_i} \frac{S_{Ci}^{a}(p_i(Q)) S_{Ci}^{a'}(p_i(Q))}{|D_i^{'}(p_i(Q))|} \right),$$

we conclude that $Pr_1^{'}(Q) > 0$ and $Pr_2^{'}(Q) < 0$. Thus, the total profit at the market 1 increases in Q while the total profit at the market 2 reduces.

Consider the case where no efficient producers exist at the market 2, that is, $S_{2C}(p_2) \equiv 0$. Then, for any Q such that $p_1(Q) \notin P_1$,

$$W'(Q) = p_2(Q)\lambda^{-1} - p_1(Q) + p_1'(Q) \cdot \sum_{a \in \overline{A}_1(p_1(Q))} \frac{S_{C1}^{a}(p_1(Q)) S_{C1}^{a'}(p_1(Q))}{|D_1{}'(p_1(Q))|} \geqslant 0,$$

because $p_2(Q)\lambda^{-1} - p_1(Q) \geqslant 0$, $p_1'(Q) > 0$. Thus, if we ignore the costs of the expansion, the social welfare increases in Q in this case. However, as we show below, the derivative typically is non-monotonous and changes in a rather sophisticated way. Consider the case with linear demand functions $D_i(p_i) = \max\left\{\overline{D}_i - p_i, 0\right\}, i = 1, 2$, and piece-wise constant marginal costs of producers: $C^a(0) = 0$, $C^{a'}(v) = c_k^a$ for $v \in (V_{k-1}^a, V_k^a)$, $k \in \overline{1, m^a}$, $V_0^a = 0$, $V_{m_a}^a = V^a$, $a \in A_1$.

Then $p_1(Q) = Q + \overline{D}_1 - S_{C1}(p_1(Q))$, $p_2(Q) = \overline{D}_2 - Q = \overline{D}_1 + \overline{D}_2 - p_1(Q) - S_{C1}(p_1(Q))$, $p_1'(Q) = \left(|\overline{A}_1(p_1(Q))| + 1\right)^{-1}$.

Proposition 11 *In this case, for any* $Q \in (0, \underline{Q})$ *such that* $p_1(Q) \notin P_1$, $W'(Q) = \overline{D}_1 + \overline{D}_2 - 2p_1(Q) - S_{C1}(p_1(Q)) + R(p_1(Q))$, *where* $R(p_1) = \frac{|\overline{A}_1(p_1)|}{|\overline{A}_1(p_1) + 1|}(p_1 - c_{av}(p_1))$, $c_{av}(p_1) = \sum_{a \in \overline{A}_1(p_1)} C^{a'}(S_{C1}^a(p_1))/|\overline{A}_1(p_1)|$*—the average marginal costs of the producers for whom the capacity constraints are not binding.*

The function $R(p_1)$ is typically discontinuous and has many jumps up and down. Consider an example where $A_1 = \{1, 2\}$, $c_1^1 = 0$, $c_2^1 = 2$, $V_1^1 = 2$, $V_2^1 = 6$, $c_1^2 = 3$, $c_2^2 = 6$, $V_1^2 = 3$, $V_2^2 = 2$, $\overline{D}_1 = 4$. Figure 8 shows the Cournot supply function, the equilibrium price $p_1(Q)$ and the graph of $R(p_1)$ for this case.

The problem of the total welfare optimization with account of the transmission expansion costs is non-standard even for the simple case under consideration. While $W'(Q)$ decreases in each interval where it exists, the jumps up of the function R may cause multiple extrema of $TW(Q)$.

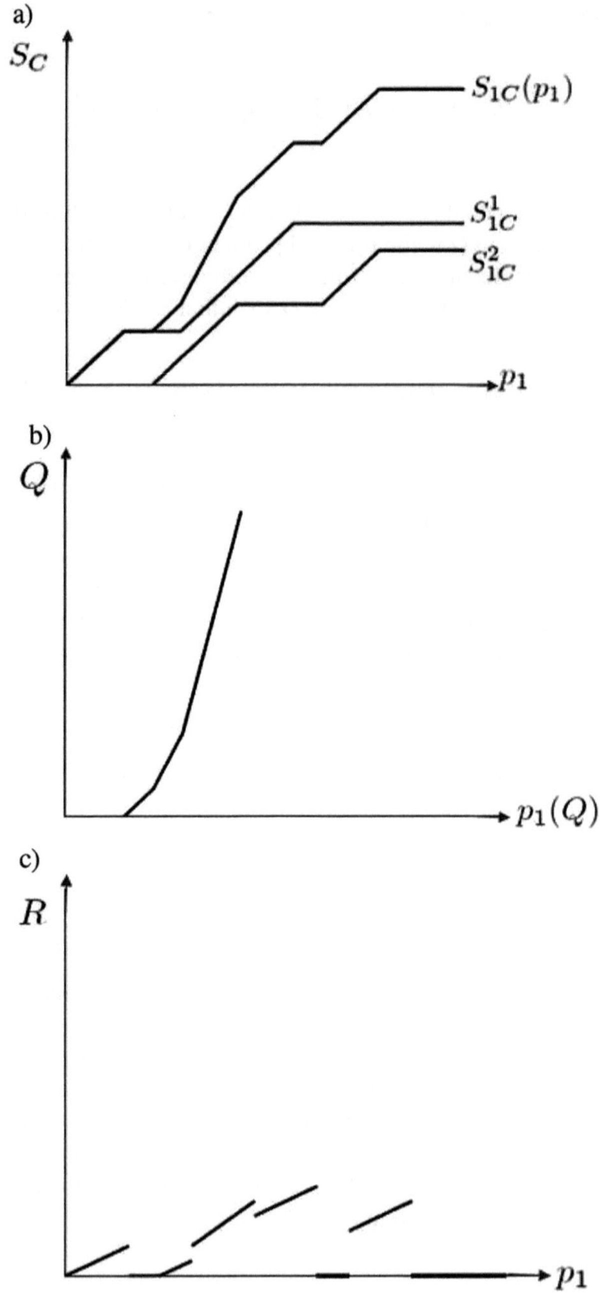

Fig. 8 Cournot supply function, the equilibrium price and the graph of R for the example with two producers

5 Conclusion

Our study shows that even the simplest network market under Cournot competition is rather sophisticated for the formal analysis. Under general conditions, there may co-exist several local equilibria of different types. Some of them (in particular, all or none of them) may be true Nash equilibria. The problem of the social welfare optimization by means of the transmission capacity increasing turns out to be multi-extremal.

Besides that, for electricity markets the deviation from the competitive equilibrium and the corresponding reduction of the consumer surplus are usually rather large at the Cournot equilibrium, because electric energy is a good with a low price elasticity. The general Cournot Oligopoly Theory for multi-node markets is rather incomplete. Daylova and Vasin (2014) provide a method for computation of NE for a chain-type market. In order to construct the model and provide the method for more general structure, it is necessary to use Kirchhoff laws for electricity markets with alternating current. Even for perfectly competitive markets, computation of the equilibrium is rather sophisticated (see Davidson et al. (2009)). So the nasty mathematical properties of the model (especially in comparison with the model for a perfectly competitive market) may be considered as one more argument in favour of economic mechanisms that support behavior close to the Walrasian equilibrium. Our previous study (Vasin et al. 2007) shows that Vickrey auction and its modifications may be a good alternative for the uniform price auction related to Cournot competition. Another promising way is the development of forward markets (see Bushnell 2005; Vasin and Daylova 2012).

Acknowledgements The research was supported by Russian Foundation for Basic Research (project No. 14-01-91163 GFEN a). We thank a referee for useful comments.

References

Amir R (1996) Market power and power markets: structural problems of Russian wholesale electricity market. EERC final report

Amir R, Lambson V (2000) On the effects of entry in Cournot markets. Rev Econ Stud 67:235–254

Bertoletti P, von Mouche P (2014) Inferior factor in Cournot oligopoly revisited. J Econ 112(1):85–90

Bushnell J (2005) Oligopoly equilibria in electricity contract markets. University of California Energy Institute: CSEM working paper, WP-148

Cornes R, Sato T (2016) Existence and uniqueness of Nash equilibrium in aggregative games: an expository treatment. Springer, Cham

Cournot A (1838) Researches into the mathematical principles of the theory of wealth. Macmillan, New York, 1897, first published in Paris in 1838

Davidson MR, Dogadushkina YV, Kreines EM, Novikova NM, Seleznev AV, Udaltsov YA, Shiryaeva LV (2009) Mathematical model of power system management in conditions of a competitive wholesale electric power (capacity) market in Russia. J Comput Syst Sci Int 48:243–253

Daylova E, Vasin A (2014) Determination of transmission capacity for a two-node market. Procedia Comput Sci 31:151–157
Hogan W (1998) Competitive electricity market design: a wholesale primer. Tech. Rep. Harvard Electricity Policy Group
Hogan W (2007) Toward a combined merchant-regulatory mechanism for electricity transmission expansion. Department of Economics, Boston University
Kreps D, Scheinkman J (1983) Quantity recommitment and Bertrand competition yield Cournot outcomes. Bell J Econ 14:326–337
Kukushkin N (1994) A fixed point theorem for decreasing mappings. Econ Lett 46:23–26
Kukushkin N (2016) Cournot tatonnement in aggregative games with monotone best responses. Springer, Cham
Moreno D, Ubeda L (2006) Capacity precommitment and price competition yield the Cournot outcome. Games Econ Behav 56(2):323–332
Newbery D (2009) Predicting market power in wholesale electricity markets. EUI working papers RSCAS
Novshek W (1985) On the existence of Cournot equilibrium. Rev Econ Stud 52:85–98
Okuguchi K (2007) Quasi-competitive Cournot oligopoly with product differentiation and symmetric firms. Rev Dev Econ 11(2):404–411
Okuguchi K (2010) Inferior factor in Cournot oligopoly. J Econ 101(2):125–131
Vasin A, Daylova E (2012) Game-theoretic model of agents' interaction on a two-stage market with a random factor. Matematicheskaya Teoriya Igr i Prilozheniya 4:3–32
Vasin A, Shamanaev A (2010) Cournot equilibria in a model of two-hub homogeneous commodity market. Comput Math Model (Springer New York) 21(1):51–69
Vasin A, Vasina P (2005) Models of supply functions competition with application to the network auctions. Final research report on the EERC project R03–1011
Vasin A, Vasina P (2006) Electricity markets analysis and design. Working paper 2006/053. Moscow New Economic School
Vasin A, Vasina P, Ruleva T (2007) On organization of markets of homogeneous goods. J Comput Syst Sci Int 46:93–106
Vasin A, Sosina Y, Weber G (2012) Evaluation of market power in local and two-node markets. International workshop networking games and management NGM-2012. Extended abstracts. KarRC RAS, pp 66–69

Kant-Nash Equilibria in a Quantity-Setting Oligopoly

Ngo Van Long

Abstract This paper studies oligopolistic equilibria when not all firms behave in a Nashian fashion. The Kantian firms act according to norms of behavior that reflect the Kantian categorical imperative. We introduce the concepts of Exclusive Kant-Nash Equilibrium and Inclusive Kant-Nash Equilibrium to study such situations. Our model formalises Sen's conjecture that in actual situations people often do not follow the selfish strategy because of the Kantian norm of behavior.

1 Introduction

In the economic theory of oligopoly theory, the dominant stream of literature focuses on issues such as existence and uniqueness of a Nash-Cournot equilibrium (see, e.g. Okuguchi, 1977; Okuguchi and Szidarovszky, 1990; Gaudet and Salant, 1991; Long and Soubeyran , 2000). There is also a literature on collusive behavior that relies mainly on the theory of repeated games (e.g. Fudenberg and Tirole, 2000). These two streams of literature share a common foundation: the idea of rationality. However, recent advances in behavioral economics show convincingly that rationality (in the narrow sense used by economists) is often violated: people may deviate from 'rationality' because they are conditioned by norms of behavior, or habit, or intuitive reactions, or misperception, and so on. (The social norm equilibrium is related to the concepts of tenable strategy blocks and settled equilibria; see Myerson and Weibull, 2015.)[1]

[1]The main idea of tenable strategy block is that "when people interact in familiar settings, social conventions usually develop so that people tend to disregard alternatives outside the convention. For rational players to usually restrict attention to a block of conventional strategies, no player should prefer to deviate from the block when others are likely to act conventionally and rationally inside the block."

N. Van Long
Department of Economics, McGill University, Montreal, QC, H3A2T7, Canada
e-mail: ngo.long@mcgill.ca

P. von Mouche, F. Quartieri (eds.), *Equilibrium Theory for Cournot Oligopolies and Related Games*, Springer Series in Game Theory,
DOI 10.1007/978-3-319-29254-0_12

A number of authors emphasize that norms of behavior can sustain co-operation. Adam Smith (1790[2002]) finds that co-operation and mutual help are incorporated in established rules of behavior, and that "*upon the tolerable observance of these duties, depends the very existence of human society, which would crumble into nothing if mankind were not generally impressed with a reverence for those important rules of conduct.*"[2] A most important norm of behavior is the categorical imperative formalised by Kant (1785). While Kant's book offers several formulations of this concept, for our purposes it seems adequate to present it as follows: 'Act as if the maxim of your action were to become through your will a general natural law.'[3] Reflecting on Kantian conduct, Sen (1977, p. 341) observes that "in actual situations people often do not follow the selfish strategy" and that "even in controlled experiments in laboratory conditions people playing the Prisoners' Dilemma frequently do the unselfish thing." He argues that possibly the person "has asked himself what type of preference he would like the other player to have, and on somewhat Kantian grounds has considered the case for himself having those preferences, or behaving *as if* he had them."[4]

In the same vein, Laffont (1975) investigates "the possibility and the consequences of collective behavior (without any direct communication between agents) that is in contradiction to the maximization of direct personal interest." Laffont (1975) considers situations where "a typical agent assumes (according to Kant's moral) that the other agents will act as he does, and he maximizes his utility function under this new constraint." Laffont considers only the case of identical agents, and does not formally model Kantian situations in which agents are heterogeneous. This task is taken up by John Roemer (2010), who defines the concept of Kantian equilibrium in games with heterogeneous agents. Recall that in the Nash equilibrium, when an agent contemplates a deviation, he assumes that other agents do not deviate. In contrast, in a Kantian equilibrium as formulated by Roemer, every agent compares his equilibrium payoff with the payoff he would obtain if he were to deviate, assuming that his deviation would be accompanied by 'similar deviations' by other agents. Roemer thinks of the Kantian equilibrium as a normative concept. The *counterfactual alternative* proposed by Roemer is that, when agent i's deviates from his current action level by a factor $\lambda > 0$, the action level of all *all other agents* will also be adjusted by the same factor λ. By definition,

---

[2]See Smith (1790[2002]), Part III, Chapter V, p. 190. There is a large literature on norms of behavior and on their origins, including the evolutionary view (see e.g. Bergstrom, 1995).

[3]See Bertrand Russell's *A History of Western Philosophy*. As explained by Russell (1945, p. 737) 'There are two sorts of imperative: the hypothetical imperative, which says "You must do so-and-so if you wish to achieve such-and-such and end"; and the categorical imperative, which says that a certain kind of action is objectively necessary, without regard to any end.'

[4]Another approach to the possible avoidance of the prisoner's dilemma outcome would be to change the payoff matrix by allowing each player to care about the other's player. This approach was suggested by Rescher (1975). However, Waymark (1978) points out that Rescher's method might turn a game without the prisoner's dilemma properties into a prisoner's dilemma game. Waymark admits that his technical note does not go beyond the narrow technical issues.

in a Kantian equilibrium, each agent finds that his payoff under such a counterfactual alternative is inferior to the equilibrium payoff.[5] Roemer (2010, p. 14) applies the Kantian equilibrium concept to a quantity setting oligopoly and proves that under certain mild assumptions, a collusive solution is a Kantian equilibrium. See also De Donder and Roemer (2014) for an interesting application.

Economists traditionally argue that the business of a firm is to earn money (i.e. to maximize its market value). According to that view, it would be inappropriate to apply the notion of Kantian behavior to firms. However, as pointed out by Tirole (2001, pp. 3–4), the view that the firm's sole responsibility is to serve the financial interests of its shareholders is not widely accepted in some circles. Tirole therefore defines corporate governance as "the design of institutions that induce or force management to internalize the welfare of stakeholders" (p. 4). This is in accord with Arrow (1973) and Sen (1993).

Both Laffont (1975) and Roemer (2010) restrict attention to equilibrium outcomes where all agents are Kantian. This is indeed a very strong restriction. In this paper, we investigate oligopolistic situations where there exists a non-empty subset of firms that behave in a non-Kantian way: they are "Nashians." Denote this subset by $\mathcal{N}$. What kind of equilibrium can we expect? In any attempt to define a concept of equilibrium when Kantian agents know that there are Nashian players, it is important to specify the counterfactual alternative. We propose two concepts of Kant-Nash equilibrium, corresponding to two different counterfactual alternatives. The first concept is called Exclusive Kant-Nash equilibrium (EKNE). The second is called Inclusive Kant-Nash Equilibrium (IKNE).[6] In the EKNE, the counterfactual alternative that a Kantian agent contemplates is that the action level of all *all other agents, except the Nashian ones*, will also be adjusted by the same factor λ. In contrast, in the IKNE, the counterfactual alternative is that the action level of all *all other agents, including the Nashian ones*, will also be adjusted by the same factor λ. While some may argue that the latter scenario is utterly unrealistic, as Kantian agents should know that Nashians would not follow suit, it is important to recall that Kantians act according to rules of behavior, which need not be 'rational' in the narrow economic sense.

In Sect. 2, we propose the model and present some formal definitions. Section 3 presents some general analytical results. Section 4 provides some examples to illustrate the general results. In Sect. 5, we generalize the equilibrium concept by proposing a Generalised Kant-Nash Equilibrium (GKNE) which contains IKNE and EKNE as special cases. In Sect. 6, we have followed the recommendation of a referee to offer a brief analysis of Kant-Nash equilibria in a price-setting oligopoly with non-homogeneous products. Section 7 offers some concluding remarks.

[5]Roemer does not claim that this counterfactual alternative outcome is what an agent should expect would necessarily materialize if he were to deviate.

[6]These concepts were first introduced by Ghosh and Long (2015) in a model of voluntary contribution to a public good.

2 The Model

We consider an oligopolistic industry consisting of m heterogeneous firms that produce a homogeneous good. Let y_i denote firm i's output. The industry output is $Q = \sum_{i=1}^{m} y_i$. The inverse demand function is $P(Q)$, where $P'(Q) < 0$. The total cost of producing y_i is $c_i(y_i)$, a convex and increasing function, with $c_i(0) = 0$. Let y denote the output vector,

$$y = (y_1, y_2, \dots, y_m).$$

Firm i's profit is

$$\pi_i(y) = P(Q)y_i - c_i(y_i).$$

Roemer defines a Kantian equilibrium as an output vector y such that any scaling up (or scaling down) of the output vector by a factor $\lambda > 0$ will (weakly) decrease the profit of all firms, that is,

$$(\text{for all } i = 1, \dots, m), \qquad 1 \in \arg\max_{\lambda > 0} \pi_i(\lambda y). \tag{1}$$

Roemer (2010) supposes that all firms are Kantian (in the sense that each firm s will maintain his current output level y_s iff condition (1) is satisfied). Assuming that the function $P(Q)$ is concave, and that all firms produce strictly positive output, Roemer shows that a Kantian equilibrium in a quantity-setting oligopoly achieves the same outcome as a collusive solution: industry profit is maximized (p. 14).

In this paper, we consider a more general case where not all firms are Kantian. We assume that there are n Nashian firms and k Kantian firms, and we let $m = n + k$. Without loss of generality, we label firms so that the first n firms are Nashian and the last k firms are Kantian. Let $\mathscr{K}$ and $\mathscr{N}$ denote respectively the set of Kantian firms and the set of Nashian firms. Let $\mathscr{M} \equiv \mathscr{K} \cup \mathscr{N}$.

A Nashian firm j is in equilibrium if and only if no deviation in y_j can improve its profit, assuming that all other firms do not change their output. What about Kantian firms? We propose two different concepts of equilibrium, corresponding to two different questions that Kantian firms ask themselves.

Definition 1 (Exclusive Kant-Nash equilibrium) An output vector

$$y = (y_1^N, \dots, y_n^N, y_{n+1}^K, \dots, y_{n+k}^K)$$

is an Exclusive Kant-Nash equilibrium iff

(i) No Nashian firm j can gain by deviating from y_j^N, assuming that all other firms in $\mathscr{M}$ maintain their output level.
(ii) For each Kantian firm i, the answer to the following question is "No": Can my profit be increased by applying a scaling factor $\lambda > 0$ to the output vector

$(y_{n+1}^K, \ldots, y_{n+k}^K)$ of all k Kantian firms? In other words, it holds for all $i = n+1, \ldots, n+k$ that

$$1 \in \arg\max_{\lambda>0} \pi_i(y_1^N, \ldots, y_n^N, \lambda y_{n+1}^K, \ldots, \lambda y_{n+k}^K) \tag{2}$$

An EKNE is 'interior' if all firms produce strictly positive outputs.

Remark 1 The game we consider is a static one. However, as a referee points out, a possible interpretation of an EKNE is that of a supergame in which firms have different discount factors. Kantian firms have a low discount factor and collude while Nashian firms have a high discount factor and play the one shot Cournot equilibrium.

Definition 2 (Inclusive Kant-Nash equilibrium) An output vector

$$y = (y_1^N, \ldots, y_n^N, y_{n+1}^K, \ldots, y_{n+k}^K)$$

is an Inclusive Kant-Nash equilibrium iff

(i) No Nashian firm j can gain by deviating from y_j^N, assuming that all other firms in $\mathscr{M}$ maintain their output level.
(ii) For each Kantian firm i, the answer to the following question is "No": Can my profit be increased by applying a scaling factor $\lambda > 0$ to the output vector $y = (y_1^N, \ldots, y_n^N, y_{n+1}^K, \ldots, y_{n+k}^K)$? In other words, it holds for all $i = n+1, \ldots, n+k$ that

$$1 \in \arg\max_{\lambda>0} \pi_i(\lambda y_1^N, \ldots, \lambda y_n^N, \lambda y_{n+1}^K, \ldots, \lambda y_{n+k}^K) \tag{3}$$

An IKNE is 'interior' if all firms produce strictly positive outputs.

To ensure that sufficient conditions are met, we adopt the following standard assumption:

Assumption A1: $P'(Q) < 0$ and

$$\frac{P''(Q)Q}{P'(Q)} > -2$$

This assumption ensures that the industry marginal revenue function is decreasing in industry output (see e.g. Long and Soubeyran, 2001). We maintain this assumption throughout the analysis. This condition, together with the assumption that cost functions are convex, ensures that the second order condition for each firm is satisfied.

3 Some General Results

In this section, we prove some results under general demand and convex cost functions.

Proposition 1 (Equivalence between EKNE and Nash equilibrium with merger) *Consider an oligopoly with k Kantian firms having identical convex cost function $c(y_i)$ and n Nashian firms having (possibly non-identical) convex cost function $c_j(y_j)$. Under these conditions, an interior EKNE is indistinguishable from a Nash equilibrium with $n + 1$ firms, where the first n firms have the cost function $c_j(y_j)$ and the $(n + 1)$th firm is a merger of k identical Kantian firms, each with cost $c(y_i)$.*

Proof Let y^M be the output of the merger of k identical firms. Let $C_M(y^M)$ denote the merger's least cost of producing a given output y^M. Clearly,

$$C_M(y^M) = kc\left(\frac{y^M}{k}\right)$$

This implies

$$C'_M(y^M) = c'\left(\frac{y^M}{k}\right)$$

The Nash equilibrium for this $n+1$ firm oligopoly is characterised by two conditions

$$P'(Q)y_j^N + P(Q) = c'_j(y_j^N), j = 1, 2, \ldots, n \tag{4}$$

$$P'(Q)y^M + P(Q) = C'(y^M) = c'\left(\frac{y^M}{k}\right) \tag{5}$$

Clearly, if we write $y^M = ky_i^K$, condition (5) is identical to the following condition

$$P'(Q)ky_i^K + P(Q) = C'(y^M) = c'\left(\frac{y^M}{k}\right) \equiv c'(y_i^K) \tag{6}$$

which, together with (4), defines an interior EKNE.

This concludes the proof. □

Proposition 2 *Assume all firms have the same convex cost function c(.). At an interior EKNE, the output of the representative Kantian is strictly smaller than (respectively, equal to) that of the representative Nashian if $k > 1$ (respectively, if $k = 1$).*

Proof We let the subscript E denote variables at an exclusive Kant-Nash equilibrium. Consider the necessary conditions for an interior EKNE:

$$P'(Q_E)y_E^N + P(Q_E) = c'(y_E^N) \tag{7}$$

$$P'(Q_E)ky_E^K + P(Q_E) = c'(y_E^K) \tag{8}$$

Let $y_E^K(k, m)$ and $y_E^N(k, m)$ denote equilibrium outputs of Kantian and Nashian firms when there are a total of m firms in the industry, and k of them are Kantian. From Eqs. (7) and (8) we see that if $k = 1$, then clearly $y_E^K(1, m) = y_E^N(1, m)$. To show that $y_E^K < y_E^N$ if $k > 1$, we construct a proof by contradiction. Let $k > 1$ and suppose that $y_E^K \geq y_E^N$. Then $c'(y_E^N) \leq c'(y_E^K)$, and subtracting (8) from (7) yields

$$\left[y_E^N - ky_E^K\right] P'(Q_E) = c'(y_E^N) - c'(y_E^K) \leq 0$$

This inequality implies $y_E^N - ky_E^K \geq 0$, which, together with $k > 1$, entails $y_E^N > y_E^K$, contradicting the supposition that $y_E^K \geq y_E^N$. This concludes the proof. □

Proposition 3 *Assume all firms have the same convex cost function $c(.)$. At an interior IKNE, the output of the representative Kantian is strictly smaller than that of the representative Nashian, for all possible values of k.*

Proof Let the subscript I denote variables at an inclusive Kant-Nash equilibrium. Consider the necessary conditions for an interior IKNE:

$$P'(Q_I)y_I^N + P(Q_I) = c'(y_I^N) \tag{9}$$

$$P'(Q_I)(ky_I^K + ny_I^N) + P(Q_I) = c'(y_I^K) \tag{10}$$

Subtracting (10) from (9) yields

$$\left[y_I^N - ky_I^K - ny_I^N\right] P'(Q_I) = c'(y_I^N) - c'(y_I^K)$$

Suppose $y_I^K \geq y_I^N$. Then $c'(y_I^N) - c'(y_I^K) \leq 0$. This inequality implies that $\left[y_I^N - ky_I^K - ny_I^N\right] P'(Q_I) \leq 0$, which is not possible. This concludes the proof. □

Proposition 4 *Assume that Kantian firms have constant marginal costs γ_i^K and Nashian firms have constant marginal costs γ_j^N. In any IKNE where some Nashian firm j^* produces a strictly positive output, all Kantian firms must produce zero output if*

$$\min_{i \in \mathcal{K}} \left\{\gamma_i^K\right\} \geq \gamma_{j^*}^N \tag{11}$$

Proof Let Q_I denote industry output under IKNE. The necessary condition for Kantian firm i to be in equilibrium is

$$P'(Q_I)Q_I + P(Q_I) - \gamma_i^K \leq 0, y_i^K \geq 0,$$
$$\left[P'(Q_I)Q_I + P(Q_I) - \gamma_i^K\right] y_{iI}^K = 0$$

For Nashian firm j^* with a strictly positive output,

$$P'(Q_I)y_{jI}^N + P(Q_I) - \gamma_j^N = 0$$

Assume that condition (11) holds. Suppose $y_{iI}^K > 0$ for some i, then

$$\gamma_i^K = P'(Q_I)Q_I + P(Q_I) < P'(Q_I)y_{j^*I}^N + P(Q_I) = \gamma_j^N$$

This inequality violates condition (11). This concludes the proof. □

Proposition 5 (Comparing aggregate outputs under IKNE and EKNE) *Assume that all Kantian firms have constant marginal cost γ^K, while Nashian firms may have non-identical constant or increasing marginal costs. Then if all Kantian firms and at least one Nashian firm j produce strictly positive outputs, the industry output under IKNE is strictly smaller than under EKNE.*

Proof Let Q_I (respectively, Q_E) denote industry output under IKNE (EKNE). Let $\phi(Q) = P'(Q)Q + P(Q)$. Note that $\phi'(Q) < 0$ by Assumption A1. Then

$$\phi(Q_I) = P'(Q_I)Q_I + P(Q_I) = \gamma^K \tag{12}$$
$$P'(Q_I)y_{jI}^N + P(Q_I) = c_j'(y_{jI}^N) \tag{13}$$
$$P'(Q_E)ky_E^K + P(Q_E) = \gamma^K \tag{14}$$

and

$$P'(Q_E)y_{jE}^N + P(Q_E) = c_j'(y_{jE}^N) \tag{15}$$

Suppose that $Q_I \geq Q_E$. Then, since $\phi(.)$ is a decreasing function, we must have $\phi(Q_I) \leq \phi(Q_E)$, hence

$$\gamma^K \leq P'(Q_E)Q_E + P(Q_E) < P'(Q_E)ky_E^K + P(Q_E) = \gamma^K$$

which is a contradiction. This concludes the proof. □

4 Examples with Linear Demand and Quadratic Cost

In what follows we consider some simple cases where explicit solutions can be obtained.

4.1 *Example 1: Interior EKNE When Demand Is Linear and Cost Is Quadratic*

In this example, we assume that $P(Q) = B - Q$, and firms have identical cost functions $c_i(y_i) = by_i + \frac{1}{2}\theta y_i^2$, where $\theta > 0, B > b > 0$. Let $y_{jE}^N(k,m)$ and $y_{iE}^K(k,m)$ denote the output of the Nashian firm j and the Kantian firm i in an industry consisting of m firms, of which k are Kantian. Let

$$Q(k,m) \equiv \sum_{i\in\mathcal{K}} y_i^K(k,m) + \sum_{j\in\mathcal{N}} y_j^N(k,m)$$

4.1.1 Equilibrium Outputs in an EKNE

If all firms are Nashian, i.e., $k = 0$ and $n = m$, we have the first order conditions

$$P'(Q)y_j + P(Q) = b + \theta y_j \text{ for all } j = 1, 2, \ldots, m.$$

Then we obtain the Nash equilibrium output, given that $k = 0$,

$$y_i^N(0,m) = \frac{P(Q(0,m)) - b}{\theta + 1} = \frac{(B-b) - my_i^N(0,m)}{\theta + 1}$$

i.e.,

$$y_i^N(0,m) = \frac{A}{\theta + m + 1}, A \equiv B - b > 0$$

And

$$Q(0,m) = \frac{mA}{\theta + m + 1}$$

If all firms are Kantian, i.e., $k = m$, then the equilibrium condition (1) is satisfied at the allocation $(y_1, y_2, \ldots, y_m)$ iff, for all $i = 1, 2, \ldots, m$,

$$\frac{d}{d\lambda}\left\{P(\lambda Q)\lambda y_i - b\lambda y_i - \frac{1}{2}(\lambda y_i)^2\right\} = 0 \text{ at } \lambda = 1$$

i.e. iff

$$P'(\lambda Q)Q\lambda y_i + P(\lambda Q)y_i - by_i - \theta\lambda(y_i)^2 = 0 \text{ at } \lambda = 1$$

After dividing by y_i, this condition reduces to

$$P'(Q)Q + P(Q) = b + \theta y_i$$

Thus the equilibrium output, given that $k = m$, satisfies

$$(\theta + m)y_i^K(m, m) = P(Q) - b = A - my_i^K(m, m)$$

Then

$$y_i^K(m, m) = \frac{A}{\theta + 2m}$$

and industry output is

$$Q(m, m) = \frac{Am}{\theta + 2m}$$

This is, of course, the same as the collusive outcome.

When $m > k \geq 1$, we have to distinguish the exclusive Kant-Nash equilibrium from the inclusive one. In this example, we consider only interior EKNE. We denote equilibrium variables in an EKNE with the subscript E.

For the Nashian firms, the first-order condition is

$$P'(Q_E)y_{jE}^N + P(Q_E) = b + \theta y_{jE}^N$$

i.e.,

$$-y_{jE}^N + A - ny_{jE}^N - ky_{jE}^K = \theta y_{jE}^N \tag{16}$$

where we have omitted the subscripts. For the Kantian firms, the equilibrium condition (2) is equivalent to

$$\frac{d}{d\lambda}\left\{P(k\lambda y_{iE}^K + ny_{jE}^N)\lambda y_{iE}^K - b\lambda y_{iE}^K - \frac{1}{2}(\lambda y_{iE})^2\right\} = 0 \text{ at } \lambda = 1$$

i.e.,

$$P'(Q_E)k(y_{iE}^K)^2 + P(Q)y_{iE}^K = by_{iE}^K + \theta(y_{iE}^K)^2$$

Upon division by y_{iE}^K, we obtain

$$-ky_{jE}^K + A - ny_{jE}^N - ky_{jE}^K = \theta y_{jE}^K \tag{17}$$

The two FOCs (16) and (17) give the system of equations

$$\begin{bmatrix} \theta + n + 1 & k \\ n & \theta + 2k \end{bmatrix} \begin{bmatrix} y_E^N \\ y_E^K \end{bmatrix} = \begin{bmatrix} A \\ A \end{bmatrix}$$

Using $n = m - k$ the determinant is

$$\Delta = (\theta + 1)(\theta + 2k) + (m - k)(\theta + k)$$

Then

$$y_E^N(k, m) = \frac{A\,(\theta + k)}{(\theta + 1)(\theta + 2k) + (m - k)(\theta + k)} \tag{18}$$

$$y_E^K(k, m) = \frac{A\,(\theta + 1)}{(\theta + 1)(\theta + 2k) + (m - k)(\theta + k)} \leq y_E^N(k, m) \tag{19}$$

with equality holding iff $k = 1$.

Industry output in the EKNE is

$$Q_E(k, m) = \frac{(m\theta + k(m - k + 1))A}{(\theta + 1)(\theta + 2k) + (m - k)(\theta + k)}$$

It is easy to show that $Q_E(k, m)$ decreases in k, for given m. This is plausible. The greater the proportion of Kantian firms in the industry, the closer is the aggregate output to the collusive output.

4.1.2 Equilibrium Profits in an EKNE

Let us compare the profits. For simplicity, let $b = 0$. Then $A = B$, and the equilibrium price is

$$P_E(k, m) = B - Q_E = \frac{A\,(\theta + 1)\,(k + \theta)}{(\theta + 1)(\theta + 2k) + (m - k)(\theta + k)}$$

The representative Nashian firm's total revenue is $P_E(k,m) \times y_E^N = (\theta + 1)\left(y_E^N\right)^2$ and its equilibrium profit is

$$\pi_E^N = \left(y_E^N\right)^2 \left[1 + \frac{\theta}{2}\right]$$

$$= \left(\frac{A\,(\theta + k)}{(\theta + 1)(\theta + 2k) + (m - k)(\theta + k)}\right)^2 \left(\frac{2 + \theta}{2}\right)$$

The total revenue of a Kantian firm is $P_E(k,m) \times y_K^N = (k+\theta)\left(y_E^K\right)^2$. Its equilibrium profit is

$$\pi_E^N = \left(y^K\right)^2 \left[k + \frac{\theta}{2}\right]$$

$$= \left(\frac{A\,(\theta + 1)}{(\theta + 1)(\theta + 2k) + (m - k)(\theta + k)}\right)^2 \left(\frac{2k + \theta}{2}\right)$$

Notice that $\pi_E^N(k,m)$ is increasing in k.

The ratio of profits is

$$\frac{\pi_E^K}{\pi_E^N} = \frac{(\theta + 1)^2}{(\theta + k)^2}\frac{(2 + \theta)}{(2k + \theta)} < 1 \text{ iff } k > 1$$

Even though each Kantian firm's profit is lower than the profit of Nashian firms in the EKNE, it is easy to see that in an industry with m firms with identical cost functions, if the ratio k/m is large enough (i.e., sufficiently close to 1), each Kantian firm earns greater profit in an EKNE than the profit that would be earned if all firms behave in the Nashian way. This inequality holds if and only if

$$((\theta + 1)(\theta + 2k) + (m - k)(\theta + k))^2\,(2+\theta) - (\theta + 1)^2\,(\theta + m + 1)^2\,(2k+\theta) < 0$$

This holds if, for example, $\theta = 0, m = 5, k = 4$. This result is consistent with that of Salant et al. (1983, p. 193), who show that in the case of constant marginal cost, i.e. $\theta = 0$, then "it is sufficient for a merger to be unprofitable that less than 80 percent of the firms to collude."

However, if marginal cost is increasing, we can show that the result stated in Salant et al. (1983) for the constant marginal cost case is too pessimistic. Let us take an example with $\theta = 3$ and $m = 20$. The above inequality becomes

$$((3 + 1)(3 + 2k) + (20 - k)(3 + k))^2\,(2+3) - (3 + 1)^2\,(3 + 20 + 1)^2\,(2k+3) < 0$$

Suppose $k = 15$. Then the inequality is satisfied, indicating the Kantian firms are better off than if all firms were Nashian (even though they constitute less than 80 %

of the firms). The same result holds with $k = 14$, 13. However, if $k = 12$ or less, the Kantian firms are worse off than if all firms were Nashian.

4.2 Example 2: Interior IKNE

Let $m > k \geq 1$, and assume that

$$\theta + 1 > n \tag{20}$$

We maintain the assumptions that $P(Q) = B - Q$, and $c_i(y_i) = by_i + \frac{1}{2}\theta y_i^2$, where $\theta > 0, B > b > 0$.

For the Nashian firms, the first-order condition at an interior IKNE is

$$P'(Q_I)y_{jI}^N + P(Q_I) = b + \theta y_{jI}^N \tag{21}$$

i.e.,

$$-y_I^N + A - ny_I^N - ky_I^K = \theta y_I^N$$

where we have omitted the subscripts. For the Kantian firms, the equilibrium condition (3) is equivalent to

$$\frac{d}{d\lambda}\left\{P(k\lambda y_i^K + n\lambda y_j^N)\lambda y_i^K - b\lambda y_i - \frac{1}{2}(\lambda y_i)^2\right\} = 0 \text{ at } \lambda = 1$$

i.e.,

$$P'(Q_I)Q_I y_{iI}^K + P(Q_I)y_{iI}^K = by_{iI}^K + \theta(y_{iI}^K)^2$$

We focus on an interior equilibrium. (This implies some restrictions on the parameters; as we shall see, assumption (20) is sufficient.) Upon division by y_{iI}^K, we obtain

$$P'(Q_I)Q_I + P(Q_I) = b + \theta y_{iI}^K \tag{22}$$

Dropping the subscript i and j, and subtracting (21) from (22), we obtain, at an interior equilibrium,

$$\theta(y_I^K - y_I^N) + \left[(n-1)y_I^N + ky_I^K\right](-P') = 0$$

The two FOCs give the system of equations

$$\begin{bmatrix} \theta + n + 1 & k \\ 2n & \theta + 2k \end{bmatrix} \begin{bmatrix} y_I^N \\ y_I^K \end{bmatrix} = \begin{bmatrix} A \\ A \end{bmatrix}$$

The determinant is

$$\Delta_I = (\theta + 1)(\theta + 2k) + n\theta$$

Then

$$y_I^N(k, m) = \frac{A(\theta + k)}{(\theta + 1)(\theta + 2k) + (m - k)\theta} \tag{23}$$

$$y_I^K(k, m) = \frac{A(\theta + 1 - m + k)}{(\theta + 1)(\theta + 2k) + (m - k)\theta} > 0 \text{ iff } \theta + 1 > m - k \tag{24}$$

Industry output is

$$Q_I(k, m) = \frac{A(k + \theta m)}{(\theta + 1)(\theta + 2k) + (m - k)\theta}$$

We can verify that industry output in an interior IKNE is smaller than that obtained in an interior EKNE:

$$Q_E(k, m) - Q_I(k, m) < 0.$$

Example 3 (Non-interior IKNE) Assume $1 \leq k < m$, and

$$\theta + 1 < n \tag{25}$$

We maintain the assumptions that $P(Q) = B - Q$, and $c_i(y_i) = by_i + \frac{1}{2}\theta y_i^2$, where $\theta > 0, B > b > 0$. We now show that, because of condition (25), the Inclusive Kant-Nash equilibrium involves $y_I^K(k, m) = 0$ and $y_I^N(k, m) = \frac{A}{\theta+n+1}$. Let us verify that the output vector $\left(\frac{A}{\theta+n+1}, \ldots, \frac{A}{\theta+n+1}, 0, 0, 0, \ldots 0\right)$ is an equilibrium. Consider the FOCs. For Nashian firms

$$P'(ny^N + 0)y^N + P(ny^N + 0) = b + \theta y^N$$

For Kantian firms, at output $y_I^K(k, m) = 0$,their profit is $\pi_i(y^N, \ldots, y^N, 0, 0, \ldots, 0) = 0$. Thus

$$1 \in \arg\max_{\lambda \geq 0} \pi_i(\lambda y^N, \ldots, \lambda y^N, \lambda \times 0, \lambda \times 0, \ldots, \lambda \times 0).$$

5 Extension 1: Generalised Kant-Nash Equilibrium

In the preceding sections, we considered two extreme formulations of Kant-Nash equilibrium. In the EKNE, when a Kantian firm contemplates whether deviate by scaling its equilibrium output by a factor $\lambda > 0$, it considers as counterfactual alternative that the action level of all *all other agents, except the Nashian ones,* will also be adjusted by the same factor λ, while it assumes that the Nashian firms do not adjust at all. In contrast, in the IKNE, it assumes that all firms, including the Nashians, adjust by the same factor λ.

It seems natural to consider a more general formulation which nests both the IKNE and the EKNE as special cases. We propose that a Kantian firm asks itself the following question:

Question G: *"If I were to scale up or scale down of my output by any non-negative factor $\lambda \neq 1$, and if all other Kantian firms in the industry were to scale up or down their effort levels by the same factor, while the Nashian firms were to scale up or down their effort levels by a factor $\mu(\lambda)$, would my profit level be (weakly) lower?"*

It is important to stress that in asking this question, the Kantian firm does not necessarily believe that Nashians would actually scale up or down their activity levels. The question only serves as a guide to find an action that satisfies 'a good rule of conduct'.

How should $\mu(\lambda)$ be specified? It seems sensible to suppose that $\mu(\lambda) \neq \lambda$ iff $\lambda \neq 1$. An operational specification would be to introduce a parameter τ, such that

$$\mu(\lambda) = (\lambda - 1)\tau + 1 \text{ where } 0 \leq \tau \leq 1 \tag{26}$$

so that $\mu'(\lambda) = \tau \leq 1$. This means that if $\lambda = 1$ (neither scaling up nor down) then $\mu = 1$ too; if $\lambda > 1$ then $\mu(\lambda) \geq 1$, and $\mu(\lambda) \leq \lambda$; if $\lambda < 1$ (scaling down), then $\mu(\lambda)$ is such that $1 \geq \mu(\lambda) \geq \lambda$.

The resulting equilibrium is called the *Generalised Kant-Nash Equilibrium* (GKNE). This specification (26) has the advantage that the IKNE and the EKNE can be seen clearly as special cases: If $\tau = 0$, we have EKNE, and if $\tau = 1$,we have IKNE.

Example of GKNE: The case of linear demand and linear marginal cost.

We continue to assume that $P(Q) = B - Q$ and $c_i(y_i) = by_i - \frac{1}{2}y_i^2$. Assume that (26) holds. Then, for the n Nashian firms, the first order condition is

$$-y_G^N + A - ny_G^N - ky_G^K = \theta y_G^N \tag{27}$$

where the subscript G denotes variables at the GKNE.

And for the each of the k Kantian firms, the FOC is

$$\frac{d}{d\lambda}\left\{P(k\lambda y_G^K + \mu(\lambda)ny_G^N)\lambda y_G^K - b\lambda y_G^K - \frac{1}{2}(\lambda y_G)^2\right\} = 0 \text{ at } \lambda = 1$$

which is equivalent to

$$P'(Q_G)\left[k(y_G^K)^2 + \tau n y_G^K\right] + P(Q_G)y_G^K = by_G^K + \theta(y_G^K)^2$$

Recall that $P' = -1$ and $P(Q_G) = B - Q_G$, and that $A \equiv B - b$, the FOC reduces to

$$-k(y_G^K)^2 - \tau n y_G^N y_G^K + Ay_G^K - (ky_G^K + ny_G^N)y_G^K = \theta(y_G^K)^2 \tag{28}$$

We now use the two FOCs (27) and (28) to solve for the equilibrium output of each type of firm in a GKNE. From Eq. (27), we obtain

$$y_G^N = \frac{A - ky_G^K}{\theta + n + 1} \tag{29}$$

Substituting (29) into (28), we obtain, after simplification,

$$y_G^K(k,m) = \frac{A(\theta + 1 - (m-k)\tau)}{\theta^2 + \theta(m+k+1) + (m-k+2)k + \tau(m-k)k} \tag{30}$$

Then

$$y_G^N(k,m) = \left(\frac{A}{n+\theta+1}\right)\frac{\left(k + \theta + m\theta + \theta^2 - 2k^2\tau + km - k^2 + 2km\tau\right)}{(\theta^2 + \theta(m+k+1) + (m-k+2)k + \tau(m-k)k)} \tag{31}$$

Obviously, Eqs. (30) and (31) reduce to (19) and (18) when $\tau = 0$, and to (24) and (23) when $\tau = 1$.

6 Extension 2: Kant-Nash Equilibrium in a Price-Setting Oligopoly with Heterogeneous Goods

In this section, we consider an oligopoly in heterogeneous goods, and study Kant-Nash equilibrium when firms set prices. We assume that there are m firms in the oligopoly, each producing a differentiated product. There is also a numeraire good produced under perfect competition. Let $\mathbf{q} = (q_1, q_2, \ldots, q_m)$ denote the representative consumer's vector of consumption of the goods produced by the m oligopolists, and z the quantity of the numeraire good she consumes. Her utility function is

$$U(\mathbf{q}, z) = z + u(\mathbf{q})$$

where

$$u(q) = AQ - \frac{1}{2}ebQ^2 - \frac{1}{2}b(1-e)\sum_{i=1}^{m} q_i^2$$

with $b > 0$, $0 < e < 1$, and

$$Q \equiv \sum_{i=1}^{m} q_i$$

The assumption that $b > 0$ implies diminishing marginal utility with respect to each good q_i. The restriction $0 < e < 1$ ensures that the consumer's necessary conditions for utility maximization are also sufficient.

The consumer maximizes utility subject to the budget constraint $z + \sum_{i=1}^{m} p_i q_i = M$ where M is her income and p_i is the price of good i.

We consider an one-shot game, where each firm i sets the price of its product, p_i. Let $\mathbf{p} = (p_1, p_2, \ldots, p_m)$ denote the price vector.

Assuming an interior solution, the first order conditions of the consumer's optimization problem are

$$p_i = A - bq_i - eb\sum_{j \neq i} q_j$$

In matrix form,

$$\begin{bmatrix} g_{11} & g_{12} & \cdot & \cdot \\ \cdot & \cdot & \cdot & \cdot \\ \cdot & \cdot & \cdot & \cdot \\ \cdot & \cdot & \cdot & g_{mm} \end{bmatrix} \begin{matrix} q_1 \\ q_2 \\ \cdot \\ q_m \end{matrix} = \begin{matrix} A - p_1 \\ A - p_2 \\ \cdot \\ A - p_m \end{matrix}$$

where G is an $m \times m$ matrix with elements $g_{ii} = 1$ and $g_{ij} = e$, for $j \neq i$. The inverse of matrix G is denoted by H with

$$h_{ii} = \frac{1 + (m-2)e}{\Delta}$$

$$h_{ij} = -\frac{e}{\Delta}$$

$$\Delta = 1 + (m-2)e - (m-1)e^2 < 0 \text{ for } 0 < e \leq 1.$$

Then the demand function for good i is

$$q_i = D^i(\mathbf{p}) = \frac{1}{b\Delta}\left\{A(1-e) - [1 + (m-2)e]p_i + e\sum_{s \neq i} p_s\right\} \tag{32}$$

with

$$D^i_{p_i}(\mathbf{p}) = -\frac{[1+(m-2)e]}{b\Delta} < 0$$

$$D^i_{p_j}(\mathbf{p}) = \frac{e}{b\Delta} > 0 \text{ for } j \neq i$$

Note that if all prices increase by $\varepsilon > 0$, then the quantity demanded, q_i , will change by $\frac{-(1-e)}{b\Delta}\varepsilon < 0$. Note also that if all firms charge the same price, the demand for each good is

$$q_i = \frac{1}{b\Delta}(1-e)(A-p)$$

and the total revenue of the industry is

$$\frac{m}{b\Delta}(1-e)(A-p)p.$$

Thus, the common price that maximizes industry revenue is $p^* = A/2$.

Assume the production cost of q_i is $c_i(q_i) \geq 0$, the profit of firm i is

$$\pi_i(p) = p_i D^i(p_1, p_{2,} \ldots, p_m) - c_i\left(D^i(p_1, p_{2,} \ldots, p_m)\right)$$

Definition PS1 (Kantian equilibrium in a price-setting oligopoly with differentiated goods): A Kantian equilibrium in a price-setting oligopoly with differentiated goods (KE-PS) is a price vector $p > 0$ such that any scaling up (or scaling down) of the output vector by a factor $\lambda > 0$ will (weakly) decrease the profit of all firms, that is,

$$(\text{for all } i = 1, 2, \ldots, m), \qquad 1 \in \arg\max_{\lambda > 0} \pi_i(p).$$

The necessary condition for an interior Kantian equilibrium in this price-setting game is

$$\frac{d}{d\lambda}\left[\lambda p_i D^i(\lambda p_1, \lambda p_{2,} \ldots, \lambda p_m) - c_i\left(D^i(\lambda p_1, \lambda p_{2,} \ldots, \lambda p_m)\right)\right] = 0 \text{ at } \lambda = 1.$$

That is,

$$D^i(\mathbf{p}) + \left[p_i - c_i'(D^i(p))\right]\sum_{s=1}^{m} D^i_{p_s}(\mathbf{p}) = 0 \tag{33}$$

It is easy to see that this condition is identical to the necessary condition for a collusive equilibrium.

For simplicity, suppose that $c^i(q_i) = 0$ for all $q_i \geq 0$. Then in a symmetric Kantian equilibrium, the necessary condition (33) reduces to

$$D(p,p,\ldots,p) + p\sum_{s=1}^{m} D_{p_s}(p,p,\ldots,p) = 0 \tag{34}$$

Using the linear demand function (32), condition (34) gives the Kantian equilibrium price p^*, which is identical to the collusive price:

$$p^* = \frac{A}{2}.$$

At the Kantian equilibrium, the profit per firm is

$$\pi^* = p^* D(\mathbf{p}^*) = \frac{(A-p^*)p^*}{b\Delta} = \frac{A^2}{4b\Delta}.$$

Definition PS2 (Exclusive Kant-Nash Equilibrium in a price-setting oligopoly with differentiated goods): An Exclusive Kant-Nash equilibrium in a price-setting oligopoly with differentiated goods (EKNE-PS) with k Kantian firms and $n = m-k$ Nashian firms is a price vector $p > 0$

$$\mathbf{p} = (p_1^N, \ldots, p_n^N, p_{n+1}^K, \ldots, p_{n+k}^K)$$

such that

(i) No Nashian firm j can gain by deviating from p_j^N, assuming that all other firms in $\mathcal{M}$ maintain their output level.
(ii) For each Kantian firm i, the answer to the following question is "No": Can my profit be increased by applying a scaling factor $\lambda > 0$ to the price vector $(p_{n+1}^K, \ldots, p_{n+k}^K)$ of all k Kantian firms? In other words, it holds that

$$(\text{for all } i = n+1, \ldots, n+k), \qquad 1 \in \arg\max_{\lambda>0} \pi_i(p_1^N, \ldots, p_n^N, \lambda p_{n+1}^K, \ldots, \lambda p_{n+k}^K)$$

Applying this definition to our example with zero marginal cost, we find that for Nashian firm i to be in equilibrium, it is necessary that

$$D^i(p_1^N, \ldots, p_n^N, p_{n+1}^K, \ldots, p_{n+k}^K) + p_i^N \frac{\partial D^i}{\partial p_i^N} = 0$$

i.e.,

$$\frac{1}{b\Delta}\left\{A(1-e) - [1+(m-2)e]p_i^N + (m-k-1)ep_j^N + kep^K\right\} - p_i^N \frac{[1+(m-2)e]}{b\Delta} = 0$$

i.e., under symmetry,

$$A(1-e)+kep^K = (2+(m+k-3)\,e)\,p^N \tag{35}$$

And, for (exclusive) Kantian firms to be in equilibrium, we require that

$$p_i^K D^i + p_i^K\left[\frac{(k-1)ep_j^K - (1+(m-2)e)p_i^K}{b\Delta}\right] = 0$$

Under symmetry, this reduces to

$$A(1-e)+(m-k)ep^N = 2p^K\,[1+(m-k-1)e] \tag{36}$$

Write the system of equations (35) and (36) as

$$\begin{bmatrix} 2+(m+k-3)e & -ek \\ -e(m-k) & 2+2e(m-k-1) \end{bmatrix}\begin{bmatrix} p^N \\ p^K \end{bmatrix} = \begin{bmatrix} A(1-e) \\ A(1-e) \end{bmatrix}$$

The determinant is

$$\Omega = (2+(m+k-3)e)\,(2+2e(m-k-1)) - e^2k(m-k) > 0 \text{ for } 0 < e < 1$$

because $2+(m+k-3)e > ek$ and $2+2e(m-k-1) > m-k$. Then

$$p^N = \frac{A(1-e)\,[2+2e(m-k-1)+ek]}{\Omega}$$

and

$$p^K = \frac{A(1-e)\,[2+(m+k-3)e+e(m-k)]}{\Omega}$$

We see that, if $k > 1$, the price set by a Kantian firm is higher than that set by a Nashian firm:

$$p^K - p^N = \frac{e(k-1)A(1-e)}{\Omega}$$

What can we say about the relative profit? The profit of a Nashian firm is

$$\begin{aligned}\pi^N(k,m) &= p^N D(p^N,\ldots,p^N,p^K,\ldots,p^K) \\ &= \frac{1}{b\Delta}p^N\,(A(1-e)-(1+(k+1)e)\,p^N + kep^K)\end{aligned}$$

and that of a Kantian firm is

$$\pi^K(k,m) = \frac{1}{b\Delta} p^K\left(A(1-e) - (1+(m-k+1)e)\,p^K + (m-k)ep^N\right)$$

As a numerical example, let $A = 1, b = 1, e = 0.5, m = 20, k = 2$. Then $\Omega = 209.5$ and $\Delta = 5.25$.

$$\Delta = 1 + (m-2)e - (m-1)e^2$$

The Nashian price is

$$p^N = \frac{\frac{1}{2}\left(2 + (18-1) + 1\right)}{209.5} = 4.7733 \times 10^{-2}$$

The Kantian price is

$$p^K = \frac{\frac{1}{2}\left(2 + (22-3)\frac{1}{2} + \frac{1}{2}(18)\right)}{209.5} = 4.8926 \times 10^{-2} > p^N$$

The profit of the representative Nashian firm is

$$\frac{1}{\Delta} p^N\left(A(1-e) - (1+(k+1)e)\,p^N + kep^K\right)$$

$$\frac{2.0506 \times 10^{-2}}{5.25} = 3.9059 \times 10^{-3}$$

The profit of the representative Kantian firm is

$$\frac{0.40566}{5.25} = 7.7269 \times 10^{-2}$$

This example shows that, in a price-setting oligopoly with heterogeneous goods, it can pay to be a Kantian (in an Exclusive Kant-Nash Equilibrium) as long as there is at least another Kantian. This is in sharp contrast to the quantity setting case. This result is not too surprising, since it is well known that price-competition is more damaging to oligopolists than quantity competition.

7 Concluding Remarks

This paper studies oligopolistic equilibria when not all firms behave in a Nashian fashion. The Kantian firms act according to norms of behavior that reflect the Kantian categorical imperative. We introduce the concepts of Exclusive Kant-Nash Equilibrium and Inclusive Kant-Nash Equilibrium to study such situations. Our model formalises Sen's conjecture that in actual situations people often do not follow the selfish strategy, because each has asked himself "what type of preference he would like the other player(s) to have, and on somewhat Kantian grounds has considered the case for himself having those preferences, or behaving *as if* he had them."

While Roemer (2010) advocates the Kantian equilibrium as a normative concept, our formulation of EKNE and IKNE is more in line with Sen (1977, p. 341) who sees Kantian behavior among some individuals as a reasonable *description* of some real world situations. Laffont (1975) seems to share both views. On the one hand, he uses the Kantian rule of behavior to explain the *fact* "that (at least in some countries) people do not leave their beer cans on beaches" (p. 431). On the other hand, he also takes a normative view, stating that one may hopefully "induce people to behave in a Kantian way", because this may be "a very good solution to some problems" (p. 431). He concludes that "we can expect to realize the positive aspects of Kantian behavior even when the social interest is, to some reasonable degree, in conflict with private interest."

While this paper's focus in on Exclusive and Inclusive Kant-Nash equilibria in the context of a quantity-setting oligopoly, it should be evident that the theory of Kant-Nash equilibrium has potentially many applications. An interesting extension would be to apply the Kant-Nash equilibrium concepts to differential games (Chiarella et al., 1984; Dockner et al., 2000; Long, 2010; Ghosh and Long, 2015).

Acknowledgements The author thanks Mardi Dungey, Scott Finley, Guy Laroque, Vinh Nguyen and Eytan Sheshinsky for helpful comments on an earlier version of this paper.

References

Arrow KJ (1973) Social responsibity and economic efficiency. Public Policy 21:303–317

Bergstrom T (1995) On the evolution of altruistic ethical rules for siblings. Am Econ Rev 85:58–81

Chiarella C, Kemp MC, Long NV, Okuguchi K (1984) On the economics of international fisheries. Int Econ Rev 25(1):85–92

De Donder P, Roemer JE (2014) An allegory of the political influence of the top 1%. Paper presented at CESifo area conference on public sector economics, Munich

Dockner E, Jorgensen S, Long NV, Sorger G (2000) Differential games, with applications to economics and management science. Cambridge University Press, Cambridge

Fudenberg D, Tirole J (2000) Game theory. MIT, Cambridge, MA

Gaudet G, Salant S (1991) Uniqueness of Cournot equilibrium: new results from old methods. Rev Econ Stud 58(2):399–404

Ghosh A, Long NV (2015) Kant's rule of behavior and Kant-Nash equilibria in games of contribution to a public good. Paper presented at the CESifo conference on public sector economics, Munich, April 2015

Kant I (1785) Grundlegung zur Metaphysik der Sitten. Translated by Mary J. Gregor: Groundwork of the metaphysics of morals. Cambridge University Press, Cambridge, UK (1998)

Laffont J-J (1975) Macroeconomic constraints, economic efficiency and ethics: an introduction to Kantian economics. Economica 42:430–437

Long NV (2010) A survey of dynamic games in economics. World Scientific, Singapore

Long NV, Soubeyran A (2000) Existence and uniqueness of Cournot equilibrium: a contraction mapping approach. Econ Lett 67:231–236

Long NV, Soubeyran A (2001) Cost manipulation games in oligopoly, with cost of manipulating. Int Econ Rev 42(2):505–533

Myerson R, Weibull J (2015) Tenable strategy blocks and settled equilibria. Econometrica 83(3):943–967

Okuguchi K (1977) Expectations and stability in oligopoly models. Lecture notes in economics and mathematical systems. Springer, Heidelberg

Okuguchi K, Szidarovszky F (1990) The theory of oligopoly with multi-product firms. Lecture notes in economics and mathematical systems. Springer, Heidelberg

Rescher N (1975) Unselfishness. University of Pittsburg Press, Pittsburg

Roemer JE (2010) Kantian equilibrium. Scand J Econ 112(1):1–24

Russell B (1945) A history of western philosophy. George Allen and Unwin, London

Salant S, Switzer S, Reynolds R (1983) Losses from horizontal merger: the effects of an exogenous change in industry structure on Cournot Nash equilibrium. Q J Econ 98(2):185–199

Sen A (1977) Rational fools: a critique of the behavioral foundation of utility. Philos Public Aff 6(4):317–344

Sen A (1993) Does business ethics make economic sense? Bus Ethics Q 3(1):45–54

Smith A (1790 [2002]) The theory of moral sentiments (revised edition, 1790). Republished in the series Cambridge Texts in the History of Philosophy, edited by Knud Haakonssen.

Tirole J (2001) Corporate Governance. Econometrica 69(1):1–35

Waymark JA (1978) Unselfishness and Prisoner's Dilemma. Philos Stud 34:417–425

Evolutionary Oligopoly Models of Commercial Fishing with Heterogeneities

Gian Italo Bischi, Fabio Lamantia and Elena Viganò

Abstract In this paper we provide an overview of some recent dynamic models of commercial fisheries. Our starting points are the papers by Okuguchi (Keio Econ Stud 35:9–17, 1998) and Okuguchi and Szidarovszky (Seoul J Econ 11(3):321–330, 1998) and Szidarovszky and Okuguchi (Seoul J Econ 13:471–476, 2000) on oligopoly competition in international fisheries, which are based on the assumptions of homogeneous fish population, homogeneous space and continuous time. Following also insightful hints given in Okuguchi (Keio Econ Stud 21:37–44, 1984) and Erjaee and Okuguchi (Keio Econ Stud 41:61–71, 2006), we describe how some heterogeneities can be introduced in fishery models. The examples provided in this paper include bioeconomic models of fisheries with several species and with different time scales, ranging from continuous-time to discrete-time dynamic models and also addressing hybrid models, where some variables evolve in continuous time whereas other ones change in (event-driven) discrete time. Fishery models where the aquatic environment is subdivided into adjacent patches, characterized by different fishing policies including, as a particular case, Marine Protected Areas, are also described. Finally, we describe a fishery models with cooperators and defectors and punishment policies to encourage cooperation. All the models described in this overview include an evolutionary mechanism, based on imitative behavior, through which fishers are allowed to change their harvesting strategy over time. Some of these models have been motivated by projects or management experiences performed in the Adriatic Sea by Italian fishing authorities.

1 Introduction

Starting from the seminal work by Smith (1969) on commercial fishing, modelled by a system of two ordinary differential equations that describe the time evolution of a fish species harvested by a time-evolving number of fishing vessels, Koji Okuguchi addressed several relevant extensions, briefly recalled below. In Okuguchi (1984)

G.I. Bischi (✉)
Department of Economics, Society, Politics, University of Urbino, Urbino, Italy
e-mail: gian.bischi@uniurb.it

P. von Mouche, F. Quartieri (eds.), *Equilibrium Theory for Cournot Oligopolies and Related Games*, Springer Series in Game Theory,
DOI 10.1007/978-3-319-29254-0_13

he considered two interacting fish populations, preys and predators, both harvested and sold at fixed prices, with landing cost proportional to the square of harvest and inversely proportional to available fish stock. In 1998 Okuguchi studied a model of international trade of a single species, harvested by two countries under imperfect competition in an open-access sea; different prices are present in each country, obtained through two linear demand functions, which depend on the total amount of fish sold in each country by both local and foreign fishers. The questions of existence and stability of Nash equilibria for this duopoly model are studied under the assumption of a continuous-time logistic growth of the fish population. These results were then extended by Szidarovszky and Okuguchi (1998) to the case of an oligopoly with $n \geq 2$ countries, and the effect of entry of a new fishing country on existence and stability of equilibria is also studied. The duopoly model of international imperfectly competitive fishery is re-considered by Erjaee and Okuguchi (2006) in a discrete-time setting, and sequences of bifurcations leading to periodic and chaotic oscillations are studied, both analytically and numerically.

Following this stream of literature, several dynamic bioeconomic models have been recently proposed, with the aim of describing new policies for sustainable exploitation of fisheries; the underlying assumptions of these works on fish growth, demand and cost functions are similar to the ones proposed in the above mentioned papers by Okuguchi and coauthors, so that the dynamic equations are still analytically tractable even if numerical methods have been used to confirm and extend the analytical results.

In this paper, we describe some of these fishery models, motivated by a research project, in which the authors have been involved, concerning the study of the effects of some harvesting policies and fishing limitations proposed by Italian authorities to regulate fisheries in the Adriatic Sea, see e.g. the book edited by Antonelli et al. (2005), as well as the papers Bischi et al. (2013a,b), dealing with a model with two fish species and a policy adopted in the Adriatic Sea that allows each fisher to fish only one with the possibility to switch to the other one at discrete-time periods, or Antonelli et al. (2006), Bischi and Lamantia (2007), Bischi et al. (2009), and Lamantia (2006), dealing with the creation of marine protected areas, or Bischi et al. (2004), dealing with different fishing strategies with the possibility to apply sanctions or taxes on the less environmental friendly fishing method.

The plan of the paper is as follows. Section 2 illustrates an evolutionary game for the exploitation of different species, where market and cost externalities help lowering harvesting pressure to the most endangered species. Section 3 describes an evolutionary model where exploiters can decide the aquatic region to harvest from, where different harvesting rules are established. Section 4 deals with a model where agents can decide whether to cooperate and harvest the amount that maximizes the utility of a coalition or to defect and maximize individual utility. Agents can switch over time between the two groups and can enforce cooperation through punishments. Finally, Sect. 5 concludes.

2 Two-Species Fisheries with Evolutionary Strategy Switching

In Bischi et al. (2013a) a dynamic model is proposed to describe a fishery with two non-interacting species where a Fishing Regulatory Authority allows each fisher to harvest just one species during a given time period. Then fishers can reconsider their harvesting decision at discrete "switching times" on the basis of accrued profits. The model is motivated by an Italian fisheries management experience in the Northern Adriatic Sea, where this kind of "self-adjusting" fishing policy has been proposed to regulate harvesting of two shellfish species, with time periods of three years between switching times. Indeed, as the harvesting cost is inversely proportional to the available fish biomass, at each switching time more fishers are attracted towards the more abundant species, thus giving the possibility to restore the biomass of the scarcer resource.

To model this context, in Bischi et al. (2013a) a population of N fishers is considered, such that, at time t, $n_1(t)$ of them harvest only fish population 1, whose biomass density is $X_1(t)$, and $n_2(t)$ harvest only fish populations 2, whose biomass density is $X_2(t)$, with $n_1(t) + n_2(t) = N$. Let $r(t) = r_1(t) = n_1(t)/N$ be the fraction of fishers that at time t are allowed to harvest species 1. Of course, the complementary fraction $r_2(t) = 1 - r(t)$ represents the fraction of fishers that harvest species 2. The dynamics of the system is modelled assuming that natural growth and harvesting of the two species occurs in continuous time, whereas discrete-time strategy switching (decision-driven) regulates how agents adaptively update their harvesting strategy period by period, according to an evolutionary mechanism based on profit-driven replicator dynamics. Denoting by $G_i(X_i)$ the specific growth function of species i and by H_i its harvesting rate, the model assumes the following form:

$$\begin{cases} \dot{X}_1(t) = X_1(t)G_1(X_1(t)) - H_1(X_1(t), X_2(t)) \\ \dot{X}_2(t) = X_2(t)G_2(X_2(t)) - H_2(X_1(t), X_2(t)) \\ r(t) = \begin{cases} r(t-s)\frac{\overline{\pi}_1^*(t)}{r(t-s)\overline{\pi}_1^*(t)+[1-r(t-s)]\overline{\pi}_2^*(t)} & \text{if } \frac{t}{s} = \left\lfloor \frac{t}{s} \right\rfloor \\ r\left(\left\lfloor \frac{t}{s} \right\rfloor s\right) & \text{otherwise} \end{cases} \end{cases} \tag{1}$$

Here $\dot{X}_i$ denotes the derivatives of biomass with respect to time, $\lfloor x \rfloor$ is the largest integer not greater than x (i.e. the floor of x), and

$$\overline{\pi}_i^*(t) = \frac{\int_{t-s}^{t} \pi_i^*(\tau)\, d\tau}{s} \quad ; i = 1, 2 \tag{2}$$

is the average profit observed by each agent at the end of every switching time of length s. So, the third equation in (1) states that at each switching time, each fisher, assumed to know the average profits of both groups during the previous period, compares the two average profits over the last fishing period. Average profit of each strategy is taken as fitness measure of that strategy, according to the paradigms of evolutionary game theory (see Weibull 1995; Hofbauer and Sigmund 1998): if $\overline{\pi}_1^*(t) > \overline{\pi}_2^*(t)$ then $r(t)$ increases, i.e. a fraction of fishers harvesting species 2 switch to harvest species 1, otherwise $r(t)$ decreases.

Following Szidarovszky and Okuguchi (1998), the harvesting rates $H_1(t)$ and $H_2(t)$ are computed according to the profit maximization problem of a Cournot oligopoly game. Assuming that the current total harvesting is wholly supplied to the market, prices are determined according to the following horizontal differentiated linear inverse demand system (see Singh and Vives 1984 for details)

$$\begin{cases} p_1 = f_1(H_1, H_2) = a_1 - b_1(H_1 + \sigma H_2) \\ p_2 = f_2(H_1, H_2) = a_2 - b_2(\sigma H_1 + H_2) \end{cases} \tag{3}$$

where a_i is the reservation price for species i, b_i represents the slope of the demand for fish i and $\sigma \in [0, 1]$ is the symmetric degree of substitutability between the two fish varieties. In particular, if $\sigma = 0$ then the two varieties are independent in demand. On the other hand, for $\sigma = 1$ they are perfect substitutes.[1] As usual in fishery oligopoly models, quadratic harvesting costs are assumed for both species, i.e. for harvesting h_i units of species i an agent faces a cost given by

$$C_i(X_i, h_i) = \gamma_i \frac{h_i^2}{X_i} \tag{4}$$

where γ_i measures the fishers' inefficiency for catching species i. This cost function is obtained and employed in Clark (1990), Okuguchi (1998), and Szidarovszky and Okuguchi (1998) and is very common in fishery models, since it captures the fact that it is easier and less expensive to catch fish if the fish population is large. Given these specifications of cost and prices functions, the profit of fisher q belonging to group i ($i = 1, 2$) when harvesting $h_{i,q}$ reads

$$\pi_{i,q} = p_i h_{i,q} - \gamma_i \frac{h_{i,q}^2}{X_i}, i = 1, 2 \tag{5}$$

Therefore, in deciding instantaneous harvesting of species i, the representative fisher q in group i solves the problem $\max_{h_{i,q}} \pi_{i,q}$. By taking the first order conditions and employing the symmetry property that players within each group are homogeneous

[1]In the context we are considering, we disregard the case $\sigma < 0$, related to varieties that are demand complementary.

(i.e. $h_{i,q} = h_{i,u}, i = 1, 2;$) we obtain for $i, j = 1, 2$ with $i \neq j$ the following harvesting quantities at a Nash equilibrium (see Bischi et al. 2013a for details) $h_i^*(X_i, X_j) =$

$$\frac{a_i X_i (b_j X_j \left[1 + N r_j\right] + 2\gamma_j) - a_j b_i N r_j X_i X_j \sigma}{(b_i X_i (1 + Nr) + 2\gamma_i)(b_j X_j (1 + N(1-r)) + 2\gamma_j) - b_i b_j N^2 (1-r) r X_i X_j \sigma^2}, \tag{6}$$

where $r_1 = r = \frac{n_1}{N}$ and $r_2 = (1-r) = \frac{n_2}{N}$ represent, respectively, the fractions of fishers that harvest species 1 and 2, respectively. Hence, the harvesting terms in (1) become $H_1 = Nrh_1^*$ and $H_2 = N(1-r)h_2^*$ respectively. Moreover, by inserting (6) into (5), we get optimal individual profits

$$\pi_i^* = \left(b_i + \frac{\gamma_i}{X_i}\right)\left(h_i^*\right)^2 \tag{7}$$

which shows that profits are always non-negative.

Different kinds of growth functions can be considered in (1), according to the biological properties of the fish population and the marine environment at hand. In Bischi et al. (2013a), it is assumed that the two populations of (shell)fish follow a logistic natural growth rate of the form

$$G_i = \rho_i\left(1 - \frac{X_i}{k_i}\right) \quad ; i = 1, 2 \tag{8}$$

where ρ_i and k_i are, respectively, the intrinsic rate of growth and the carrying capacity of species i.

In order to investigate the model proposed by using both analytical and numerical methods, Bischi et al. (2013a) studies some benchmark cases. For example, in the limiting case $s \to 0$, i.e. with fishers changing their fishing strategy continuously, one gets $\overline{\pi}_i^*(t) = \pi_i^*(t)$, and the last equation in (1) can be replaced with

$$\begin{aligned} \dot{r}(t) &= r(t)\left[\pi_1^*(t) - \left(r(t)\pi_1^*(t) + (1-r(t))\pi_2^*(t)\right)\right] \\ &= r(t)(1-r(t))\left[\pi_1^*(t) - \pi_2^*(t)\right] \end{aligned} \tag{9}$$

which is the well-known replicator equation in continuous time (see again Weibull 1995; Hofbauer and Sigmund 1998), stating that $\dot{r}(t) > 0$ [<] if $\pi_1^*(t) > \pi_2^*(t)$ [<]. In this case, the model assumes the simpler form of a nonlinear 3-dimensional system of ordinary differential equations. In order to obtain analytical results, the case of constant prices is considered (i.e. $b_1 = b_2 = 0$). The dynamical model (1) assumes the following simplified form:

$$\begin{cases} \dot{X}_1 = X_1\rho_1\left(1 - \frac{X_1}{k_1}\right) - Nr\frac{a_1 X_1}{2\gamma_1} \\ \dot{X}_2 = X_2\rho_2\left(1 - \frac{X_2}{k_2}\right) - N(1-r)\frac{a_2 X_2}{2\gamma_2} \\ \dot{r} = r(1-r)\left[\frac{a_1^2 X_1}{4\gamma_1} - \frac{a_2^2 X_2}{4\gamma_2}\right] \end{cases} \tag{10}$$

Note that in the invariant subspaces defined by $r = 0$ and $r = 1$, the first two differential equations in (10) are uncoupled. The following propositions (proved in Bischi et al. 2013a) describe the steady states of the model and their local stability properties. To keep the notation short, in these propositions it is useful to define the aggregate parameters

$$\alpha_1(r) = \rho_1 - \frac{a_1 N r}{2\gamma_1} \text{ and } \alpha_2(r) = \rho_2 - \frac{a_2 N(1-r)}{2\gamma_2} \tag{11}$$

Proposition 1 (Boundary equilibria and their stability) *For the model (10) the following statements hold:*

- *the total extinction fixed points $E_r^0 = (0, 0, r)$, where $r \in [0, 1]$, are unstable non-hyperbolic nodes provided that $\alpha_i(r) > 0$, $i = 1, 2$;*
- *the equilibria with harvesting of only one species are given by:*
  - *$E_1^0 = (k_1, 0, 0)$ and $E_2^0 = (0, k_2, 1)$ (extinction of the harvested species), which are saddle points;*
  - *$E_1^1 = \left(0, k_2\left(1 - \frac{Na_2}{2\gamma_2\rho_2}\right), 0\right)$, if $Na_2 < 2\gamma_2\rho_2$, and $E_2^1 = \left(k_1\left(1 - \frac{Na_1}{2\gamma_1\rho_1}\right), 0, 1\right)$, if $Na_1 < 2\gamma_1\rho_1$ (extinction of the non-harvested species), which are saddle points;*
  - *$E_1^2 = \left(k_1, k_2\left(1 - \frac{Na_2}{2\gamma_2\rho_2}\right), 0\right)$ if $Na_2 < 2\gamma_2\rho_2$ (both viable species with no harvesting of species 1), which is a stable node provided that*

$$a_1 < \sqrt{\frac{a_2^2 k_2 \gamma_1 (2\gamma_2\rho_2 - Na_2)}{2k_1\gamma_2^2\rho_2}} \tag{12}$$

  and a saddle point if the reverse inequality in (12) holds;
  - *$E_2^2 = \left(k_1\left(1 - \frac{Na_1}{2\gamma_1\rho_1}\right), k_2, 1\right)$ if$Na_1 < 2\gamma_1\rho_1$ (both viable species with no harvesting of species 2) which is a stable node provided that*

$$a_2 < \sqrt{\frac{a_1^2 k_1 \gamma_2 (2\gamma_1\rho_1 - Na_1)}{2k_2\gamma_1^2\rho_1}} \tag{13}$$

  and a saddle point if the reverse inequality in (13) holds.

The next proposition characterizes an equilibrium with $r^* \in (0, 1)$, which is the case in which each species is always harvested by some fishers, of course with the restriction that each agent is allowed to fish just one species at each time.

Proposition 2 (Interior equilibrium and its stability) For the model (10) the following statements hold:

- *There exists a unique interior equilibrium $E^* = \left(X_1^*, X_2^*, r^*\right)$ with $r^* \in (0, 1)$, where*

$$X_i^* = \frac{a_j^2 k_1 k_2 \gamma_i (2a_2\gamma_1\rho_1 + 2a_1\gamma_2\rho_2 - a_1 a_2 N)}{2\left(a_2^3 k_2 \gamma_1^2 \rho_1 + a_1^3 k_1 \gamma_2^2 \rho_2\right)}, i = 1,2; i \neq j \tag{14}$$

$$r^* = \frac{\gamma_1\rho_1\left(a_2^3 k_2 N \gamma_1 - 2a_2^2 k_2 \gamma_1\gamma_2\rho_2 + 2a_1^2 k_1 \gamma_2^2 \rho_2\right)}{N\left(a_2^3 k_2 \gamma_1^2 \rho_1 + a_1^3 k_1 \gamma_2^2 \rho_2\right)}$$

- *Equilibrium biomass levels* $X_i^* > 0, i = 1,2$ *iff* $\alpha_i(r^*) > 0$, *with* $0 < r^* < 1$*; this occurs in the following cases:*
  - *case 1: If* $\alpha_1(1) > 0$ *and* $\alpha_2(0) > 0$ *then the carrying capacity* k_1 *must satisfy*

$$\hat{k}_1 = \frac{a_2^2 k_2 \gamma_1 (2\gamma_2\rho_2 - Na_2)}{2a_1^2\gamma_2^2\rho_2} < k_1 < \frac{2a_2^2 k_2^2 \gamma_1^2 \rho_1}{2a_1^2\gamma_1\gamma_2\rho_1 - Na_1^3\gamma_2} = \overline{k}_1;$$

  - *case 2: If* $\alpha_1(1) > 0$ *and* $\alpha_2(0) < 0$ *then the carrying capacity* k_1 *must satisfy* $0 < k_1 < \overline{k}_1$;
  - *case 3: If* $\alpha_1(1) < 0$ *and* $\alpha_2(0) > 0$ *then the carrying capacity* k_1 *must satisfy* $k_1 > \hat{k}_1$;
  - *case 4: If* $\alpha_1(1) < 0$ *and* $\alpha_2(0) < 0$ *then it must be* $\frac{2\gamma_1\rho_1}{N} < a_1 < \frac{2a_2\gamma_1\rho_1}{Na_2 - 2\gamma_2\rho_2}$, *and at* $a_1 = \frac{2a_2\gamma_1\rho_1}{Na_2 - 2\gamma_2\rho_2}$ *it is* $X_1^* = X_2^* = 0$;
  - *finally, if* $k_1 = \hat{k}_1$ *then it is* $r^* = 0$ *and* $E^* = E_1^2 = \left(k_1, k_2\left(1 - N\frac{a_2}{2\gamma_2\rho_2}\right), 0\right)$ *whereas if* $k_1 = \overline{k}_1$ *then it is* $r^* = 1$ *and* $E^* = E_2^2 = \left(k_1\left(1 - N\frac{a_1}{2\gamma_1\rho_1}\right), k_2, 1\right)$.
- *If the equilibrium* $E^* = \left(X_1^*, X_2^*, r^*\right)$ *involves positive biomasses, then it is stable under the replicator dynamics in continuous time.*

The cases discussed in the previous proposition help to understand how the interior equilibrium appears or disappears through contacts with border equilibria. In particular, the last two cases indicate that the interior equilibrium can have a contact with the extinction equilibrium E_r^0 or with the border equilibria E_1^2 and E_2^2. It is also interesting to notice that, by Proposition 1, a single species, say species 1, does not become extinct provided that $\frac{2\gamma_1\rho_1}{a_1 N} > r$, whereas by Proposition 2, the condition $r^* < 1$ leads to $\frac{2\gamma_1\rho_1}{a_1 N} < 1 + \frac{2a_2^2 k_2 \gamma_1^2 \rho_1}{Na_1^3 k_1 \gamma_2}$. Therefore, if

$$\frac{2\gamma_1\rho_1}{a_1 N} \in \left(r, 1 + \frac{2a_2^2 k_2 \gamma_1^2 \rho_1}{Na_1^3 k_1 \gamma_2}\right) \tag{15}$$

then species 1 will survive both with an exogenous fixed r or with a r^* to which the continuous-time switching mechanism converges. However, if the fixed $r = \overline{r} \in (0,1)$ is such that

$$\frac{2\gamma_1\rho_1}{a_1 N} \in (0, r) \tag{16}$$

i.e. too much harvesting pressure is imposed on species 1, then this fixed fraction of exploiters $\overline{r}$ will lead the resource to extinction, whereas an endogenous r could avoid the occurrence of extinction for species 1 (the same reasoning applies, of course, to species 2).

For the sake of comparison, the analytical results on the coexistence of both species can be summarized as follows:

Corollary 1 *If* $k_1, k_2 > 0$ *and* $\frac{2\rho_1\gamma_1}{Na_1} + \frac{2\rho_2\gamma_2}{Na_2} > 1$, *the model with continuous replicator dynamics (10) converges to one of the following fixed points with coexistence of the two species:*

- E_1^2 *if* $\frac{2\gamma_2\rho_2}{a_2N} \in \left(1 + \frac{2a_1^2k_1\gamma_2^2\rho_2}{Na_2^3k_2\gamma_1}, +\infty\right)$;
- E_2^2 *if* $\frac{2\gamma_1\rho_1}{a_1N} \in \left(1 + \frac{2a_2^2k_2\gamma_1^2\rho_1}{Na_1^3k_1\gamma_2}, +\infty\right)$;
- E^* *whenever* $\frac{2\gamma_1\rho_1}{a_1N} < 1 + \frac{2a_2^2k_2\gamma_1^2\rho_1}{Na_1^3k_1\gamma_2}$ *and* $\frac{2\gamma_2\rho_2}{a_2N} < 1 + \frac{2a_1^2k_1\gamma_2^2\rho_2}{Na_2^3k_2\gamma_1}$.

In the model (10) with the last differential equation dropped and the fraction r *exogenously fixed to* $\overline{r}$, *if* $k_1, k_2 > 0$ *and* $\frac{2\rho_1\gamma_1}{Na_1} + \frac{2\rho_2\gamma_2}{Na_2} > 1$, *any* $\overline{r} \in \left(1 - \frac{2\rho_2\gamma_2}{Na_2}, \frac{2\rho_1\gamma_1}{Na_1}\right)$ *ensures the coexistence of both species.*

In short, if there is a coexistence equilibrium for the model (10) then there is at least an $\overline{r}$ such that the model with an exogenously fixed r also converges to a coexistence equilibrium. On the contrary, if there exists an $\overline{r}$ such that the model with exogenous r converges to a coexistence equilibrium, then the model with continuous replicator dynamics also converges to a coexistence equilibrium.

The simpler specification considered above constitutes a useful benchmark for a deeper understanding of the full-fledged hybrid model. In fact, an equilibrium point for the system with continuous replicator (9) is also a fixed point for the hybrid system (1), although the converse is not necessarily true. This follows from the fact that the first and the second dynamic equations in the two specifications are identical, and the replicator dynamics in discrete and in continuous time share the same equilibrium conditions: $r(t) = r(t-s)$ for $r = 0$, $r = 1$ or $\pi_1^*(t) = \pi_2^*(t)$. In fact, if instantaneous profits are identical in equilibrium, then the average profits of the two strategies over the non-switching time interval of length s are also identical. Nonetheless, we can have an equilibrium point such that the average profits of the two strategies over the interval s are equal, even though instantaneous profits are not equal over the interval. In the case of the hybrid model (1), $r(t)$ becomes a piecewise-constant function, like an endogenously driven bang-bang variable, whose discontinuous jumps occur at discrete times and lead to sudden switches among different dynamic scenarios, which is typical behavior of hybrid systems.[2]

[2]A comparison of models with endogenous switching among different regimes and with different time-scales is carried out in Bischi et al. (2014). A related hybrid evolutionary model for the adoption of technologies with different efficiency and environmental impact is considered in Lamantia and Radi (2015).

Numerical simulations performed in Bischi et al. (2013a) allow to investigate the role played by s (the switching time) as well as the effects of non-constant prices, i.e. demand functions (3) with slope $b_i \neq 0$, $i = 1, 2$, on the dynamics of the more realistic hybrid model (1). Such analysis gives evidence of possible advantages of profit-driven self-regulated harvesting strategy choices over other practices, both from the point of view of biomass levels (i.e. biological sustainability) and wealth (economic profitability). Indeed, for proper sets of parameters, numerical simulations show that this kind of myopic and adaptive self-regulation may ensure a virtuous trade-off between profit maximization and resource conservation, driven by cost externalities and market pressure.

Other similar models can be proposed, as useful exercises, to describe situations with interacting species. In this case the two growth functions in (1) assume the more general forms $G_1(X_1, X_2)$ and $G_2(X_1, X_2)$, thus giving a further source of interdependence among the dynamic variables. Even if it is quite difficult to harvest only a single species when two fish populations interact in the same environment, the simulation of such a situation may provide useful theoretical information on the understanding of the trade-off between species interactions and endogenous evolutionary processes based on economic forces. For example, following the idea contained in Okuguchi (1984), in Bischi et al. (2013b) the case of two fish species which interact through a prey-predator relationship is considered. The time evolution of the fish biomasses is modelled by a system of differential equations known as Rosenzweig-MacArthur prey-predator model (see e.g. Rosenzweig and MacArthur 1963; Holling 1965; De Feo and Rinaldi 1997) with extra mortality terms both in prey and predator equations due to harvesting:

$$\begin{cases} \dot{X}_1 = X_1 G_1(X_1, X_2) - Nr(t)h_1^* (X_1, X_2) \\ \quad\; = X_1 \left[\rho \left(1 - \frac{X_1}{K}\right) - \frac{\beta X_2}{\alpha + X_1}\right] - Nr(t)h_1^* (X_1, X_2) \\ \dot{X}_2 = X_2 G_2(X_1, X_2) - N(1 - r(t))h_2^* (X_1, X_2) \\ \quad\; = X_2 \left(\frac{\eta \beta X_1}{\alpha + X_1} - d\right) - N(1 - r(t))h_2^* (X_1, X_2) \end{cases}$$

Also in this case, a simpler benchmark model is considered by assuming constant prices and a given fraction $r \in [0, 1]$ as an exogenous parameter. This leads to the following two-dimensional dynamical system:

$$\begin{cases} \dot{X}_1 = \rho X_1 \left(1 - \frac{X_1}{K}\right) - \frac{\beta X_1 X_2}{\alpha + X_1} - rN \frac{a_1 X_1}{2\gamma_1} \\ \dot{X}_2 = X_2 \left(\frac{\eta \beta X_1}{\alpha + X_1} - d\right) - (1 - r) N \frac{a_2 X_2}{2\gamma_2} \end{cases} \tag{17}$$

The following characterization of equilibrium points holds (see Bischi et al. 2013b):

Proposition 3 *The dynamical system (17) has three non-negative equilibrium points, given by* $S_0 = (0,0)$, $S_1^r = \left(\frac{K(2\rho\gamma_1 - rNa_1)}{2\rho\gamma_1}, 0\right)$ and $S_2^r = (X_1^r, X_2^r)$, with $X_1^r = \frac{\alpha\left(d+(1-r)N\frac{a_2}{2\gamma_2}\right)}{\eta\beta - d-(1-r)N\frac{a_2}{2\gamma_2}}$, $X_2^r = \frac{(\alpha+X_1^r)}{\beta}\left[\rho - \frac{\rho X_1^r}{K} - rN\frac{a_1}{2\gamma_1}\right]$.

- *The Equilibrium* S_1^r *is positive if* $2\rho\gamma_1 > rNa_1$, *and* S_2^r *is positive provided that* $\eta\beta > d + (1-r)N\frac{a_2}{2\gamma_2}$ *and* $X_1^r < \frac{K(2\rho\gamma_1 - rNa_1)}{2\rho\gamma_1}$.
- S_2^r *becomes stable through a transcritical bifurcation at which* S_1^r *and* S_2^r *exchange stability, and it loses stability through a supercritical Hopf bifurcation; the analytical expressions for bifurcations curves are given by*

$$K = K_T^r = \frac{2\rho\gamma_1\alpha\left(d + (1-r)\frac{a_2N}{2\gamma_2}\right)}{(2\rho\gamma_1 - Na_1)\left(\eta\beta - d - (1-r)\frac{a_2N}{2\gamma_2}\right)} \textit{(Transcritical bifurcation curve)} \tag{18}$$

$$K = K_H^r = K_T^r + \frac{2\rho\gamma_1\alpha\eta\beta}{(2\rho\gamma_1 - Na_1)\left(\eta\beta - d - (1-r)\frac{a_2N}{2\gamma_2}\right)} \textit{(Hopf bifurcation curve)} \tag{19}$$

The endogenous evolutionary dynamics of the fraction $r(t)$ can be introduced through two steps: first the case of continuous-time replicator dynamics is considered, modelled by (9); then the more general hybrid model obtained through discrete dynamics of $r(t)$ according to the third equation in (1) is addressed. The analytical and numerical results in Bischi et al. (2013b) show that this type of evolutionary mechanism may lead, also in this case, to a good compromise between profit maximization and resource conservation thanks to an evolutionary self-regulation based on cost and price externalities. In fact, the reduction of biomass of one species leads to increasing landing costs and it consequently favours the endogenous switching to the more abundant species; moreover, severe overfishing of one species causes decreasing prices and consequently decreasing profits.

The numerical results given in Bischi et al. (2013b) show that the endogenous switching mechanism, where fishers decide the variety to catch on the basis of their profits, leads to a high probability of coexistence of the two species provided that the dynamics of unexploited species converge to the stable coexistence equilibrium. In this case, the endogenous switching mechanism significantly reduce the negative effects of exploitation. Another surprising characteristic of this endogenous switching is the reduction of the "oscillatory effect" due to oversupply of food. In fact, it is well known that, in a food-chain population model, the presence of self-sustained oscillations means oversupply of nutrients. In De Feo and Rinaldi (1997) some practical rules are given to reduce oscillations caused by overabundance of food at the bottom of the food chain. So, the exercise carried out in Bischi et al. (2013b) offers glimpse into the interesting properties of myopic and adaptive harvesting mechanisms driven by endogenous evolutionary processes. However this is just a starting point for further and deeper analysis. There are several aspects of the model that deserve to be explored more deeply. For example, the variable r, i.e. the fraction of fishers harvesting a given fish stock, is assumed to range in the interval $[0, 1]$, where 0 and 1 are always equilibria. When r converges to 0 or 1, one of the two species is no longer harvested and consequently it is not available in the market. This could be a reasonable practice only if the two species of fish are perfect substitute in consumers tastes (corresponding to the case $\sigma = 1$ in (3)); otherwise consumers may be heavily penalized in such equilibria. This may be avoided, for example, by introducing constraints on the dynamics of r. We refer to Bischi et al. (2013b) for further comments on this point.

3 Discrete-Time Fishery Models with Marine Protected Areas

Even if dynamic models in ecology have been traditionally formulated in continuous-time, discrete-time population dynamics have received a great amount of attention not only for the complex and intriguing time evolutions that they can produce even in the simplest systems, but also for relevant biological motivations that have been proposed to explain their usefulness in ecologic modeling. In fact, several authors stress that discrete-time population models should be used whenever reproduction happens at given breeding seasons, as several animal species successfully mate only during certain times of the year, thus giving non-overlapping generations (see e.g. Geritz and Kisdi 2004; May 1975). So, more and more discrete-time population models have been proposed in the literature (see e.g. Getz and Haight 1989; Cushing et al. 2004). A comparison between continuous-time and discrete-time dynamics in imperfectly competitive international commercial fisheries is given in Erjaee and Okuguchi (2006). A review of elementary fishery models in discrete time is given in Antonelli et al. (2005), and their applications in the modelling of fishing regulation by the creation of marine protected areas is discussed in Antonelli et al. (2006), Bischi and Lamantia (2007), Bischi et al. (2009), and Sanchirico and Wilen (2001). International agreements for the creation of Marine Protected Areas (MPA) exist, and an increasing interest for this kind of regulation is witnessed by the rich literature, conferences and congresses devoted to it (see e.g. Bohnsack 1993; Clark 1996; Sumaila 1998; Holland 2002; Sumaila and Charles 2002). As explained in Antonelli et al. (2006), the dynamic fishery models proposed in Bischi and Lamantia (2007) and Bischi et al. (2009), which will be described in this section, have been motivated by the project for the creation of a MPA in an international sea zone in the Adriatic Sea, in a region called "Fossa di Pomo". In order to avoid strong opposition of fishers, this MPA is not intended as a reserve area (i.e. a no fishing zone) but it should be a region with carefully regulated fishing activity, for example where a limited constant-effort fishing is allowed. This may give rise to the possible fishers' choice whether to harvest under an imposed limited effort, compulsory inside a MPA with abundance of fish stock, or without imposed limits (e.g. under oligopolistic competition), possible outside the MPA, i.e. in a zone with (likely) depleted fish stock. Of course, this choice is strongly influenced by the level of imposed fishing effort inside the MPA. On order to model this kind of fishery, in Bischi et al. (2009) an aquatic environment is assumed to be divided into two adjacent patches, characterized by different, and in some way complementary, fishing policies: an imposed *constant fishing effort* in region 1 and *oligopolistic competition* in region 2, where fishers decide their harvesting on the basis of a boundedly rational "profit incursion" attitude. Moreover, given the total number of fishers, an adaptive switching mechanism is introduced, based on imitation dynamics (see e.g. Bischi et al. 2003; Vega-Redondo 1996), that endogenously regulates the fraction of fishers that, at each time period, decide to operate in one of the two regions on the basis of observed profits.

Clearly the division in two different patches is a virtual one, in the sense that no physical boundaries exist, and fish can move among regions so that the quantity of biomass in one region depends not only on harvesting and biological growth in that region, but also on the stock and catch in the neighboring regions. The consequence is an interdependence which, at each time period, can increase the population of the zone where the fish density is smaller.

Let $X(t)$ be the total quantity of fish biomass in the water basin at time period t. We suppose that the water basin is virtually divided in two patches, labelled as patch 1 and patch 2. We denote by X_1 and X_2 the quantity of biomass in the different regions, with $X = X_1 + X_2$. We can imagine that a central authority draws a virtual line on the water basin so that it can impose a specific harvesting policy for agents operating in one of the two areas. Assuming that fish can migrate between the different patches according to a linear diffusion mechanism, the biological evolution of the resource in each patch is modelled by the following two-dimensional dynamical system:

$$\begin{cases} X_1(t+1) = X_1(t)\, G(X(t)) - \sigma[X_1(t) - X_2(t)] - H_1(t) \\ X_2(t+1) = X_2(t)\, G(X(t)) + \sigma[X_1(t) - X_2(t)] - H_2(t) \end{cases} \tag{20}$$

where $\sigma > 0$ is a diffusion coefficient, and $H_i(t)$ represents the quantity of fish harvested in time period t in region i.

A subdivision of a fishing region into patches, with these two kinds of fishing policies, is also considered in Bischi and Lamantia (2007). However, the division modelled in that paper is markedly different, because two independent growth functions are considered, in the sense that the fish population of each region is assumed to interact only with individuals of the same region. Instead, in this case the growth function in each patch involves interaction with both subpopulations, i.e. summing up the two Eqs. (20) we have

$$(X_1 + X_2)(t+1) = (X_1(t) + X_2(t))G(X_1(t) + X_2(t)) - H(t)$$

where $H(t) = H_1(t) + H_2(t)$ is the total harvested quantity.

In the following, a logistic growth function is considered, with intrinsic growth rate α and carrying capacity $K = \alpha/\beta$, i.e.

$$X(t+1) = X(t)G(X(t)) = X(t)(1 + \alpha - \beta X(t))$$

In order to derive the two different harvesting functions in patch 1 and patch 2 let us assume that in patch 1 n_1 fishers are allowed to harvest with an imposed constant fishing effort, whereas in patch 2 n_2 fishers are engaged in a Cournot oligopolistic competition. Let $N = n_1 + n_2$ be the total number of fishers. We denote by $r = n_2/N$ the fraction of fishers that decide to harvest outside the MPA, choosing the quantity to harvest according to a profit-increasing Cournot oligopoly strategy, as specified below. Of course $(1 - r)$ represents the fraction of agents fishing in region 1, i.e. inside the MPA.

If E is the individual fishing effort imposed by a regulator in the MPA, denoted as patch 1, then

$$H_1(X_1) = qEN(1 - r)X_1.$$

where q is the "catchability" coefficient, linked to the adopted technology. The idea of a constant fishing effort is very common in the literature on mathematical bioeconomic modelling of fisheries (see e.g. Clark 1990 and references therein) and constitutes the most employed method of control in fisheries.

In patch 2 fishers engage in a Cournot oligopolistic competition, i.e. they decide, at each time period, their harvesting quantity according to profit-increasing behavior taking into account economic externalities. Here we assume the existence of a unique market for the harvested resource determined by the linear inverse demand function

$$p(t) = a - bH(t) \tag{21}$$

where a and b are positive parameters. Notice that we implicitly assumed that the total harvested quantity $H(t)$ is entirely sold in the market.

The profit at time t of a representative fisher operating in zone 1 is given by

$$\begin{aligned} \pi_1(t) &= qEX_1(t)(a - bH(t)) - \delta E \\ &= qEX_1(t)\left\{a - b\left[ENq(1 - r(t))X_1(t) + Nr(t)x(t)\right]\right\} - \delta E \end{aligned} \tag{22}$$

where δ is the unitary cost of effort, $H_1(t) = N[1 - r(t)]\,qEX_1(t)$ and $H_2 = Nr(t)x(t)$ represent the total harvesting in patch 1 and 2 respectively, $r(t)$ is the fraction of agents operating outside the MPA in period t, $x(t)$ is their individual catch, whose determination is given below.

The cost function of a representative fisher that chooses to fish in zone 2, for harvesting a quantity x when the fish stock is X_2, is given by

$$C(x, X_2) = c + \gamma \frac{x^2}{X_2}$$

where γ reflects the inefficiency for harvesting outside the MPA and c is a fixed cost, which includes also the effort for solving the harvesting decision problem. The quantity of resource harvested by a single oligopolist, $x(t)$, is determined in the following way. Assuming that all fishers are homogeneous, we can write the time t profit of a representative fisher working in zone 2 as

$$\pi_2(t) = x(t)\,(a - bH) - \gamma \frac{x(t)^2}{X_2(t)} - c \tag{23}$$

Due to the complexity of the underlying model, it could be difficult for an oligopolist to compute the profit maximizing level of harvesting. For this reason, we assume that oligopolists act in a "boundedly rational" way, in the sense that they update their harvesting plans in the direction of increasing profits, as specified by the so called profit-gradient dynamics (see Flam 1993; Furth 1986; Bischi and Naimzada 2000). Thus, the quantity $x(t+1)$ of a representative oligopolist as a function of $r(t), x(t), X_1(t)$ and $X_2(t)$ can be written as

$$\begin{aligned} x(t+1) &= x(t) + k \cdot \frac{\partial \pi_2(t)}{\partial x(t)} = \\ &= x(t) + k \cdot \left\{ a - bN\left[2r(t)x(t) + Eq(1-r(t))X_1(t)\right] - \frac{2x(t)\gamma}{X_2(t)} \right\} \end{aligned} \tag{24}$$

where $k > 0$ is the speed of adjustment to the direction of the profit gradient.

Agents can decide at any time period the patch to fish in and, consequently, the harvesting strategy, i.e. whether to harvest the resource according to free competition outside the MPA, following a Cournot oligopolistic game, or through (controlled) constant effort landing inside the MPA. Since agents do not have full knowledge of the underlying structure of the game, the switching mechanism is assumed to be governed by an imitative behavior.[3] Again, a replicator dynamics may be used (see e.g. Weibull 1995; Vega-Redondo 1996; Hofbauer and Sigmund 1998) where at each time period any fisher samples an agent that has chosen a different strategy in the past, and switches to the other strategy if the profit of the sampled agent has been greater than his own. However, in Bischi et al. (2009) the probability to change behavior is modelled according to a different kind of imitative behavior, more similar to word of mouth dynamics (see Bischi et al. 2003), based on the following switching probabilities

$$p_{O\to E} = (1-r)\mathbb{P}(\pi_1 > \pi_2)\,;\, p_{O\to O} = 1 - p_{O\to E}$$

$$p_{E\to O} = r\mathbb{P}(\pi_2 > \pi_1)\,;\, p_{E\to E} = 1 - p_{E\to O}$$

[3] A non-evolutionary version of this model has been addressed in Lamantia and Sbragia (2006).

where "$p_{O\to E}$" is the probability to change from strategy "O" ("Oligopoly") to strategy "E" (constant effort). This probability is given by the product between the probability to meet an agent harvesting with constant effort, i.e. $(1-r)$, and the probability that the profit of that agent is greater than the profit for a representative oligopolist, that is $\mathbb{P}(\pi_1 > \pi_2)$; a similar definition holds for the other quantities, with an obvious meaning of the symbols. The cumulative distribution (probability) function $\mathbb{P}(\pi_1 > \pi_2)=\Psi(\pi_1 - \pi_2)$ is monotone increasing with $\Psi(-\infty) = 0$ and $\Psi(+\infty) = 1$. Of course, $\mathbb{P}(\pi_2 \geq \pi_1) = 1 - \Psi(\pi_1 - \pi_2)$.

All in all, the dynamic (deterministic mean field) equation describing how the expected fraction of oligopolists evolves is given by

$$\begin{aligned} r(t+1) &= r(t) + [1-r(t)]\,p_{E\to O} - r(t)p_{O\to E} \\ &= r(t) + [1-r(t)]\,r(t)\,[1-\Psi(\pi_1(t)-\pi_2(t))] - r(t)(1-r(t))\Psi(\pi_1(t)-\pi_2(t)) \\ &= r(t)\,[1-(1-r(t))M(\pi_1(t)-\pi_2(t))] \end{aligned} \tag{25}$$

where $M(.) = 2\Psi(.) - 1$. Equation (25) can be interpreted as a balance equation, saying that the fraction of agents that at time $t+1$ operates outside the MPA, $r(t+1)$, is increased by the expected fraction of agents that decide to exit the MPA, $[1-r(t)]\,p_{E\to O}$, and decreased by agents entering the MPA, i.e. $r(t)p_{O\to E}$.

The main properties that the function M inherits from the commonly used cumulative distributions Ψ are: (1) $M(0) = 0$; (2) $\lim_{z\to-\infty} M(z) = -1$; (3) $\lim_{z\to+\infty} M(z) = 1$; (4) M is an odd function and (5) M is convex in $(-\infty, 0)$ and concave in $(0, +\infty)$.

In the following, we explicitly assume the functional form

$$M(z) = \frac{2}{\pi}\arctan\left(\frac{\lambda\pi}{2}z\right)$$

where the parameter $\lambda = M'(0) = \frac{1}{2}\Psi'(0) > 0$ represents the agents' propensity to strategy switching.

All in all, the dynamical system obtained by considering the natural resources evolution in Eqs. (20) with $H_1(t) = N\,[1-r(t)]\,qEX_1(t)$ and $H_2(t) = Nr(t)x(t)$, individual oligopolistic harvesting in (24), and expected fraction of oligopolistic competitors (25), is described by the following map in $\mathbb{R}^4$ with dynamic variables X_1, X_2, x and r:

$$T: \begin{cases} X_1(t+1) = X_1\,(t)\,G(X_1(t)+X_2(t)) - \sigma[X_1(t)-X_2(t)] - N\,[1-r(t)]\,qEX_1(t) \\ X_2(t+1) = X_2\,(t)\,G(X_1(t)+X_2(t)) + \sigma[X_1(t)-X_2(t)] - Nr(t)x(t) \\ x(t+1) = \max\left[0,\ x(t) + k\cdot\left\{a - bN\,[2r(t)x(t) + qE(1-r(t))X_1(t)] - \frac{2x(t)\gamma}{X_2(t)}\right\}\right] \\ r(t+1) = r(t)\left\{1-(1-r(t))\frac{2}{\pi}\arctan\left[\frac{\lambda\pi}{2}\left(\pi_1(t)-\pi_2(t)\right)\right]\right\} \end{cases} \tag{26}$$

It is important to note that the max operator prevents that oligopolists can choose negative quantity of harvesting. The sets $r = 0$ and $r = 1$ are dynamic invariant, in the sense that if $r(t) = 0$ (1) then also $r(t+1) = 0$ (1). This corresponds to the obvious statement that when a strategy is not performed at a given time it can not be imitated. When $r = 0$, i.e. all agents exert constant effort, the model reduces to a two-dimensional map, as the equations for x and r in (26) are redundant.

A steady state of the dynamical system (26) is a point $S = (X_1^*, X_2^*, x^*, r^*)$ satisfying the following nonlinear system of equations

$$\begin{cases} X_1\left[\alpha-\beta\left(X_1+X_2\right)\right]-\sigma\left(X_1-X_2\right)-N\left(1-r\right)qEX_1=0 & (1)\\ X_2\left[\alpha-\beta\left(X_1+X_2\right)\right]+\sigma[X_1-X_2]-Nrx=0 & (2)\\ aX_2-bNX_2\left[2rx+qE(1-r)X_1\right]-2x\gamma=0 & (3)\\ r(1-r)\left(\pi_1-\pi_2\right)=0 & (4) \end{cases} \tag{27}$$

where π_1 and π_2 are given in (22) and (23) respectively. When no biomass is present in each patch, i.e. $X_1 = X_2 = 0$, from (27, 2) we get $x = 0$ and (27, 4) reduces to $r(1-r)(c-\delta E) = 0$. In this case equilibria are given by $S_1^0 = (0,0,0,0)$ and $S_2^0 = (0,0,0,1)$. When $c = \delta E$, any level $r^* \in [0,1]$ guarantees an equilibrium of the type $S_3^0 = (0,0,0,r^*)$. It is easy to prove (see Bischi et al. 2009) that all these boundary equilibria S_1^0, S_2^0, S_3^0 are unstable for any parameters' values. Indeed, all these cases represent extinction equilibria as they involve no biomass in each patch.

Let us now consider the case $r = 0$, i.e. all agents harvesting inside the MPA. Obviously Eq. (27, 4) is satisfied and from (27, 3) we get that the harvesting equilibrium level of each representative agent is given by $x_0^* = \frac{(a-bNqEX_1)X_2}{2\gamma}$. Note that as all agents harvest in this case inside the MPA, the equilibrium level x_0^* has no practical relevance whatsoever. Hence system (27) reduces now to

$$\begin{cases} X_1\left[\alpha-\beta\left(X_1+X_2\right)\right]-\sigma\left(X_1-X_2\right)-NqEX_1=0\\ X_2\left[\alpha-\beta\left(X_1+X_2\right)\right]+\sigma[X_1-X_2]=0\ . \end{cases}$$

From the second equation, for $X_2 \neq \frac{\sigma}{\beta}$, we obtain $X_1 = g(X_2) = \frac{X_2(\alpha-\sigma-\beta X_2)}{\beta X_2-\sigma} > 0$ that substituted back in the first equation gives the equilibrium level in the variable X_2 as a zero of the single variable function

$$f(X_2) = \frac{X_2\left[(2X_2\beta-\alpha)(\alpha-2\sigma)\sigma+ENq(\sigma-X_2\beta)(\alpha-\sigma-X_2\beta)\right]}{(\sigma-X_2\beta)^2}$$

Clearly, it results that $f(0) = 0$ and $\lim_{X_2\to+\infty} f(X_2) = +\infty$; moreover, f has a vertical asymptote at $X_2 = \frac{\sigma}{\beta}$, as $\lim_{X_2\to\frac{\sigma}{\beta}} f(X_2) = -\infty$. By continuity of f in the interval $(\frac{\sigma}{\beta}, +\infty)$, at least one $X_2^* > \frac{\sigma}{\beta}$ exists such that $f(X_2^*) = 0$, hence defining the equilibrium point $S_0^* = (g(X_2^*), X_2^*, x^*, 0)$. Note that condition $f'(0) > 0$, equivalent to $\frac{\alpha}{\sigma} < 1 + \frac{1}{1+ENq}$, implies the existence of another equilibrium point such that $X_2^* \in \left(0, \frac{\sigma}{\beta}\right)$.

The analysis of the case $r = 1$, i.e. with all agents harvesting outside the MPA, leads to a similar exercise. Again Eq. (27, 4) is satisfied and the harvesting equilibrium level of each representative agent is $x_1^* = \frac{aX_2}{2(bNX)} > 0$. The existence of equilibria can be studied similarly to the previous case. Under the likely assumption that $\alpha > 2\sigma$ it is possible to prove analytically that no equilibrium exists with $X_1^* < \frac{\sigma}{\beta}$ and at least one equilibrium $S_1^* = (X_1^*, g(X_1^*), x_1^*, 1)$ always exists with $X_1^* > \frac{\sigma}{\beta}$. Moreover, it is easy to verify that condition $a < \frac{4b\gamma\sigma(\alpha-2)}{\beta}$ ensures uniqueness of equilibrium S_1^* under the case $r = 1$.

Now we tackle the problem of existence of interior equilibria with $r \in (0, 1)$ and with $X_1, X_2 > 0$. As before, we can find equilibria in terms of a two-dimensional system of equations. In fact, from (27, 1) we can write

$$r(X_1, X_2) = \frac{ENq - \alpha + \beta(X_1 + X_2) - \sigma\,(X_1 - X_2)}{ENq}.$$

By adding together (27, 1) and (27, 2) and substituting back the expressions for r and $1 - r$, an expression for $x(X_1, X_2) = \frac{[(\alpha-\beta(X_1+X_2))X_2+\sigma(X_1-X_2)]EqX_1}{X_1[ENq-\alpha+\beta(X_1+X_2)]-\sigma(X_1-X_2)}$ is obtained. By employing the previous expressions of $r(X_1, X_2)$ and $x(X_1, X_2)$ in (27, 3) and (27, 4), the following system of equations is obtained, in which only the variables X_1 and X_2 appear, which characterize interior equilibria with $r \in (0, 1)$ and $X_1, X_2 > 0$:

$$\begin{cases} a - bNqE(1-r)X_1 - 2\left(bNr + \frac{\gamma}{X_2}\right)x = 0 \\ qEX_1\left[a - b\left(qENX_1\,(1-r)\right)\right] - \delta E + c = x\left[a + bNqEX_1(2r-1) - \left(bNr + \frac{\gamma}{X_2}\right)x\right] \end{cases}$$

where the last equation can be referred to as the locus of points where profits inside and outside the MPA are equal. Note that any solution (X_1^*, X_2^*) only define a feasible equilibrium point when $r(X_1^*, X_2^*) \in [0, 1]$ and $x(X_1^*, X_2^*) > 0$.

This system can be solved numerically. An easy situation is depicted in Fig. 1a, with parameters $\alpha = 2$; $\beta = 0.2$; $\sigma = 0.6$; $E = 1$; $N = 80$; $q = 0.1$; $a = 2$; $b = 0.3$; $\gamma = 0.5$; $\delta = 0.5$; $c = 0.5$, where the unique positive equilibrium with $(X_1^*, X_2^*) \simeq (6.7341, 4.6296)$ is represented as the intersection point of two curves defined by the previous system. It can be easily checked that the corresponding equilibrium level of r and x are also feasible. More involved equilibrium curves are depicted in Fig. 1b, where parameters are set as $\alpha = 3$; $\beta = 0.4$; $\sigma = 0.3$; $E = 2$; $N = 80$; $q = 0.1$; $a = 2$; $b = 0.3$; $\gamma = 0.5$; $\delta = 0.5$; $c = 0.5$. In this second example, four equilibria, A, B, C, D, are obtained (also in these cases the corresponding equilibrium level of r and x are feasible).

As stability analysis for equilibria of the type S_0^* and S_1^*, as well as for interior equilibria with $r \in (0, 1)$, is not an easy task, only some numerical simulations are proposed in Bischi et al. (2009), from which the following economic interpretations are obtained. When the regulator is able to enforce low effort levels in the MPA, then

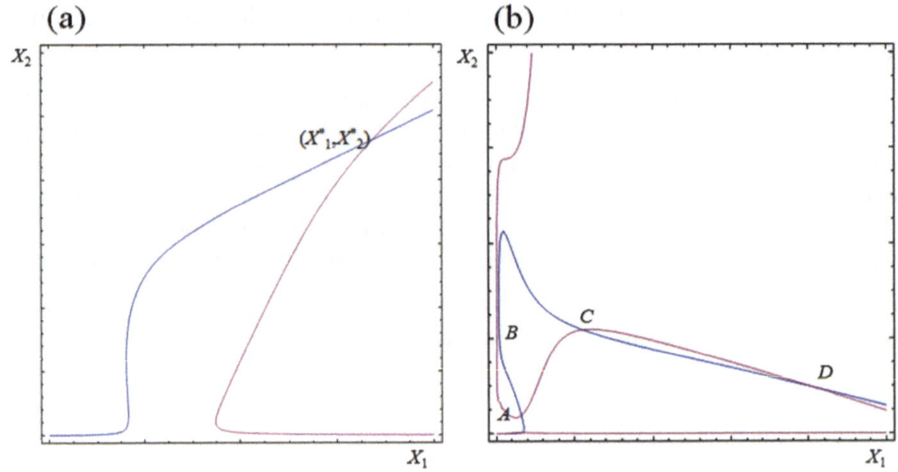

Fig. 1 (**a**) Unique interior equilibrium of system (27) as intersection point of the depicted curves; (**b**) multiple interior equilibria of system (27)

the system is characterized by stability in all relevant quantities: biomass in each patch, harvesting and fraction of oligopolists. This is a consequence of imposing low harvesting levels in the first patch so that many agents are better off by operating in patch 2. Indeed, patch 1 serves as a restocking area, since harvesting levels are kept at a minimum. However, if the regulator let agents increase their catches in patch 1, then a larger fraction of fishers tends to switch to the first patch. As a consequence, also individual harvesting in patch 2 is increased, in order to compensate higher harvesting (and profit) from patch 1. This mechanism becomes more complicated when a Neimark-Sacker bifurcation of the interior equilibrium occurs, after which persistent endogenous oscillations in state variables (biomasses, harvesting, and fraction of oligopolists) are observed. These effects are more evident with higher agents' propensity to switch strategies, with consequent more persistent oscillations in biomass levels and fractions of oligopolists. Similar results can be obtained, for a given level of imposed efforts, by varying the cost of effort δ. In particular, we observe that the higher the cost of effort is, the more stable the system appears. Moreover, the oscillatory behavior encountered with low costs of effort disappears as δ is increased over a given value.

The effect of changing c, the fixed costs of a representative oligopolist, is also considered. This fixed cost can also be interpreted as a fixed tax imposed to each oligopolist at each time period. In this case, as before, there is a range of fixed costs ensuring stability of the steady state but now loss of stability of the steady state can happen through flip (period doubling) bifurcations, and when c is above a given level, exploiting patch 2 gives rise to sure losses so that almost all agents operate in patch 1. Hence patch 2 tends to become a no-take area. In this scenario we have that $r \to 0$ as time passes and the third equation in (26), representing harvesting of a representative oligopolist, does not influence biomass levels in practice.

To summarize, the model with two different kinds of fishing strategies reveals a much broader array of dynamical behavior with respect to those obtained in the system with a homogeneous fishing environment: besides steady state behavior, periodic, quasi-periodic and chaotic dynamics emerge, and qualitative changes (or bifurcations) are detected as some parameters of the model are varied. The main conclusions are related to the sustainability of exploitation in the whole region. In fact, sustainability is favored by reducing the harvesting levels in one of the two patches, which hence serves as a restocking area.

The important role of marine protected areas can be deduced from the results shown in Bischi et al. (2009), where it is shown that low levels of effort tend to stabilize the system around equilibrium levels, whereas large variability of state variables are present when effort is over some threshold values. This effect is more evident as agents' propensity to switch strategy is increased. Stability of the state variable is important as it implies also stability in profits. In any case, as fishing effort is increased more variability in profits for oligopolists than for constant-effort exploiters is observed; hence oligopolistic behavior can be very risky when the MPA is not properly enforced. This point suggests to keep regulated effort at low levels not only to prevent overfishing inside the MPA, but also to reduce the variability of oligopolists' profits. Moreover the regulator can influence the fishing level outside the MPA by properly tuning the parameter c, that is interpretable as a tax on oligopolists.

A peculiar feature that can be deduced from the numerical results shown in Bischi et al. (2009) is the tendency to produce oscillatory time patterns, with periodic, quasi-periodic or chaotic oscillations. In a real system this property may induce some difficulties in the analysis of empirical data when policy makers try to understand the trend of available fish stock after the creation of a MPA. In other words, the benefits of the creation of a MPA can be properly evaluated only if data is observed for a long period after the MPA is established. Of course, the model suggested in Bischi et al. (2009) provides a very simplified and stylized representation of the underlying real system. In fact, the presence of many important factors have been neglected, such as the presence of different species of fish as well as the existence of different age (or size) classes. This point is especially important in the particular case that motivated the paper Bischi et al. (2009), the project of a MPA in the region known as Fossa di Pomo in the central Adriatic Sea. In fact, the importance of this region is related to the presence of young (hence small) fishes (it is denoted as a "nursery"). So, the limitation of fishing activity in that region has the important effect of increasing the number of big size fishes around it. This effect, as well as other boundary effects around a MPA, has not been dealt with in Bischi et al. (2009). Moreover, also the effects of the mobility of fishes different from linear diffusion due to density differences have been neglected. In other words, the model proposed in Bischi et al. (2009) constitutes a first step towards more advanced models, which can be obtained by including more realistic assumptions. Of course, the inclusion of such assumptions has the consequence of yielding a more complicated mathematical structure to the model, and the usual trade-off between mathematical simplicity and a realistic description of real systems always constitutes a difficult and challenging issue.

4 Heterogeneous Harvesting Rates with Punishment

In this section we illustrate a dynamic model proposed in Bischi et al. (2004) to describe an open access fishery exploited by a population of interacting agents that sell the harvested resource in the same market. This population of harvesters is divided into two groups: cooperators, that decide their harvesting policy by maximizing the overall profit of their group, and defectors, that just maximize their own profit. Defectors are punished by sanctions (or taxes or social disapproval) imposed by cooperators. An evolutionary mechanism, based on the replicator dynamics, is introduced to model the time changes in the proportions of defectors and cooperators within the population, and this leads to the qualitative study of a two-dimensional nonlinear dynamical system that describes the time evolution of the resource stock and the population share between cooperators and defectors.

Let us assume that a population of n agents harvests from a common property renewable resource stock X, and sells the harvested resource at a price p determined by the total harvested quantity according to a given linear demand function

$$p = a - bH \tag{28}$$

where a and b are positive constants. H is the total harvesting, which is entirely supplied and sold in the market:

$$H = \sum_{i=1}^{ns} x_c^i + \sum_{i=1}^{n(1-s)} x_d^i$$

where $0 \leq s \leq 1$ is the fraction of "cooperators" that form a coalition (a cooperative venture) and consequently each of them decides to harvest a quantity x_c^i, $i = 1, \ldots, ns$, that maximizes the overall profit of the coalition. Agents in remaining fraction $(1 - s)$ behave as "selfish" profit maximizers and are denoted as "defectors" (with respect to the socially more desirable cooperative behavior). Each of them harvest a quantity x_d^i, $i = 1, \ldots, n(1 - s)$, that maximize his/her own profit.

Player's i cost function for harvesting a quantity x when a fish stock X is present is given by the usual expression

$$C(x, X) = \gamma \frac{x^2}{X}. \tag{29}$$

Following Sethi and Somanathan (1996), an extra-cost is considered, due to the presence of social norms that are intended to punish fishers that behave as defectors, i.e. self-interested profit maximizers without any care of social optimum. Like in Sethi and Somanathan (1996) cooperators are entrusted to punish defectors by applying sanctions. This may be done directly by exerting a direct punishment, such as social disapproval damage or physical damage or destruction of equipment, as observed in less developed societies, or by alerting authorities so that they can impose sanctions according to the laws in force. Such punishment is costly for the defectors, the cost being $ns\xi$, where ξ is the amount of the sanction and ns represents the probability that a defector is notified by a cooperator. However, in general, this kind of punishment is also costly for the cooperators, the cost being proportional to the number of defectors. Let $n(1-s)\psi$ be this extra cost for cooperators (of course $\xi > \psi$, and we shall often consider $\psi = 0$ in the following). All in all, the profit of the i-th cooperator is

$$\pi_c^i = x_c^i (a - bH) - \gamma_c \frac{\left(x_c^i\right)^2}{X} - n(1-s)\psi \tag{30}$$

where γ_c measures the fishing inefficiency of cooperators and $n(1-s)\psi$ represents the cost that cooperators have to face in order to punish defectors. The profit of the i-th defector is

$$\pi_d^i = x_d^i (a - bH) - \gamma_d \frac{\left(x_d^i\right)^2}{X} - ns\xi \tag{31}$$

where γ_d measures the fishing inefficiency of defectors and $ns\xi$ represents the punishment that defectors have to bear for causing the negative externality in the community.

Each cooperator determines x_c^i by solving the optimization problem

$$\max_{x_c^i} \pi^V = \max_{x_c^i} \sum_{i=1}^{ns} \pi_c^i \tag{32}$$

where π^V, which is a concave function in the variables x_c^i, denotes the total profit of the cooperative venture. Assuming interior optimum, the first order conditions give a system of linear equations in the unknowns x_c^i

$$\frac{\partial \pi^V}{\partial x_c^i} = a - 2b \sum_{k=1}^{ns} x_c^k - b \sum_{k=1}^{n(1-s)} x_d^k - \frac{2\gamma_c}{X} x_c^i = 0 \tag{33}$$

Each defector determines x_d^i by solving the optimization problem

$$\max_{x_d^i} \pi_d^i \tag{34}$$

Assuming, again, interior optimum, the first order conditions give a system of linear equations in the unknowns x_d^i

$$\frac{\partial \pi_d^i}{\partial x_d^i} = a - b\sum_{k=1}^{sn} x_c^k - b\sum_{k=1}^{(1-s)n} x_d^k - bx_d^i - \frac{2\gamma_d}{X}x_d^i = 0 \qquad (35)$$

Equations (33) and (35) constitute a linear system of n equations with n unknowns. However, it is straightforward to see that any cooperator faces the same optimization problem, and analogously for defectors. In fact, if we denote by $x_c^{TOT} = \sum_{k=1}^{sn} x_c^k$ the total harvest of cooperators and by $x_d^{TOT} = \sum_{k=1}^{(1-s)n} x_d^k$ the total harvest of defectors, from (33) we get

$$x_c^i = \frac{X}{2\gamma_c}\left(a - 2bx_c^{TOT} - bx_d^{TOT}\right) \quad \forall i = 1, \ldots, sn$$

and from (35) we get

$$x_d^i = \frac{X}{bX + 2\gamma_d}\left(a - b(x_c^{TOT} + x_d^{TOT})\right) \quad \forall i = 1, \ldots, (1-s)n$$

Thus, denoting the optimal harvesting decision of each cooperator by x_c and the optimal harvesting decision of each defector by x_d, these quantities can be obtained by solving the two linear equations

$$\begin{cases} a - 2\left(bsn + \frac{\gamma_c}{X}\right)x_c - b\,(1-s)\,nx_d = 0 \\ a - bsnx_c - \left[b\,(1 + (1-s)n) + \frac{2\gamma_d}{X}\right]x_d = 0 \end{cases}$$

from which the two linear *reaction functions* are obtained

$$\begin{aligned} x_c = h_c\,(x_d) &= \frac{aX}{2\,(bsnX + \gamma_c)} - \frac{b\,(1-s)\,nX}{2\,(bsnX + \gamma_c)}x_d \\ x_d = h_d\,(x_c) &= \frac{aX}{b\,(1 + (1-s)n)\,X + 2\gamma_d} - \frac{bsnX}{b\,(1 + (1-s)n)\,X + 2\gamma_d}x_c \end{aligned} \qquad (36)$$

These reaction functions allow one to compute, respectively, the optimal harvesting decision of a "representative cooperator", given the harvesting decision of a representative defector, and the optimal harvesting decision of a "representative defector", given the harvesting decision of a representative cooperator. These two reaction functions always intersect in a unique point $\left(x_d^*, x_c^*\right)$, whose coordinates are given by

$$x_d^*\,(X, s) = aX\frac{bsnX + 2\gamma_c}{b^2 sn\,(n\,(1-s) + 2)\,X^2 + 2b\,(2ns\gamma_d + \gamma_c + \gamma_c n(1-s))\,X + 4\gamma_c\gamma_d} \qquad (37)$$

and

$$x_c^*(X,s) = \frac{X}{2}\frac{a - b(1-s)\,nx_d^*(X,s)}{bsnX + \gamma_c}. \tag{38}$$

For each $s \in [0,1]$ both x_d^* and x_c^* are increasing functions of X that vanish at $X = 0$, where they have slopes $\frac{\partial x_d^*(X,s)}{\partial X}|_{X=0} = \frac{a}{2\gamma_d}$ and $\frac{\partial x_c^*(X,s)}{\partial X}|_{X=0} = \frac{a}{2\gamma_c}$, respectively. Moreover, both x_d^* and x_c^* tend to saturate as $X \to +\infty$, at the values

$$x_d^*(+\infty,s) = \frac{a}{b\left[(1-s)\,n+2\right]} \quad \text{and} \quad x_c^*(+\infty,s) = \frac{a}{bsn\left[(1-s)\,n+2\right]}$$

respectively. The gap between x_d^* and x_c^*, for large values of the fish stock, increases with increasing prices and with the number ns of cooperators, being

$$x_d^*(+\infty,s) - x_c^*(+\infty,s) = \frac{a}{b\left[(1-s)\,n+2\right]}\left(1 - \frac{1}{sn}\right)$$

The total harvesting at the Nash equilibrium between the two groups, given by

$$H^*(X,s) = n\left[sx_c^*(X,s) + (1-s)x_d^*(X,s)\right] \tag{39}$$

is an increasing and concave function with respect to X, such that $H^*(0,s) = 0$,

$$\frac{\partial H^*(X,s)}{\partial X}|_{X=0} = \frac{na}{2}\left(\frac{s}{\gamma_c} + \frac{1-s}{\gamma_d}\right)$$

and for $X \to \infty$ it saturates at the value

$$H^*(+\infty,s) = \frac{a}{b}\left(1 - \frac{1}{2+(1-s)\,n}\right)$$

The properties of the harvesting function H^* are similar to the ones assumed in Sethi and Somanathan (1996), but in this case the harvesting function is derived from an explicit underlying optimization problem. This implies that the effects of the economic and biologic parameters can be unequivocally studied. First of all we notice that for $sn \geq 2$ (i.e. if at least two cooperators exist) it is $x_c^* < x_d^*$ for large values of fish stock, whereas for small values of X the comparison between x_d^* and x_c^* depends on the respective cost parameters. It is also worth to notice that both x_d^* and x_c^* increase if the selling price increases, i.e. a increases and/or b decreases in the demand function (28). Notice also that $H^* < a/b$ for each $X > 0$, so that prices are always positive. Moreover, if s is decreased, i.e. the number of defectors is increased, then $H^*(+\infty,s)$ increases. In other words, as expected, in the presence of abundant resource the total harvest is greater if the number of defectors increases. In the limiting case $s = 0$ (all defectors) we have $H^*(+\infty,0) = \frac{a}{b}\left(1 - \frac{1}{2+n}\right)$, and in the opposite limiting case $s = 1$ (all cooperators, equivalent to the case of a sole owner) we get $H^*(+\infty, 1) = a/2b$, the monopolist optimal harvesting.

These results allow us to compute, at the Nash equilibrium, the profit of a representative defector and that of a representative cooperator, given by

$$\pi_d^* = x_d^* \left(a - bH^*\right) - \gamma_d \frac{\left(x_d^*\right)^2}{X} - ns\xi, \; \pi_c^* = x_c^* \left(a - bH^*\right) - \gamma_c \frac{\left(x_c^*\right)^2}{X} - n(1-s)\psi \tag{40}$$

respectively. Plugging the expressions of x_c^*, x_d^* and H^* into (40), π_d^* and π_c^* and can be rewritten as:

$$\pi_d^* = \frac{a^2X(bnsX + 2\gamma_c)^2 \, (bX + \gamma_d)}{\left[bX\,(bn\,(-n(1-s)-2))\,sX - 2bX\,(n(1-s)+1)\,\gamma_c - 4\,(bnsX + \gamma_c)\,\gamma_d\right]^2} - ns\xi \tag{41}$$

and

$$\pi_c^* = \frac{a^2X(bnsX + \gamma_c)\,(bX + 2\gamma_d)^2}{\left[bX\,(bn\,(-n(1-s)-2))\,sX - 2bX\,(n(1-s)+1)\,\gamma_c - 4\,(bnsX + \gamma_c)\,\gamma_d\right]^2} - (1-s)n\psi \tag{42}$$

respectively. These expressions show that, at the Nash equilibrium, the profits of defectors (cooperators) are positive provided that the applied sanctions (the costs to apply sanctions) are not too heavy. For example, under the assumption $\psi = 0$ and $\xi > 0$, we have π_c^* always positive and π_d^* positive or negative depending on the sanctions applied and on the number of cooperators.

We now relax the assumptions of a fixed fish biomass X and of a fixed share s between cooperators and defectors. For the fish stock at time period t, denoted by $X(t)$, a logistic growth function in discrete time is assumed:

$$X(t+1) = F(X(t)) = X(t)\,(1 + \alpha - \beta X(t)) - H^*(X(t), s) \tag{43}$$

Moreover, an evolutionary mechanism is introduced to describe how, at each time period, the dynamics of $s(t)$ is updated. For this purpose, the replicator dynamics in discrete time is employed, given by

$$s\,(t+1) = s\,(t)\,\frac{\pi_c^*(t)}{\overline{\pi}} \tag{44}$$

where

$$\overline{\pi}(t) = s\pi_c^*(t) + (1-s)\,\pi_d^*(t) \tag{45}$$

represents the average profit observed at time t (see e.g. Hofbauer and Sigmund 1998, 2003; Weibull 1995).

Of course, (44) states that $s(t+1)$ will be greater than $s(t)$ if $\pi_c^*(t) > \overline{\pi}(t)$ whereas $s(t)$ will decrease if $\pi_c^*(t) < \overline{\pi}(t)$. As $\pi_c^*(t) > \overline{\pi}(t)$ if and only if $\pi_c^*(t) > \pi_d^*(t)$, it follows that the population share related to the better performing strategy at time period t increases in the next period.

Both $\pi_c^*(t)$ and $\overline{\pi}(t)$ depend on $s(t)$, as well as $X(t)$, so the difference Eqs. (43) and (44) define a two-dimensional discrete dynamical system in the dynamic variables $X(t)$ and $s(t)$, obtained by the iteration of a map of the plane $T : (X(t), s(t)) \to (X(t+1), s(t+1))$ given by

$$T : \begin{cases} X(t+1) = X(t)\,(1+\alpha-\beta X(t)) - H^*\,(X(t), s(t)) \\ s\,(t+1) = s\,(t)\,\frac{\pi_c^*(X(t),s(t))}{s\pi_c^*(X(t),s(t))+(1-s)\pi_d^*(X(t),s(t))} \end{cases} \tag{46}$$

where $H^*\,(X, s)$ is given by (39), with x_d^* and x_c^* defined in (37) and (38) respectively, and π_d^*, π_c^*, are given in (41) and (42) respectively.

It is straightforward to see that if $\pi_c^*(t) > 0$ and $\pi_d^*(t) > 0$ then $s(t) \in (0, 1)$ implying $s(t+1) \in (0, 1)$ as well. However, negative profits may arise if the parameters ξ and/or ψ are positive. As the influence of these parameters is always related to the difference $\xi - \psi$, without loss of generality we shall assume $\psi = 0$, and whenever $\pi_d^*(t) < 0$ we shall assume $s(t+1) = 1$ (instead of the meaningless $s(t+1) > 1$, as obtained by simply applying (46)).

It is important to notice that if $s(t) = 0$ then $s(t+1) = 0$ for each $t \geq 0$, and if $s(t) = 1$ then $s(t+1) = 1$ for each $t \geq 0$, i.e. the two boundary lines $s = 0$ and $s = 1$ are trapping lines, on which the dynamics are governed by one-dimensional unimodal maps, given by the restrictions of the two-dimensional map (46) to them. These two cases correspond to particular benchmark cases, where we have all cooperators and all defectors respectively, i.e. the cases considered in Szidarovszky and Okuguchi (1998, 2000) respectively, see also Bischi et al. (2005) on this point. The properties of these one-dimensional dynamical systems can be easily obtained. For example, the dynamics along the invariant edge $s = 0$, where all players are defectors, are governed by the one-dimensional map (43) with

$$H^*(X, 0) = nx_d^*(X, 0) = \frac{naX}{b(n+1)X + 2\gamma_d}$$

So the steady states are $X = 0$ and the solutions (if any) of the equation

$$b\beta\,(n+1)\,X^2 + (2\beta\gamma_d - \alpha b\,(n+1))\,X + na - 2\alpha\gamma_d = 0 \tag{47}$$

On the other invariant edge $s = 1$, where all players are cooperators, the dynamics are governed by (43) with

$$H^*(X, 1) = nx_c^*(X, 1) = \frac{naX}{2(bnX + \gamma_c)}$$

So the fixed points are $X = 0$ and the solutions (if any) of the equation

$$2b\beta nX^2 + 2(\beta\gamma_c - \alpha bn)X + na - 2\alpha\gamma_c = 0 \tag{48}$$

The analysis is the same as in Szidarovszky and Okuguchi (2000) or in Bischi et al. (2005).

As usual, the starting point for the qualitative analysis of a nonlinear dynamical system is the localization of the steady states and the study of their local stability. The steady states of the two-dimensional dynamical system (46) are the fixed points of the map T, solutions of the system $T(X,s) = (X,s)$. It is straightforward to see that two *corner equilibria* always exist, given by $E_0 = (0,0)$ and $E_1 = (0,1)$, characterized by extinction of the resource. Other boundary equilibria may exist along the invariant lines $s = 0$ and $s = 1$, given by the solutions, if any, of Eqs. (47) and (48) respectively. If two equilibria with positive fish stock exist both on the invariant edge $s = 0$ and on invariant edge $s = 1$, say $X_2(0)$, $X_1(0)$ and $X_2(1)$, $X_1(1)$ respectively, then, on the basis on the arguments of Sect. 3, the following relation must hold: $X_2(1) < X_2(0) < X_1(0) < X_1(1)$. A necessary condition for the existence of two positive equilibria along $s = 0$ is that two positive equilibria exist along $s = 1$. However, it may happen that two positive equilibria exist along $s = 1$ and no positive equilibria exist along $s = 0$.

The stability of these equilibria with respect to the one-dimensional dynamics trapped inside the invariant edges can be easily deduced from the discussion on the one-dimensional dynamics. However, the stability with respect to perturbations transverse to the invariant edges, i.e. what happens if a few defectors appear starting from a situation with all cooperators, or what happens if a few cooperators appear starting from a situation with all defectors, is more interesting: are such small mutations eliminated by the evolutionary dynamics, so that the original benchmark case is restored (case of transverse stability) or do these behaviors grow up thus causing an irreversible departure from the original benchmark case?

An answer to these questions requires the study of the local stability of the boundary equilibria, i.e. the localization, in the complex plane, of the eigenvalues of the Jacobian matrix of (46) computed at the boundary steady states. This is not difficult in principle, as eigenvalues are always real because the Jacobian matrix of (46) is a triangular matrix along the invariant edges. This implies that we can only have nodes or saddle equilibria on the boundaries. However, the expressions of the eigenvalues are quite involved, and the obtained stability conditions are not easily interpreted.

Interior equilibria, i.e. steady states characterized by coexistence of cooperators and defectors, are obtained solving the nonlinear system

$$\begin{cases} X(\alpha - \beta X) = H^*(X,s) \\ \pi_c^*(X,s) = \pi_d^*(X,s) \end{cases} \tag{49}$$

with $0 < s < 1$. The set of points of the plane (X, s) that satisfy the first equation represent locus of points that give one-period stationary resource stock, i.e. $X(t+1) = X(t)$. This set of points may be formed by two branches, say $X_1(s)$ and $X_2(s)$, with $X_2(s) < X_1(s)$ for each s. Moreover, $X_2(s)$ is a decreasing function and $X_1(s)$ is increasing, so the branch $X_1(s)$ has positive slope and the branch $X_2(s)$ has negative slope in the plane (X, s). The intersection of the branch $X_1(s)$ $(X_2(s))$ with the invariant edge $s = 1$ gives the X coordinate of the boundary equilibrium $X_1(1)$ $(X_2(1))$, and the same holds for the intersections (if any) with the other invariant edge $s = 0$. However, it may happen that the two branches intersect $s = 1$ but have no intersections with $s = 0$, because they may merge for $s > 0$. The knowledge of these curves give us the following information: starting from a given point (X, s), a one-step iteration of (46) generates a new point $(X', s') = T$, rank-1 image of (X, s) by T, with $X' > X$ if (X, s) is in the strip between the curves $X_2(s)$ and $X_1(s)$ (or between the axis $X = 0$ and the curve $X_1(s)$ if only the branch $X_1(s)$ exists) and with $X' < X$ if (X, s) is on the left of the curves $X_2(s)$ (provided it exists) or on the right of the curve $X_1(s)$.

A similar reasoning can be applied to the set of points that satisfy the second Eq. (49), which represents the locus of points that give one-period stationary population share, i.e. $s(t+1) = s(t)$. A numerical solution of the equation $\pi_c^*(X, s) = \pi_d^*(X, s)$ for different sets of parameters gives rise to decreasing curves in the plane (X, s). The points above the curve $\pi_c^*(X, s) = \pi_d^*(X, s)$ are points where $\pi_c^*(X, s) < \pi_d^*(X, s)$, hence the rank-1 images $(X', s') = T(X, s)$ of points (X, s) above the curve have $s' < s$. Of course, the points below the curve are characterized by $\pi_c^*(X, s) > \pi_d^*(X, s)$, hence $(X', s') = T(X, s)$ are such that $s' > s$. These arguments allow us to obtain a global qualitative picture of the dynamic behavior of the dynamical system (46).

Any intersection between the curve $\pi_c^*(X, s) = \pi_d^*(X, s)$ and one of the curves $X_i(s)$, $i = 1, 2$, represents a steady state of the dynamical system. The steady states along the branch $X_2(s)$ cannot be stable, because all the points of that branch behave as repelling points along the X direction. Instead, steady states located along the branch $X_1(s)$ are candidates to be stable equilibria.

The results given in Bischi et al. (2004) can be summarized as follows. First, the presence of many cooperators always leads to a relatively high level of the resource stock, hence wealth in the long run, whereas many defectors can cause a severe depletion of the resource, incrementing also the basin of attraction of steady states with extinction of the resource.[4] Second, from the point of view of a regulator, not only the cooperative behavior can be supported by increasing the level of sanctions, or lowering the prices, but also it is possible to reach steady states characterized by the presence of only cooperators. These equilibria may be transversely stable, i.e. even if a defector enters the market, this behavior does not spread over the population. Otherwise, low sanctions and/or high prices could lead to steady states

[4]See Bischi and Lamantia (2005) for the topological definition and for an overview on nonlinear dynamical models in discrete time.

with only defectors, and the potential depletion (even extinction) of the resource. In this case, even if the system starts from an initial condition with only cooperators, the non-cooperative behavior will prevail if only one agent decides to defect.

5 Conclusions

In this paper, we have described in details three evolutionary models of commercial harvesting of renewable resources. In particular, we have addressed the presence of heterogeneity in the fish populations to be harvested, in the rules established in different aquatic regions or in the behaviors of exploiters, more or less oriented towards cooperation. In most cases, these assumptions are motivated by real cases or by authorities' proposals for harvesting policies to regulate fisheries in the Italian Adriatic Sea. In any case, the underlying economic structure, the spirit of the analysis and the guiding principles are certainly gained from Prof. Okuguchi's inspiring works on commercial fisheries.

References

Antonelli G, Bischi GI, Viganò E (2005) La sostenibilità nel settore della pesca. Franco Angeli, Milano

Antonelli G, Bischi GI, Lamantia F, Viganò E (2006) Il ruolo delle riserve marine nella gestione sostenibile della pesca. Costi, benefici, modelli. In: Castellini A et al. (eds) L'apporto degli economisti agrari alla ricerca scientifica nel settore ittico. Franco Angeli, Milano

Bischi GI, Lamantia F (2005) Coexisting attractors and complex basins in discrete-time economic models. In: Lines M (ed) Nonlinear dynamical systems in economics. CISM lecture notes, vol 476, Springer, Wien-New York

Bischi GI, Lamantia F (2007) Harvesting dynamics with protected and unprotected areas. J Econ Behav Organ 62:348–370

Bischi GI, Naimzada A (2000) Global analysis of a duopoly game with bounded rationality. Adv Dyn Games Appl 5:361–385

Bischi GI, Dawid H, Kopel M (2003) Spillover effects and the evolution of firm clusters. J Econ Behav Organ 50:47–75

Bischi GI, Lamantia F, Sbragia L (2004) Competition and cooperation in natural resources exploitation: an evolutionary game approach. In: Carraro C, Fragnelli V (eds) Game practice and the environment. Edward Elgar Publishing, Cheltenham Glos (UK)

Bischi GI, Kopel M, Szidarovszky F (2005) Expectation-stock dynamics in multi-agent fisheries. Ann Oper Res 137(1):299–329

Bischi GI, Lamantia F, Sbragia L (2009) Strategic interaction and imitation dynamics in patch differentiated exploitation of fisheries. Ecol Complex 6:353–362

Bischi GI, Lamantia F, Radi D (2013a) Multi-species exploitation with evolutionary switching of harvesting strategies. Nat Resour Model 26(4):546–571

Bischi GI, Lamantia F, Radi D (2013b) A prey-predator model with endogeneous harvesting strategy switching. Appl Math Comput 219:10123–10142

Bischi GI, Lamantia F, Tramontana F (2014) Sliding and oscillations in fisheries with on-off harvesting and different switching times. Commun Nonlinear Sci Numer Simul 19:216–229

Bohnsack JA (1993) Marine reserves: they enhance fisheries, reduce conflicts, and protect resources. Oceanus 36:63–71

Clark CW (1990) Mathematical bioeconomics, the optimal management of renewable resources, 2nd edn. Wiley, New York

Clark CW (1996) Marine reserves and the precautionary management of fisheries. Ecol Appl 6:369–370

Cushing JM, Levarge S, Chitnis N, Henson SM (2004) Some discrete competition models and the competitive exclusion principle. J Differ Equ Appl 10(13–15):1139–1151

De Feo O, Rinaldi S (1997) Yield and dynamics of tritrophic food chains. Am Nat 150:328–345

Erjaee GH, Okuguchi K (2006) Bifurcation and stability in imperfectly competitive international commercial fishing. Keio Econ Stud 41:61–71

Flam SD (1993) Oligopolistic competition: from stability to Chaos. In: Gori F, Geronazzo L, Galeotti M (eds) Nonlinear dynamics in economics and social sciences. Lecture notes in economics and mathematical systems, vol 399. Springer, Wien-New York

Furth D (1986) Stability and instability in oligopoly. J Econ Theory 40:197–228

Geritz SAH, Kisdi E (2004) On the mechanistic underpinning of discrete-time population models with complex dynamics. J Theor Biol 228:261–269

Getz WM, Haight RG (1989) Population harvesting: demographic models of fish, forest, and animal resources. Princeton University Press, Princeton

Hofbauer J, Sigmund K (1998) Evolutionary games and population dynamics. Cambridge University Press, Cambridge

Hofbauer J, Sigmund K (2003) Evolutionary game dynamics. Bull Am Math Soc 40(4):479–519

Holland DS (2002) Integrating Marine protected areas into models for fishery assessment and management. Nat Resour Model 15(3):369–386

Holling CS (1965) The functional response of predators to prey density and its role in mimicry and population regulation. Mem Entomol Soc Can 45:1–60

Lamantia F (2006) Variable effort management of renewable natural resources. Chaos Solitons Fractals 29(3):771–782

Lamantia F, Radi D (2015) Exploitation of renewable resources with differentiated technologies: an evolutionary analysis. Math Comput Simul 108:155–174

Lamantia F, Sbragia L (2006) A dynamic model of patch differentiated exploitation of fisheries. Pure Math Appl 16(4):411–427

May RM (1975) Biological populations obeying difference equations: stable points, stable cycles, and chaos. J Theor Biol 51(2):511–524

Okuguchi K (1984) Commercial fishing with predator-prey interaction. Keio Econ Stud 21:37–44

Okuguchi K (1998) Long-run fish stock and imperfectly competitive commercial fishing. Keio Econ Stud 35:9–17

Rosenzweig ML, MacArthur RH (1963) Graphical representation and stability conditions of predator-prey interactions. Am Nat 97:209–223

Sanchirico JN, Wilen JE (2001) A bioeconomic model of marine reserve creation. J Environ Econ Manage 42:257–276

Sethi R, Somanathan E (1996) The evolution of social norms in common property resource use. Am Econ Rev 86:766–788

Singh N, Vives X (1984) Price and quantity competition in a differentiated duopoly. Rand J Econ 15(4):546–554

Smith VL (1969) On models of commercial fishing. J Pol Econ 77(2):181–198

Sumaila UR (1998) Protected marine reserves as fisheries management tools: a bioeconomic analysis. Fish Resour 37:287–296

Sumaila UR, Charles AT (2002) Economic models of marine protected areas: an introduction. Nat Resour Model 15:261–272

Szidarovszky F, Okuguchi K (1998) An oligopoly model of commercial fishing. Seoul J Econ 11(3):321–330

Szidarovszky F, Okuguchi K (2000) A dynamic model of international fishing. Seoul J Econ 13:471–476

Vega-Redondo F (1996) Evolution, games and economic behaviour. Oxford University Press, Oxford

Weibull JW (1995) Evolutionary game theory. MIT, Cambridge MA

Existence, Uniqueness, and Comparative Statics in Contests

Martin Kaae Jensen

Abstract Many important games are aggregative games which allows for robust comparative statics analysis even when the games do not exhibit strategic complements or substitutes (Acemoglu and Jensen, Games Econ Behav 81:27–49, 2013). This paper establishes such comparative statics results for contests improving upon existing results by (i) allowing payoff functions to be discontinuous at the origin, and (ii) allowing for asymmetric rent-seeking contests and patent races. A leading example where (i) is relevant is the classical Tullock contest (Tullock G (1980) Efficient rent-seeking. In: Buchanan JM, Tollison RD, Tullock G (eds) Toward a theory of the rent-seeking society. Texas A&M Press, College Station, pp 97–112). The paper also studies existence and uniqueness of equilibria extending the results of Szidarovszky and Okuguchi (Games Econ Behav 18:135–140, 1997) and Cornes and Hartley (Econ Theory 26:923–946, 2005) to patent races.

1 Introduction

In contests, agents exert effort to increase their chance of winning a monetary prize, capturing a patent, or defeating their enemies. The study of contests dates back to Loury (1979) and Tullock (1980) and forms a large and active research area. Contests have been used to analyze R&D rivalry (patent races), various types of rent-seeking behavior, financial institutions, market share competition, conflicts, and a number of other strategic interactions. For a long list of applications and further references see Skaperdas (1996), Nti (1997), and Corchón (2007). For recent contributions to the literature see Kolmar and Wagener (2013) and Hirai and Szidarovszky (2013) .

Szidarovszky and Okuguchi (1997) prove the existence of a unique pure strategy Nash equilibrium in the standard case where one of the contenders always wins the prize. Exploiting methods from the literature on aggregative games, Cornes and Hartley (2005) give a very simple proof of the same result (see also von Mouche

M.K. Jensen
Department of Economics, University of Leicester, Leicester LE1 7RH, UK
e-mail: mj182@le.ac.uk

P. von Mouche, F. Quartieri (eds.), *Equilibrium Theory for Cournot Oligopolies and Related Games*, Springer Series in Game Theory,
DOI 10.1007/978-3-319-29254-0_14

and Yamazaki (2015) for a treatment of uniqueness in aggregative games). Nti (1997) derives comparative statics results for symmetric contests, allowing for the possibility that the agents win the prize with probability less than 1. This possibility has merit since a contests where no agent necessarily wins is equivalent to a patent race (Loury 1979).[1] Acemoglu and Jensen (2013) provide comparative statics results that apply to asymmetric contests and asymmetric patent races but their results, at least when taken at face value, require payoff functions to be differentiable which rules out situations where the contest success function is not continuous at the origin. Just like Cornes and Hartley (2005) and Acemoglu and Jensen (2013) explicitly exploit the aggregative games structure of contests and patent races.[2]

The existing literature leaves open two question both of which are addressed in this paper.

- Firstly, does the existence and uniqueness result of Szidarovszky and Okuguchi (1997) and Cornes and Hartley (2005) extend to situations where there is a positive probability that no agent wins the prize. In particular, does this result extend to patent races in the spirit of Loury (1979)?
- Secondly, can comparative statics results similar to those of Acemoglu and Jensen (2013) be established when contest success functions are discontinuous at the origin?

The answer is in the affirmative in both cases: the comparative statics results in Acemoglu and Jensen continue to hold when contest success functions are discontinuous; and the existence and uniqueness of standard contests continues to hold when there is a positive probability that no agent wins the prize (in particular, in patent races).

The structure of the paper is as follows: In Sect. 2 the model and basic assumptions are introduced and discussed. Section 3 derives backward reply functions and proves existence and uniqueness of equilibrium. Finally, Sect. 4 contains the comparative statics results.

2 Models of Contests and Patent Races

Consider a contest with $I \in \mathbb{N}$ agents fighting for a prize. Agent $i \in \mathcal{I}$'s payoff function is,

$$\pi_i(s_i, s_{-i}) = V_i \cdot \frac{h_i(s_i)}{R + \sum_{j=1}^{I} h_j(s_j)} - c_i(s_i), \tag{1}$$

[1]This relationship is further elaborated on in Sect. 2, see also Nti (1997).

[2]For a general discussion of aggregative games see Jensen (2010).

where $s_i \in S_i = [0, \bar{s}_i] \subseteq \mathbb{R}_+$denotes *effort*, $c_i : \mathbb{R}_+ \to \mathbb{R}_+$ is the agent's *cost function*, and $V_i > 0$ his individual *valuation* of the prize.[3] The functions $h_1, \ldots, h_I : \mathbb{R}_+ \to \mathbb{R}_+$ and the constant $R \geq 0$ determine the *contest success functions* $\frac{h_i(s_i)}{R+\sum_{j=1}^{I} h_j(s_j)}$, $i \in \mathscr{I}$, which map a given profile of efforts $s = (s_1, \ldots, s_I)$ into the agents' winning probabilities.

Two classical references are Tullock (1980) and Loury (1979). Loury considers the case of a symmetric patent race with $h_i(s_i) = h(s_i)$, h a concave and strictly increasing function, $c_i(s_i) = Rs_i$ where $R > 0$ is the interest rate which also affects the agents' probability of winning the race (Loury (1979), p. 399; see also the paragraph prior to Theorem 2 in this paper). Patent races may be viewed as contests where a non-strategic agency (*the house* in gambling terminology) has a certain probability of winning the prize. Indeed when $R > 0$, the probability that *no* contender wins the prize is $1 - \frac{\sum_i h_i(s_i)}{R+\sum_i h_i(s_i)} > 0$ and this is then also the probability that the house wins. In contrast, contests a la Tullock have $R = 0$ and $c_i(s_i) = s_i$. Tullock considers symmetric rent-seeking games with $h_i(s_i) = s_i^k$, $k \in (0, 1]$, $i = 1, 2$. The more general specification of contests with risk-neutral agents has again $c_i(s_i) = s_i$, but allows $h_i(s_i)$ to be arbitrary concave and strictly increasing functions. For an axiomatization of various contest success functions including that of Tullock (1980) and the logit function of Hirshleifer (1989) returned to below, see Skaperdas (1996). Here we allow for all of the previous cases and their asymmetric extensions as long as the following standard assumption is satisfied:

Assumption 1 *For all $i \in \mathscr{I}$: $h_i : \mathbb{R}_+ \to \mathbb{R}_+$ and $c_i : \mathbb{R}_+ \to \mathbb{R}_+$ are twice continuously differentiable and strictly increasing. Furthermore, zero effort implies zero likelihood of winning the prize, i.e., $h_i(0) = 0$.*

Note that under Assumption 1, payoffs are *not* well-defined when $R = 0$ and $s = (0, \ldots, 0)$. The contest literature's convention, which we follow here, is to set the probability of winning the price equal to $\frac{1}{I}$ in this situation, and so define $\pi_i(0) = V_i \frac{1}{I} - c_i(0)$ when $R = 0$. If all agents $j \in \mathscr{I} \backslash \{I\}$ choose $s_j = 0$, it is clear from the fact that c_i is continuous that it is not optimal for agent i to also choose $s_i = 0$. Hence $s = (0, \ldots, 0)$ cannot be a Nash equilibrium under Assumption 1 when $R = 0$. In order to also rule out the trivial equilibrium $s = (0, \ldots, 0)$ when $R > 0$, we impose the following boundary condition[4]:

Assumption 2 *For all $i \in \mathscr{I}$, $V_i h_i'(0) > R c_i'(0)$.*

[3]Note that strategy sets are assumed to be compact. This is for technical convenience. If one insists on having $S_i = \mathbb{R}_+$, one can instead assume, for example, that for all $s_{-i} \in \mathbb{R}_+^{I-1}$, $\pi_i(\cdot, s_{-i})$ strictly decreases for $s_i \geq \bar{s}_i$.

[4]Note that Assumption 2 is implied by Assumption 1 when $R = 0$ (specifically, from the fact that h_i is strictly increasing).

To see that $s = (0, \ldots, 0)$ cannot be an equilibrium when $R > 0$ and Assumption 2 holds, consider the marginal payoff of agent i:

$$D_{s_i}\pi_i(s_i, s_{-i}) = V_i h_i'(s_i) \cdot \frac{R + \sum_{j=1}^{I} h_j(s_j) - h_i(s_i)}{\left(R + \sum_{j=1}^{I} h_j(s_j)\right)^2} - c_i'(s_i). \tag{2}$$

Thus $D_{s_i}\pi_i(0, \ldots, 0) = V_i h_i'(0)\frac{1}{R} - c_i'(0)$ which under Assumption 2 is strictly positive. Hence it is *not* optimal for agent i to choose $s_i = 0$ when everyone else also chooses to exert zero effort.

The key assumption for our results imposes the curvature condition of Acemoglu and Jensen (2013).

Assumption 3 *For all* $i \in \mathscr{I}$:

$$\frac{h_i''(s_i)}{h_i'(s_i)} \leq \frac{c_i''(s_i)}{c_i'(s_i)} \text{ for all } s_i \in S_i. \tag{3}$$

Note that if c_i is linear, Assumption 3 says that h_i must be concave. If h_i is linear, Assumption 3 requires c_i to be convex. More generally, (3) is equivalent to assuming that $c_i \circ h_i^{-1}$ is convex.[5] In the setting of Tullock (1980) mentioned above we have,

$$\pi_i(s_i, s_{-i}) = V \cdot \frac{s_i^k}{\sum_{j=1}^{I} s_j^k} - s_i, \tag{4}$$

Hence (3) holds since $k \leq 1$. In logit contests,

$$\pi_i(s_i, s_{-i}) = V \cdot \frac{h(s_i)}{\sum_{j=1}^{I} h(s_j)} - s_i, \tag{5}$$

(3) holds because h is concave. But in general, concavity of h_i and convexity of c_i are not necessary for (3) to be satisfied. Hirshleifer (1989) proposes taking $h_i(s_i) = e^{ks_i}$ $(k > 0)$ in the logit case which defines a *convex* function h_i. In this situation, (3) will be satisfied if, for example, $c_i(s_i) = e^{ls_i}$ where $l \geq k$.

3 Backward Replies, Existence and Uniqueness

The analysis in this paper is based on backward reply functions (Selten 1970; Novshek 1985; Kukushkin 1994). For a strategy profile s, define the *aggregate* as $Q = \sum_i h_i(s_i)$. Given this definition, we may write the marginal payoff of agent i

[5] Use that $\frac{\partial^2 c_i \circ h_i^{-1}(s_i)}{\partial s_i^2} = \frac{c_i''(h_i^{-1}(s_i))(h_i'(h_i^{-1}(s_i)))^{-1} h_i'(h_i^{-1}(s_i)) - h_i''(h_i^{-1}(s_i))(h_i'(h_i^{-1}(s_i)))^{-1} c_i'(h_i^{-1}(s_i))}{(h_i'(h_i^{-1}(s_i)))^2}$.

(see (2)) as a function of Q and the agent's own strategy s_i:

$$\Psi_i(s_i, Q) \equiv V_i h_i'(s_i) \cdot \frac{R + Q - h_i(s_i)}{(R+Q)^2} - c_i'(s_i), \tag{6}$$

When the payoff function is pseudo-concave in s_i, it is clear that s_i is an interior optimizer for agent i if and only if $\Psi_i(s_i, Q) = 0$ and $Q = \sum_j h_j(s_j)$. We may, however, also think of $Q \in \mathbb{R}$ as an independent variable and given Q look for solutions to $\Psi_i(s_i, Q) = 0$. Any such solution is called a *backward reply*.

Consider now the following definition from Acemoglu and Jensen (2013):

Definition 1 (Uniform Local Solvability) Agent $i \in \mathcal{I}$ satisfies the *uniform local solvability condition* if for all $s_i \in S_i$ and $Q > 0$: $\Psi_i(s_i, Q) = 0 \Rightarrow D_{s_i}\Psi_i(s_i, Q) < 0$.

As shown under slightly more general conditions in Acemoglu and Jensen (2013), the payoff functions in (1) satisfy the uniform local solvability condition when Assumption 3 holds. To see why, calculate,

$$D_{s_i}\Psi_i(s_i, Q) = V_i h_i''(s_i) \cdot \frac{R + Q - h_i(s_i)}{(R+Q)^2} - c_i''(s_i) - V_i \frac{(h_i'(s_i))^2}{(R+Q)^2}.$$

When $\Psi_i(s_i, Q) = 0$, this derivative reduces to,

$$D_{s_i}\Psi_i = \frac{h_i''(s_i)}{h_i'(s_i)} c_i'(s_i) - c_i''(s_i) - V_i \cdot \frac{(h_i'(s_i))^2}{(R+Q)^2}.$$

Dividing both sides by $c_i'(s_i) > 0$, we immediately see $D_{s_i}\Psi_i < 0$ when Assumption 3 holds.

Uniform local solvability requires the function $\Psi_i(s_i, Q)$ to be strictly decreasing in s_i at any zero, i.e., at any s_i with $\Psi_i(s_i, Q) = 0$. When Ψ_i is continuous—which it is under Assumption 1—there can then exist at most one backward reply for any given $Q > 0$. We may therefore define the *backward reply function* $b_i : \{Q \in \mathbb{R} : \exists s_i \in S_i \text{ s.t. } \Psi_i(s_i, Q) = 0\} \to S_i$ by means of:

$$s_i = b_i(Q) \Leftrightarrow \Psi_i(s_i, Q) = 0. \tag{7}$$

Given the individual backward reply functions $(b_i)_{i \in \mathcal{I}}$ we can then define the *aggregate backward reply function*:

$$z(Q) = \sum_i h_i(b_i(Q)). \tag{8}$$

In general, z will not be monotonic in a contest, i.e., it will be neither everywhere increasing or everywhere decreasing in Q.[6] It will, however, be a continuous function that "begins above and ends below the 45° line" in the sense of the following lemma:

Lemma 1 *Let Assumptions 1, 2 and 3 be satisfied. Then there exist $Q_H > Q_L > 0$ such that*

$$z(Q) > Q \textit{ for all } Q \in (0, Q_L) \textit{ and } z(Q) < Q \textit{ for all } Q > Q_H.$$

Furthermore, z is well-defined and continuous on the interval $[Q_H, Q_L]$.

Proof First note that by the implicit function theorem, if z is well-defined at a point Q, then z is well-defined and continuous in a neighborhood of Q. Furthermore, if z is well-defined at points Q^a and Q^b with $Q^a < Q^b$, then it will be well-defined on the interval $[Q^a, Q^b]$ (for the proof of the latter statement the reader is referred to the proof of Theorem 6 in Acemoglu and Jensen (2013)). From these two observations, all claims of the Lemma follow if Q_H and Q_L with $z(Q) > Q$ for all $Q \in (0, Q_L)$ and $z(Q) < Q$ for all $Q > Q_H$ can be shown to exist. Q_H is easy since existence of such an element is a direct consequence of the compactness of the strategy sets.

If $R > 0$, $z(0)$ is well-defined. To see this, use that $\Psi_i(0, 0) > 0$ together with the fact that $\Psi_i(\cdot, 0)$ is continuous in s_i and negative for sufficiently high s_i. When $R = 0$, a slight modification of the same argument shows that $z(Q)$ is well-defined for any $Q > 0$ sufficiently close to 0. So in all cases, $z(Q)$ is well-defined for all Q in a neighborhood of 0. It remains only to be shown then, that $z(Q) > Q$ for Q sufficiently close to 0.

For $Q > 0$, let $\tilde{s}_i = \frac{h_i(s_i)}{Q}$. Since h_i is strictly increasing by Assumption 1, we may define $\tilde{b}_i$ by,

$$\tilde{s}_i = \tilde{b}_i(Q) \Leftrightarrow V_i \frac{Q(1 - \tilde{s}_i)}{Q^2} - \frac{c_i'(h_i^{-1}(Q\tilde{s}_i))}{h_i'(h_i^{-1}(Q\tilde{s}_i))} = 0\,. \tag{9}$$

If there exists $Q_L > 0$ such that $\tilde{b}_i(Q) \geq 1$ for all $Q < Q_L$ we are clearly done since then $z(Q) = \sum_i h_i(b_i(Q)) \geq IQ > Q$ whenever $Q > 0$. So consider from now on the case where this is not so. There will then exist a sequence (Q^n) with $Q^n \downarrow 0$ so that for all n, $\tilde{b}_i(Q^n) < 1$. Consider a convergent subsequence $\lim_{m \to \infty} \tilde{b}_i(Q^{n_m}) = \hat{s}_i$. Evidently $\hat{s}_i \leq 1$. Under Assumptions 1, 2 and 3, $\frac{c_i'(h_i^{-1}(z))}{h_i'(h_i^{-1}(z))}$ is

[6]See statement 3 of Theorem 2 where conditions for z to be locally increasing/decreasing play an important role.

increasing in z and therefore $\lim_{z\downarrow 0} \frac{c_i'(h_i^{-1}(Q\tilde{s}_i))}{h_i'(h_i^{-1}(Q\tilde{s}_i))} < +\infty$. Hence we must have $\hat{s}_i = 1$ in order for (9) not to be violated for some m. So in fact $\tilde{b}_i(Q^{n_m}) \uparrow 1$. Since this applies to the limit point of any convergent subsequence, we have $\tilde{b}_i(Q) \uparrow 1$ as $Q \downarrow 0$ (in particular, $\tilde{b}_i(\cdot)$ is continuous from above when we set $\tilde{b}_i(0) = 1$). Hence for any $\epsilon > 0$ we can find a Q_L so that $Q < Q_L \Rightarrow \tilde{b}_i(Q) \geq 1 - \epsilon$. Equivalently $h_i(b_i(Q)) \geq Q - \epsilon Q$ and so $z(Q) = \sum_i h_i(b_i(Q)) \geq IQ - \epsilon IQ = (I - \epsilon)Q > Q$.

Definition 2 An equilibrium aggregate is a fixed point of the aggregate backward reply function, i.e., $Q^* \geq 0$ such that

$$z(Q^*) = Q^*.$$

If Q^* is an equilibrium aggregate, then $(b_i(Q^*))_{i\in\mathcal{I}}$ is a Nash equilibrium for the original game. Conversely, if $(s_i^*)_{i\in\mathcal{I}}$ is a Nash equilibrium, then $\sum_i h_i(s_i^*)$ is an equilibrium aggregate. Thus there is a one-to-one correspondence between equilibrium aggregates and Nash equilibria of the contest. Since as mentioned above, the aggregate backward reply function is generally not monotonic, it is not obvious in light of what has been said so far that the equilibrium aggregate (and therefore the Nash equilibrium) will be unique. This is however the case, as the following result (discussed after the proof) shows.

Theorem 1 *Let Assumptions 1, 2 and 3 be satisfied. Then there exists a unique pure strategy Nash equilibrium* $s^* = (s_1^*, \ldots, s_I^*)$, *and this equilibrium is non-trivial, i.e.,* $s^* \neq (0, \ldots, 0)$.

Proof Existence follows from Lemma 1 and the mean value theorem. That $s^* \neq (0, \ldots, 0)$ whether the equilibrium s^* is unique or not was established in-text in the paragraphs before and after Eq. (2). To prove uniqueness, we exploit the share function of Cornes and Hartley (2005). Let $\tilde{s}_i = \frac{h_i(s_i)}{Q}$, substitute into (6) and set equal to zero:

$$V_i \cdot \frac{R + Q - Q\tilde{s}_i}{(R+Q)^2} - \frac{c_i'\left(h_i^{-1}(Q\tilde{s}_i)\right)}{h_i'(h_i^{-1}(Q\tilde{s}_i))} = 0 \tag{10}$$

Write $\tilde{s}_i = \tilde{b}_i(Q)$ if and only if the previous equation is satisfied. Our objective is to show for all i, $\tilde{b}_i$ is non-increasing in Q in the neighborhood of any equilibrium aggregate $Q = Q^*$, while for at least one i, $\tilde{b}_i$ must in be strictly decreasing in Q at Q^*. Since it then follows that $\sum_j \tilde{b}_j(Q)$ is strictly decreasing at any equilibrium aggregate Q^*, i.e., when $\sum_j \tilde{b}_j(Q^*) = 1$, there can be at most one equilibrium aggregate. Equivalently, there is at most one Q^* such that $z(Q^*) = \sum_j b_i(Q^*) = Q^*$. Since Q^* uniquely determines the equilibrium efforts $s_i^* = b_i(Q^*), i = 1, \ldots, I$, this implies the conclusion of the Theorem.

That $\tilde{b}_i$ is non-increasing implies that if Q and $\tilde{s}_i$ satisfy (10), then we cannot increase Q and weakly increase $\tilde{s}_i$ and still satisfy (10). To see that this is indeed true, note first that $\frac{c_i'(h_i^{-1}(Q\tilde{s}_i))}{h_i'(h_i^{-1}(Q\tilde{s}_i))}$ is non-decreasing in $Q\tilde{s}_i$ under Assumption 3 and therefore weakly increases when Q and/or $\tilde{s}_i$ are increased. Next consider

$$\frac{R+Q-Q\tilde{s}_i}{(R+Q)^2} = \frac{R+Q(1-\tilde{s}_i)}{(R+Q)^2} \tag{11}$$

Clearly, this term is strictly decreasing in $\tilde{s}_i$ (here and below we need only consider $\tilde{s}_i \leq 1$ and $Q > 0$ where the latter is because $Q = 0$ cannot be an equilibrium aggregate). To see that (11) is non-increasing in Q under the condition of the Theorem differentiate w.r.t. Q:

$$\frac{(1-\tilde{s}_i)(R+Q)^2 - 2(R+Q)(R+Q(1-\tilde{s}_i))}{(R+Q)^4}$$

So (11) is non-increasing in Q if and only if:

$$\begin{aligned}(1-\tilde{s}_i)(R+Q)^2 - 2(R+Q)(R+Q(1-\tilde{s}_i)) \leq 0 \Leftrightarrow \\ (1-\tilde{s}_i)(R+Q) - 2(R+Q(1-\tilde{s}_i)) \leq 0 \Leftrightarrow \\ (1-\tilde{s}_i)(R-Q) - 2R \leq 0\end{aligned}$$

This inequality clearly holds if $R = 0$. In fact, since for at least one agent we must have $\tilde{s}_i < 1$ in equilibrium, the inequality must be strict for at least one agent i which implies that $\tilde{b}_i$ is strictly decreasing for that agent. If $R > 0$ there are two cases. Either $Q \geq R$, in which case $(1-\tilde{s}_i)(R-Q) - 2R \leq -2R < 0$. If $Q < R$, $(1-\tilde{s}_i)(R-Q) - 2R < (R-Q) - 2R = -R-Q < 0$. To sum up, in all cases $\tilde{b}_i(Q)$ is non-increasing when Q is an equilibrium aggregate and it will be strictly decreasing for at least one agent i. This finishes the proof.

As discussed in the Introduction, Szidarovszky and Okuguchi (1997) and Cornes and Hartley (2005) establish all of the conclusion of Theorem 1 in the case where $R = 0$. The proof of uniqueness in Theorem 1 follows the basic strategy of Cornes and Hartley (2005) in using *share functions* which are very effective in dealing with uniqueness issues in aggregative games. The basic idea is to consider $\frac{z(Q)}{Q}$ (the share function) and showing that this function must be decreasing which, in turn, implies that there can be at most one equilibrium aggregate and so at most one Nash equilibrium.

4 Comparative Statics

Contests and patent races are neither games of strategic complements or strategic substitutes. They are, however, *aggregative* which provides enough structure to develop robust comparative statics results.[7] The most general results available in the literature for the class of games considered in this paper are due to Nti (1997) and Acemoglu and Jensen (2013). Nti's results apply to symmetric equilibria in symmetric games (in particular, agents must be identical) where $c_i(s_i) = s_i$ for all i. Acemoglu and Jensen (2013) provide results for asymmetric contests and rent seeking games, but their setting requires payoff functions to be everywhere differentiable. As discussed in Sect. 2, this is violated when $R = 0$ at the point $s = (0, \ldots, 0)$. The Theorem that follows next can be viewed as an improved version of Proposition 3 in Acemoglu and Jensen (2013) which allow for the discontinuous payoffs of many contests. The critical "fix" is in this connection Lemma 1.

Definition 3 There is a positive shock to agent $i \in \mathcal{I}$, if the agent's payoff function $\pi_i(s) = V_i \frac{h_i(s_i)}{R+\sum_j h_j(s_j)} - c_i(s_i)$ changes to $\tilde{\pi}_i(s) = \tilde{V}_i \frac{\tilde{h}_i(s_i)}{\tilde{R}+\sum_j \tilde{h}_j(s_j)} - \tilde{c}_i(s_i)$ where $D_{s_i}\tilde{\pi}_i(s) \geq D_{s_i}\pi_i(s)$ for all $s \in S$.

A simple example of a positive shock to agent i is an increase in his valuation of the price from V_i to $\tilde{V}_i$. Indeed, one clearly has

$$\tilde{V}_i \frac{\partial \frac{h_i(s_i)}{R+\sum_j h_j(s_j)}}{\partial s_i} - c_i'(s_i) \geq V_i \frac{\partial \frac{h_i(s_i)}{R+\sum_j h_j(s_j)}}{\partial s_i} - c_i'(s_i)$$

If c_i is replaced with $\tilde{c}_i$ where $\tilde{c}_i'(s_i) \leq c_i'(s_i)$, *i.e.*, if the agent experiences a pointwise decrease in marginal costs, then this is also a positive shock to agent i. A decrease in R will be a positive shock for all agents $i \in \mathcal{I}$ both when the cost function is held fixed and when R enters the costs function as in Loury (1979), $c_i(s_i) = Rs_i$.[8] It is also possible to consider changes in the functions h_i but the issue will not be pursued here.

Theorem 2 *Consider a contest with payoff functions (1) and suppose that Assumptions 1, 2 and 3 are satisfied. Then:*

1. *The equilibrium aggregate is increasing in any positive shock to one or more of the agents (e.g., a decrease in R, an increase in V_i for one or more agents, or a pointwise decrease in marginal costs $c_i'(s_i)$ for one or more agents).*

---

[7]For the state-of-the-art in games that are not aggregative and also do not exhibit strategic complements see Roy and Sabarwal (2010) and Monaco and Sabarwal (2015).

[8]Note that the payoff of an agent in Loury is in fact $\frac{V}{R}\frac{h_i(s_i)}{R+\sum_j h_j(s_j)} - s_i$ (see equation (7) on page 399 of Loury (1979)). Multiplying with R—which does not alter the optimal decision—leads to the functional form considered here.

2. *Entry of an additional agent increases the equilibrium aggregate.*
3. *If at the corresponding equilibrium* s^* *with aggregate* $Q^* = \sum_i h_i(s_i^*)$*, agent i is an absolute favorite to win in the sense that,*

$$\frac{h_i\left(s_i^*\right)}{\left(R + \sum_i h_i(s_i^*)\right)} \geq \frac{1}{2}, \tag{12}$$

then agent i's equilibrium effort increases in any of the situations of 1 and 2. Conversely, if,

$$\frac{h_i\left(s_i^*\right)}{\left(R + \sum_i h_i(s_i^*)\right)} < \frac{1}{2}, \tag{13}$$

then the changes in parts 1 and 2 decrease agent i's effort provided that the shock does not affect this agent directly (e.g., corresponding to a decrease in another agent's marginal costs, or entry of an additional agent).

Proof Taking Lemma 1 into account, both 1 and 2 are direct consequences of the results in Acemoglu and Jensen (2013) because we may restrict Q to an interval $[Q_L - \epsilon, Q_H + \epsilon]$, $\epsilon > 0$ and $Q_L - \epsilon > 0$ where the discontinuity at the origin is irrelevant. The proofs will not be reproduced here. 3. Consider first the case where the shock does not affect agent i. We already know that Q increases in equilibrium. Applying the implicit function theorem to $\Psi_i(s_i, Q) = 0$ at an equilibrium, we get that $\frac{ds_i}{dQ} = -\frac{D_Q\Psi_i(s_i^*,Q^*)}{D_{s_i}\Psi_i(s_i^*,Q^*)}$. Since $D_{s_i}\Psi_i(s_i^*, Q^*) < 0$ by the uniform local solvability condition and Q is increasing in the positive shock, the first claim follows immediately when $D_Q\Psi_i(s_i^*, Q^*) \geq 0$. Since,

$$D_Q\Psi_i(s_i, Q) \equiv V_i h_i'(s_i) \cdot [-(R + Q)^{-2} + 2h_i(s_i)(R + Q)^{-3}] \tag{14}$$

It is seen that $D_Q\Psi_i \geq 0$ if and only if:

$$2h_i(s_i)(R + Q)^{-3} \geq (R + Q)^{-2} \Leftrightarrow h_i(s_i) \geq \frac{1}{2}(R + Q)$$

which in turn is equivalent to the condition of the theorem.

When $D_Q\Psi_i < 0$, the increase in Q^* will lead to a decrease in agent i's strategy by the exact same argument.

If the shock affects agent i, it will in the first of the two cases lead to a further increase in s_i and so the conclusion that s_i increases remains the same. In the second case, this is not so and it is therefore necessary to assume that the shock does not affect agent i directly. □

Note that when $R = 0$, condition (12) simply says that the agent will win the contest with at least 50 % probability. In the two-agent setting, Dixit (1987) refers to such an agent as a *favorite*. When $R > 0$, the interpretation remains the same in two-agent contests with a house that wins with probability $\frac{R}{R+\sum_i h_i(s_i^*)}$. With more than two agents, (12) says that the agent must be the favorite to win when pitched again everyone else including the house. This is what is meant by the term *absolute favorite*. So what 3 of Theorem 2 tells us is that an absolute favorite will actively defend her position by increasing her effort given any change in the game that leads to an overall increase in effort (and hence would lead to a decrease in her probability of winning if her effort remained the same). In particular, she increases her effort if another agent enters the game or another agent experiences a decrease in marginal costs, or another agent's valuation of the prize increases. Conversely, an agent who is *not* more likely to win than everybody else combined is "cowed" and lowers her effort and probability of winning the prize when, for example, another agent's marginal costs decrease or another agent's valuation of the prize increases.

5 Conclusion

This paper contributes to the literature on contests by establishing existence, uniqueness, and comparative statics results for asymmetric contests and patent races. The results exploit the fact that contests are aggregative games, and mathematically the proofs follow closely Cornes and Hartley (2005) and Acemoglu and Jensen (2013).

Acemoglu and Jensen (2013) establish comparative statics results for aggregative games that either exhibit strategic substitutes or satisfy a local solvability condition (see Definitions 7–8 in Acemoglu and Jensen (2013)). Contests are not games of strategic substitutes but they do satisfy the uniform local solvability condition (Definition 1 in the current paper or Definition 8 in Acemoglu and Jensen (2013)). While Acemoglu and Jensen assume that payoff functions are differentiable, the current results allow contest success functions to be discontinuous at the origin. As is clear from this paper's proofs, however, the extension is straight-forward. There is no doubt that one can similarly extend Acemoglu and Jensen (2013) to other situations where payoff functions are not everywhere differentiable such as Cournot oligopolies with discontinuities at the origin.[9] Another extension of Acemoglu and Jensen (2013) more closely related to the topic of the current paper would establish comparative statics results for contents in which the value of the prize is a function of aggregate effort (Hirai and Szidarovszky 2013). One could also use the weaker version of the local solvability condition in Section 4.1.1 of Acemoglu and Jensen (2013) to derive comparative statics results for contests under weaker conditions

[9] Some Cournot models of course fit directly into the set-up of the current paper. Thus with inverse demand $P(Q) = \frac{1}{R+Q}$ and cost c_i, firm i's profit function is of the form (1) with $h_i(s_i) = s_i$ and $V_i = 1$. Note, however, that this Cournot model is not a game of strategic substitutes (as one may show, firm i's optimal output is decreasing in the opponents' aggregate output if and only if $s_i \leq R + \sum_{j \neq i} s_j$).

than the curvature condition imposed here (Assumption 3). In particular, one can dispense with any differentiability conditions.

As for the existence and uniqueness results, the current paper's extension of Cornes and Hartley (2005) is again fairly straight-forward. Further extensions of the uniqueness part are likely to meet with limited success however. In fact, it is doubtful whether one can even establish uniqueness under reasonable conditions when contests are defined with general aggregators as in Acemoglu and Jensen (2013), Section 5.2.

Acknowledgements I would like to thank Pierre von Mouche, Federico Quartieri, and an anonymous referee for helpful comments. All remaining errors are my responsibility.

References

Acemoglu D, Jensen MK (2013) Aggregate comparative statics. Games Econ Behav 81:27–49

Corchón L (2007) The theory of contests: a survey. Rev Econ Des 11:69–100

Cornes R, Hartley R (2005) Asymmetric contests with general technologies. Econ Theory 26:923–946

Dixit A (1987) Strategic behavior in contests. Am Econ Rev 77:891–898

Hirai S, Szidarovszky F (2013) Existence and uniqueness of equilibrium in asymmetric contests with endogenous prizes. Int Game Theory Rev 15:1–9

Hirshleifer J (1989) Conflict and rent-seeking success functions: ratio vs. difference models of relative success. Public Choice 63:101–112

Jensen MK (2010) Aggregative games and best-reply potentials. Econ Theory 43:45–66

Kolmar M, Wagener A (2013) Inefficiency as a strategic device in group contests against dominant opponents. Econ Inq 51:2083–2095

Kukushkin NS (1994) A fixed-point theorem for decreasing mappings. Econ Lett 46:23–26

Loury GC (1979) Market structure and innovation. Q J Econ 93:395–410

Monaco AJ, Sabarwal T (2015) Games with strategic complements and substitutes. Econ Theory (forthcoming Economic Theory)

Novshek W (1985) On the existence of Cournot equilibrium. Rev Econ Stud 52:85–98

Nti KO (1997) Comparative statics of contests and rent-seeking games. Int Econ Rev 38:43–59

Roy S, Sabarwal T (2010) Monotone comparative statics for games with strategic substitutes. J Math Econ 46:793–806

Selten R (1970) Preispolitik der Mehrproduktenunternehmung in der Statischen Theorie. Springer, Berlin

Skaperdas S (1996) Contest success functions. Econ Theory 7:283–290

Szidarovszky F, Okuguchi K (1997) On the existence and uniqueness of pure Nash equilibrium in rent-seeking games. Games Econ Behav 18:135–140

Tullock G (1980) Efficient rent-seeking. In: Buchanan JM, Tollison RD, Tullock G (eds) Toward a theory of the rent-seeking society. Texas A&M Press, College Station, pp 97–112

von Mouche PHM, Yamazaki T (2015) Sufficient and necessary conditions for equilibrium uniqueness in aggregative games. J Nonlinear Convex Anal 16:353–364

Two-Group Contests with Communication Within and Between Groups

Federico Quartieri and Ryusuke Shinohara

Abstract We consider a two-group contest game with weakest-link social composition functions and convex cost functions and prove the existence of a unique group-proof Nash equilibrium. Such a refinement of the Nash equilibrium prescribes the same communication possibilities as those required by a coalition-proof Nash equilibrium—in the precise sense of Bernheim et al. (J Econ Theory 42:1–12, 1987)—only among the contenders of the same group and between the two groups. We show how a fictitious game with "most inefficient fictitious contenders" can be constructed to prove the existence of a unique group-proof Nash equilibrium of the original two-group contest game. An example evidences that cautious arguments on the (twice) differentiability of cost functions must be used in such a construction.

1 Introduction

One of first general Nash equilibrium uniqueness results for contests was postulated by Szidarovszky and Okuguchi (1997).[1] That result concerns contests among single contenders which are not ex ante grouped in any way. The assumption that contenders are not ex ante partitioned into groups is of some importance for Nash equilibrium uniqueness, in that contests among groups of contenders can well possess a multiplicity of Nash equilibria under standard conditions on "group contest success functions" and on contenders' cost functions (notwithstanding that,

[1]Subsequently various extensions have obtained: for contests among single contenders see, e.g., Cornes and Hartley (2005) and Yamazaki (2008).

F. Quartieri (✉)
Dipartimento di scienze economiche e statistiche, Università degli studi di Napoli Federico II, Naples, Italy
e-mail: federico.quartieri@unina.it

P. von Mouche, F. Quartieri (eds.), *Equilibrium Theory for Cournot Oligopolies and Related Games*, Springer Series in Game Theory,
DOI 10.1007/978-3-319-29254-0_15

some Nash equilibrium uniqueness results can be proved for some classes of contests among groups).[2]

The above mentioned multiplicity of Nash equilibria is evident from the following example. Consider a two-group contest described by the game

$$\Gamma = \left(N, (S_i)_{i\in N}, (u_i)_{i\in N}\right)$$

where the set

$$N = \{1, 2, 3, 4\}$$

of players is partitioned into two groups, say

$$G_1 = \{1, 2\} \text{ and } G_2 = \{3, 4\},$$

and where each player $i \in N$ can choose a level of effort out of the strategy set

$$S_i = \mathbb{R}_+$$

obtaining a payoff defined by the function

$$u_i(s) = \begin{cases} \dfrac{s_1 + s_2}{s_1 + s_2 + s_3 + s_4} v_i - s_i & \text{if } s_1 + s_2 + s_3 + s_4 \neq 0 \\ w & \text{otherwise} \end{cases}$$

when $i \in G_1$ and by the function

$$u_i(s) = \begin{cases} \dfrac{s_3 + s_4}{s_1 + s_2 + s_3 + s_4} v_i - s_i & \text{if } s_1 + s_2 + s_3 + s_4 \neq 0 \\ w & \text{otherwise} \end{cases}$$

when $i \in G_2$, with

$$v_i = 1 \text{ for all } i \in N \text{ and } w \in \mathbb{R}$$

(typically w is assumed to be either $1/2$ or 0, but this fact is not important in the present example). It is easily verified that, in this illustrative basic example of a two-group contest,

$$\{s \in S_1 \times S_2 \times S_3 \times S_4 : s_1 + s_2 = 1/4 = s_3 + s_4\}$$

is an infinite set of Nash equilibria for Γ.

[2]The literature on contests among groups probably originated from Katz et al. (1990) and Baik (1993) (see also Corchón (2007) for an introduction to this type of contests).

In the previous two-group contest the probability with which the "public-good prize" v_i is won by i's group, call such group $G_{\bar{i}}$, is determined by means of a perfect substitutes "group impact function"—i.e., $\sum_{l \in G_{\bar{i}}} s_l$—that coincides with a "social composition function" which assigns to each group the sum of the efforts of the contenders of that group.[3] As pointed out in Hirshleifer (1983)—but see also Cornes (1993)—such a social composition function is only one of the possible many: other social composition functions can be considered legitimate in various economic situations (see Hirshleifer (1983) and Cornes (1993) for some examples). Building on this point raised by Hirshleifer, a recent strand of the literature on contests among groups has considered group impact functions that employ social composition functions other from that consisting of the summation: for instance, the weakest-link—i.e., $\min \{s_l : l \in G_{\bar{i}}\}$—and the best-shot social composition function—i.e., $\max \{s_l : l \in G_{\bar{i}}\}$.[4]

In a recent article, Lee (2012) considered a specific contest among groups where, within each group, contenders can be ordered in terms of "efficiency" and where the group impact function is a weakest-link social composition function. For that situation Lee (2012) singled out a certain Nash equilibrium and called it a coalition-proof Nash equilibrium. The Nash equilibrium selected in Lee (2012) is in a one-to-one correspondence with the unique Nash equilibrium of a contest among single contenders (i.e., of a contest without groups) whose players are the "most inefficient" contenders of each group of the original contest among groups. Thus Lee (2012) created a correspondence between the unique Nash equilibrium for a certain class of contests among single contenders and an equilibrium selection from the set of Nash equilibria for a certain class of contests among groups of contenders.

Despite Lee (2012)'s terminology, the notion of coalition-proofness employed in that work—which is not defined therein—does not seem anything similar to that elaborated—and theoretically motivated—in Bernheim et al. (1987)[5]: see the proof of part (c) of Proposition 1 in Lee (2012) and the definition of a coalition-proof Nash equilibrium in Bernheim et al. (1987). In the present work we want to reconsider the solution proposed by Lee (2012) and provide a well-defined argument in terms of an internally consistent coalitional logic. To do this we shall introduce a variant of the original definition of Bernheim et al. (1987), which will be called a "group-proof Nash equilibrium", that allows for communication possibilities—i.e., deviation possibilities—equal to those required for a coalition-proof Nash equilibrium only within a group and among the entire groups, excluding for the

[3]In the literature, the probability of winning the prize for a group is the value attained by the so-called group contest success function. In the example, when all group impact levels are positive, Group 1 (resp. 2) contest success function is $(x, y) \mapsto \frac{x}{x+y}$ (resp. $(x, y) \mapsto \frac{y}{x+y}$) where x is Group 1 impact level $\sum_{l \in G_1} s_l$ and y is Group 2 impact level $\sum_{l \in G_2} s_l$.

[4]On this literature, see also Münster (2007), Baik (2008), Lee (2012), Chowdhury et al. (2013), Kolmar and Rommeswinkel (2013), Choi et al. (2015) and Chowdhury and Topolyan (2015).

[5]Also Chowdhury and Topolyan (2015), like Lee (2012), consider—but define—a notion of "coalition-proof" equilibrium which is considerably different from that of Bernheim et al. (1987).

possibility of communication among (and hence of deviation of) coalitions that are the union of proper subsets of different groups.[6] This certainly simplifies our—not so short—analysis, but the idea that in contests among groups there can exist very limited communication possibilities among contenders of different groups can be quite natural and reasonable,[7] and should not be viewed as a way-out from the analytical complications inhering in Bernheim et al. (1987)'s concept of a coalition-proof Nash equilibrium.

The purpose of this work is to show the uniqueness, in terms of group-proofness, of the solution proposed by Lee (2012). Though we shall restrict our analysis to contests with only two groups and to logit group impact functions, we shall allow for groups where most inefficient contenders need not exist. Therefore, in our case, one cannot generally establish any one-to-one correspondence between a two-group contest game and a two-person contest game whose players are the two "most inefficient contenders" of the two groups of the original two-group contest game. However, we shall show how "most inefficient fictitious contenders" can be constructed such that the group-proof Nash equilibrium of the original two-group contest game can be put in a one-to-one correspondence with the unique Nash equilibrium of a two-person contest game (between single contenders) whose players are these fictitious agents.

As it will be clear from a simple example, even under smooth[8] assumptions on the cost functions of the contenders of the original two-group contest games we consider, the most inefficient fictitious contenders need not possess payoff functions that are twice differentiable on the interior of their domains. This fact might cause some problem because an almost ubiquitous assumption in the Nash equilibrium uniqueness results of the literature on contests among single contenders is that payoff functions are (at least) twice differentiable on the interior of their domain. One of the few exceptions is an article by Yamazaki (2008)—a former student of Prof. Okuguchi—where the author provides a Nash equilibrium uniqueness theorem that allows for non-differentiable payoff functions.[9] The present work, that will make use of that theorem, bears witness of the benefit of such a line of research.

[6]Also Chowdhury and Topolyan (2015)—like us with group-proof Nash equilibria—do not consider coalitional deviations of unions of proper subsets of contenders of different groups; however the equilibrium refinement they consider is essentially different from a group-proof Nash equilibrium because of a lack of conceptual consistency in the credibility requirement for (sub)coalitional deviations of their "coalition-proof" equilibrium.

[7]The idea of limited communication possibilities among certain coalitions of players is not new and is not ours: at least in the case of coalition-proof equilibria, this idea should be attributed to Milgrom and Roberts (1996).

[8]I.e., infinitely many times differentiable.

[9]It must be remarked that the novelties in Yamazaki (2008) do not consist solely on this lack of differentiability assumptions.

2 Preliminary Definitions

2.1 Games

A **game** Γ is a triple $(N, (S_i)_{i\in N}, (u_i)_{i\in N})$ where N is a nonempty set of players and, for all $i \in N$, S_i is a nonempty set of strategies and $u_i : \prod_{i\in N} S_i \to \mathbb{R}$ is player i's payoff function. Henceforth N is assumed to be finite and $\prod_{i\in N} S_i$ is denoted by S_N. Let Γ be a game: a joint strategy $s \in S_N$ for Γ **strongly** Γ**-Pareto dominates** a joint strategy $z \in S_N$ for Γ iff $u_i(z) < u_i(s)$ for all $i \in N$. Let $D \in 2^N \setminus \{\emptyset, N\}$ be a coalition and $s \in S_N$ a joint strategy: the **game induced by** D **at** s **from** Γ is the game

$$\Gamma|_{s,D} = (D, (S_i)_{i\in D}, (\overline{u}_i)_{i\in D}),$$

$$\text{with } \overline{u}_i : \prod_{i\in D} S_i \to \mathbb{R}, \overline{u}_i : z \mapsto u_i(z, (s_l)_{l\in N\setminus D}),$$

where $(z, (s_l)_{l\in N\setminus D})$ denotes the tuple $s^* \in S_N$ such that $z = \left(s_l^*\right)_{l\in D}$ and $(s_l)_{l\in N\setminus D} = \left(s_l^*\right)_{l\in N\setminus D}$. A game Γ is said to be **socially bipartite** iff $|N| \geq 2$ and N is associated with a (fixed) partition $\{G_1, G_2\}$ of itself (thus, by the definition of a partition, $G_1 \neq \emptyset \neq G_2 = N\setminus G_1$).

2.2 Equilibrium Notions

The following two definitions are classical equilibrium notions in game theory. In particular, the second is the recursive definition of a coalition-proof Nash equilibrium provided in Bernheim et al. (1987): the reader is referred to the original article for its theoretical motivations.[10]

Definition (Nash equilibrium) Let Γ be a game. A joint strategy $s^* \in S_N$ is a **Nash equilibrium** for Γ iff $u_i(s^*) \geq u_i\left(s_i, s_{-i}^*\right)$ for all $s_i \in S_i$, for all $i \in N$.

Definition (CP-Nash equilibrium) Let Γ be a game. When $|N| = 1$, a joint strategy $s \in S_N$ is a **coalition-proof Nash** (CP-Nash) **equilibrium** for Γ iff there does not exist another joint strategy for Γ that strongly Γ-Pareto dominates it.[11] Let

[10]For sake of simplicity the variant of the original definition of a coalition-proof Nash equilibrium that involves the weak Pareto dominance will not be considered here. For more on the difference between the two concepts see, for instance, Konishi et al. (1999), Shinohara (2005), Quartieri (2013), and Quartieri and Shinohara (2015).

[11]Equivalently, putting $N = \{i\}$ in this single player case, iff $u_i(s) \geq u_i(z)$ for all $z \in S_N$.

$|N| \geq 2$ and assume that a coalition-proof Nash equilibrium has been defined for all games with fewer than $|N|$ players. Then:

- A joint strategy $s \in S_N$ is a **self-enforcing strategy** for Γ iff it is a coalition-proof Nash equilibrium for $\Gamma|_{s,D}$, for all $D \in 2^N \setminus \{\emptyset, N\}$.
- A joint strategy $s \in S_N$ is a coalition-proof Nash equilibrium for Γ iff it is a self-enforcing strategy for Γ and there does not exist another self-enforcing strategy for Γ that strongly Γ-Pareto dominates s.

Milgrom and Roberts (1996) explain why it can be of importance, in some game-theoretic contexts, to allow for non-symmetric coalitional deviations. Their basic point is that not all possible coalitions can actually have the same possibility to communicate and coordinate deviations. (The reader is referred to their Introduction for a longer discussion.) However, Milgrom and Roberts (1996) do not provide specific economic examples where this asymmetry in the possibility to coordinate deviations is natural. We maintain that contests among groups—that will be defined later—can be the case in a natural way. Indeed, it is not difficult to image that in many situations of conflict the communication possibilities among contenders of a same group (e.g., the players of a sport team or the countries of some military or economic alliance) are much greater than among contenders of different groups.

Building on the previous motivations, we provide the following variant of a coalition-proof Nash equilibrium which we call a "group-proof Nash equilibrium". Since our results concern only two-group contests, the definition below is specialized for such a particular two-group case: the reader might easily extend such a definition to games (and contests) with $g \geq 2$ groups.

Definition (GP-Nash equilibrium) Let Γ be a socially bipartite game.

- A joint strategy $s \in S_N$ is a **group-self-enforcing strategy** for Γ iff $(s_i)_{i \in G_1}$ is a CP-Nash equilibrium for $\Gamma|_{s,G_1}$ and $(s_i)_{i \in G_2}$ is a CP-Nash equilibrium for $\Gamma|_{s,G_2}$.[12]
- A joint strategy $s \in S_N$ is a **group-proof Nash** (GP-Nash) **equilibrium** for Γ iff s is a group-self-enforcing strategy for Γ and there does not exist another group-self-enforcing strategy for Γ that strongly Γ-Pareto dominates s.

Remark 1 Let Γ be a socially bipartite game. A CP-Nash equilibrium for Γ is a Nash equilibrium for Γ (indeed, this fact is true even when Γ is not socially bipartite). A GP-Nash equilibrium for Γ is a Nash equilibrium for Γ. Also, it is easily seen that if a joint strategy is self-enforcing for Γ then it is also group-self-enforcing for Γ. Thus the set of self-enforcing strategies for Γ, say SE_Γ, is contained in the set of group-self-enforcing strategies for Γ, say GSE_Γ; hence

$$SE_\Gamma \subseteq GSE_\Gamma.$$

[12]Clearly, $\Gamma|_{s,G_1}$ is the game induced by the deviating coalition G_1 at s and $\Gamma|_{s,G_2}$ is the game induced by the deviating coalition G_2 at s.

However, one cannot generally establish any inclusion relation between the set of CP-Nash equilibria for Γ and that of GP-Nash equilibria for Γ: the set of Γ-Pareto maximal elements of GSE_{Γ} need not include (or be included by) the set of Γ-Pareto maximal elements of SE_{Γ}.

3 The Contest Game and the Associated Fictitious Game

Here below we shall introduce the class of "contest games" that will be employed in the rest of this article.[13] One of the main features of the games in such a class is that they involve several players partitioned into two sets, called groups. The assumption that only two groups are involved is of course restrictive, even though many "real-world" situations of conflict among groups in fact involve only two groups (but, of course, many can be thought of where more than two groups are involved). Another main feature is the fact that the probability of winning the prize is related only to (a function of) the "group effort" and not to individual efforts; the valuation of the prize—which is a public good prize for the contenders of the group to which the prize is assigned—is private and may differ across contenders of a same group.

3.1 *The Contest Game*

We say that a **contest game** $\mathscr{C}$ is a game

$$(N, (S_i)_{i \in N}, (u_i)_{i \in N})$$

where N is a finite set $\{1, \ldots, n\}$ of $n > 1$ players partitioned into two sets

$$G_1 \text{ and } G_2,$$

so that $G_1 \neq \emptyset \neq G_2 = N \backslash G_1$. The sets G_1 and G_2 are also called **groups** (and hence a contest game is a socially bipartite game by definition). Each strategy set is assumed to be equal to the nonnegative real line, and hence

$$S_i = \mathbb{R}_+$$

for each $i \in N$; each player i's payoff function is assumed to be defined by

$$u_i : s \mapsto \begin{cases} \frac{\tilde{\mu}_i(s)}{\mu_1(s)+\mu_2(s)} v_i - c_i(s_i) & \text{if } \mu_1(s) + \mu_2(s) > 0 \\ w_i - c_i(0) & \text{otherwise} \end{cases}$$

[13]Of course, under the same terminology many other classes of games could be called "contest games" and our terminological choice is due only to expositional convenience.

where

$$\mu_1(s) = \min\{s_l : l \in G_1\}, \; \mu_2(s) = \min\{s_l : l \in G_2\},$$
$$\tilde{\mu}_i(s) = \mu_1(s) \text{ if } i \in G_1, \quad \tilde{\mu}_i(s) = \mu_2(s) \text{ if } i \in G_2,$$

and[14]

$$w_i = 0 < v_i,$$

and where

$$c_i : \mathbb{R}_+ \to \mathbb{R}_+$$

is convex, strictly increasing, differentiable and vanishing at 0 (so $w_i - c_i(0) = 0$). Probably, a more appropriate term for $\mathscr{C}$ would be "two-group contest game"; however in the sequel we shall refer to contests only when considering this type of games and hence we prefer to avoid the "two-group" qualification for terminological simplicity. No confusion should arise.

Remark 2 When we say that a convex and increasing function $c_i : \mathbb{R}_+ \to \mathbb{R}_+$ is differentiable we mean to say that c_i has a differentiable extension[15] to some open real interval that includes $\mathbb{R}_+$.

Remark 3 When $\mu_1(s) + \mu_2(s) > 0$, the value

$$\frac{\tilde{\mu}_i(s)}{\mu_1(s) + \mu_2(s)}$$

(i.e., the value attained by the contest success function of the group to which i belongs) can be interpreted as the probability of winning the (public good) prize for i's group and v_i as the value of the prize for player i. Thus the **probability of winning** depends only on the two least efforts of the two groups of players. As usual $c_i(s_i)$ can be interpreted as the **cost of the effort** s_i for i.

3.2 *The Associated Fictitious Game*

The fictitious game we construct just below is indeed a contest (between two single contenders) in the general sense of the literature; however we shall not refer to it as a fictitious contest—but just as to a fictitious game—for sake of terminological simplicity (and, hopefully, for the reader's convenience).

[14]The assumption that $w_i = 0$—which is clearly more restrictive than the assumption $0 \le w_i < v_i$—is important for our results; we impose it here for expositional convenience.

[15]And hence, by the assumption of convexity, even a continuously differentiable extension.

The **fictitious game $\mathscr{F}$ associated with a contest game** (as defined in Sect. 3.1) is a game

$$(\{1,2\}, (X,Y), (\upsilon_1, \upsilon_2))$$

where $\{1,2\}$ is the set of players (also called **fictitious players**) and

$$X = \mathbb{R}_+ \text{ and } Y = \mathbb{R}_+$$

are, respectively, player 1's and player 2's strategy set and where players' payoff functions are defined by[16]

$$\upsilon_1(x,y) = \begin{cases} \frac{x}{x+y} - \gamma_1(x) & \text{if } x+y>0 \\ 0 & \text{otherwise} \end{cases}$$

and

$$\upsilon_2(x,y) = \begin{cases} \frac{x}{x+y} - \gamma_2(x) & \text{if } x+y>0 \\ 0 & \text{otherwise,} \end{cases}$$

where $\gamma_1 : \mathbb{R}_+ \to \mathbb{R}_+$

$$\gamma_1 : x \mapsto \int_0^x \max\{v_i^{-1} \cdot D^+ c_i(t) : i \in G_1\} dt$$

and $\gamma_2 : \mathbb{R}_+ \to \mathbb{R}_+$

$$\gamma_2 : y \mapsto \int_0^y \max\{v_i^{-1} \cdot D^+ c_i(t) : i \in G_2\} dt$$

are convex, strictly increasing, differentiable and vanishing at 0.[17] (See the Appendix for a proof.) Note that the fictitious game satisfies all conditions assumed in Yamazaki (2008) and that γ_1 and γ_2 are not in general twice differentiable: see, in particular, Lemma A1 and Remark A1 in the Appendix. Yamazaki (2008)—though employing graphical arguments—proves a theorem that implies the following Result.

---

[16]A caveat: the greek vowel υ (which, e.g., is used to denote fictitious player 1's utility function $\upsilon_1 : \mathbb{R}^2_+ \to \mathbb{R}$) should not be confused with the graphically similar latin consonant v (which, e.g., is used to denote contender i's evaluation $v_i \in \mathbb{R}_{++}$ of the prize).

[17]In the previous integrals we have written $D^+ c_i(t)$ instead of $Dc_i(t)$ because only the right-hand derivative exists at 0, while $Dc_i(t)$—which in our case is equal to $D^+ c_i(t)$ for all $t > 0$—is not defined at 0.

Result (Yamazaki, 2008) *Let $\mathscr{C}$ be a contest game and $\mathscr{F}$ be the associated fictitious game. Then $\mathscr{F}$ has exactly one Nash equilibrium, say e. In particular $e \in \mathbb{R}^2_{++}$ (i.e., both fictitious players are active in equilibrium).*[18]

We remark the following fact[19]: when $x \in \mathbb{R}_{++}$ we have

$$D\gamma_1(x) = \max\{v_i^{-1} \cdot D^+ c_i(x) : i \in G_1\}$$

and when $(x, y) \in \mathbb{R}_{++} \times \mathbb{R}_+$ we have

$$\begin{aligned} D\upsilon_1(\cdot, y)(x) &= \frac{y}{(x+y)^2} - D\gamma_1(x) \\ &= \frac{y}{(x+y)^2} - \max\{v_i^{-1} \cdot Dc_i(x) : i \in G_1\} \\ &= \min\left\{v_i^{-1} \cdot D^- u_i(\cdot, z_{-i})(z_i) : i \in G_1\right\} \end{aligned} \tag{1}$$

where $z \in S_N$ is the joint strategy for the contest game where, for all $i \in N$,

$$z_i = \begin{cases} x \text{ if } i \in G_1, \\ y \text{ if } i \in G_2. \end{cases}$$

An analogous remark holds for $D\gamma_2$.

4 A Uniqueness Result for Group-Proof Nash Equilibria

Theorem 1 *Consider a contest game $\mathscr{C}$ and the associated fictitious game $\mathscr{F}$.*

(i) There exists a unique GP-Nash equilibrium e for $\mathscr{C}$. In that equilibrium all players are active; in particular

$$e_i = x \textit{ for all } i \in G_1 \textit{ and } e_i = y \textit{ for all } i \in G_2$$

where (x, y) is the unique Nash equilibrium for $\mathscr{F}$.

(ii) If the unique GP-Nash equilibrium e for $\mathscr{C}$ is a self-enforcing strategy for $\mathscr{C}$ then e is also the unique CP-Nash equilibrium for $\mathscr{C}$.

[18]The last sentence of the Result is a consequence of the discontinuity of payoff functions and can be readily verified by the reader. The important part is the uniqueness result.

[19]Clearly, when $x > 0$ the derivative $Dc_i(x)$ is well-defined and we can replace $D^+c_i(x)$ with $Dc_i(x)$ in the equality that defines $D\gamma_1(x)$.

Proof

(i) Yamazaki (2008) guarantees that the associated fictitious game $\mathscr{F}$ has a unique Nash equilibrium, say (x, y). Construct s^* as in Theorem A4 in the Appendix. We must have that $s^* \in \mathbb{R}^n_{++}$, where $s^* = e$. By Theorems A3 and A4 in the Appendix, s^* is the only group-self-enforcing strategy for $\mathscr{C}$ in $\mathbb{R}^n_{++}$. Thus we are done if we prove that every group-self-enforcing strategy for $\mathscr{C}$ where at least one contender in N implements the strategy 0 is strongly $\mathscr{C}$-Pareto dominated by s^*. The proof of this fact is as follows.

For every $s \in S_N$ such that $\min\{s_l : l \in G_1\} = 0 = \min\{s_l : l \in G_2\}$ we have $u_l(s) = 0$ for all $l \in N$: thus s^* strongly $\mathscr{C}$-Pareto dominates every such s by part (xii) of Lemma A2. Now the proof continues by contradiction: suppose there exists a group-self-enforcing strategy s^{**} for $\mathscr{C}$ that strongly $\mathscr{C}$-Pareto dominates s^* and a contender l in one of the two groups, say G_1 (the proof is the same when G_2 is chosen), such that $s_l^{**} = 0$. Then

$$\min\left\{s_l^{**} : l \in G_2\right\} > 0$$

for the reasons explained just above. As $\left(s_i^{**}\right)_{i \in G_1}$ is a Nash equilibrium for the induced game (note that such an induced game satisfies all conditions of Lemma A2 in the Appendix when k is put equal to $\min\{s_l : l \in G_2\}$), we must have that $\left(s_i^{**}\right)_{i \in G_1}$ is a $|G_1|$-tuple of zeros by part (i) of Lemma A2. Thus, the game induced by group G_2 at s^{**} satisfies all conditions of Theorem A2 in the Appendix when k is put equal to $\min\{s_l : l \in G_1\} = 0$. But then, by parts (i) and (ii) of that Theorem A2, we have that $\left(s_i^{**}\right)_{i \in G_2}$ cannot be a CP-Nash equilibrium for the game induced by group G_2 at s^{**} and hence s^{**} cannot be a group-self-enforcing strategy for the contest game $\mathscr{C}$.

(ii) In part (i) we have shown that e is a group-self-enforcing strategy for $\mathscr{C}$ and in particular the unique group-self-enforcing strategy for $\mathscr{C}$ that belongs to $\mathbb{R}^n_{++}$; thus, if e is self-enforcing then it must be the unique self-enforcing strategy for $\mathscr{C}$ that belongs to $\mathbb{R}^n_{++}$. We have shown also that e strongly $\mathscr{C}$-Pareto dominates every (self-enforcing) strategy s for $\mathscr{C}$ such that $\min\{s_l : l \in G_1\} = 0 = \min\{s_l : l \in G_2\}$. Thus, if e is a self-enforcing strategy for $\mathscr{C}$ then it is also the unique CP-Nash equilibrium for $\mathscr{C}$. □

The main issue of this paper was to prove the existence of a unique GP-Nash equilibrium for a contest game $\mathscr{C}$: we accomplished such purpose with part (i) of Theorem 1. Our contribution does not clarify whether the unique GP-Nash equilibrium for a contest game $\mathscr{C}$ is also a CP-Nash equilibrium for $\mathscr{C}$ (and we cannot exclude that a contest game does not posses any CP-Nash equilibrium). However, with part (ii) of Theorem 1, we have clarified the following fact: to prove the existence of a unique CP-Nash equilibrium for a contest game $\mathscr{C}$—if any—it suffices to prove that the unique GP-Nash equilibrium for $\mathscr{C}$ is self-enforcing.

5 On the Assumptions on the Boundaries

Proposition 1 below clarifies that in order to guarantee the existence of a unique GP-Nash equilibrium—and hence of at most one CP-Nash equilibrium—that coincides with the Nash equilibrium proposed by Lee (2012) one cannot impose the usual assumption that the prize is equally shared in probability among the groups when at least one contender in each group is inactive: i.e., we must impose that $w_i \neq v_i/2$ (where 2 is the number of groups). As we said, Lee (2012) does not define the notion of coalition-proofness he employees, but certainly the proofs provided in Lee (2012) are by no means sufficient to ensure the existence of a unique CP-Nash equilibrium in the sense of Bernheim et al. (1987); therefore Proposition 1 should not be understood as a counter-example to part (c) of Proposition 1 in Lee (2012).

Henceforth in this Sect. 5 we consider the following contest game $\mathscr{C}$ with

$$n = 4, G_1 = \{1, 2\}, G_2 = \{3, 4\}$$

and with

$$v_i = 1 \text{ and } c_i = \mathrm{id}_{\mathbb{R}_+} \text{ for all } i \in N,$$

and where, for all $i \in N$, we define $u_i(s)$ as in our Sect. 3.1 if $\min(s_l)_{l \in G_1} + \min(s_l)_{l \in G_2} > 0$ but we put

$$u_i(s) = 1/2 \text{ if } \min(s_l)_{l \in G_1} + \min(s_l)_{l \in G_2} = 0.$$

The contest game just defined is a specific case of the example illustrated in Sect. 3.1 of Lee (2012) when—using Lee (2012)'s notation—$v_{ik} = 1$ for all ik. Lee (2012) proposes[20] the following solution

$$s^L := (1/4, 1/4, 1/4, 1/4)$$

as the unique coalition-proof Nash equilibrium for $\mathscr{C}$. About s^L we can prove the following fact.

Proposition 1 *The joint strategy s^L is neither a GP-Nash equilibrium nor a CP-Nash equilibrium for $\mathscr{C}$ because the strategy where all players are inactive is self-enforcing (and hence group-self-enforcing) for $\mathscr{C}$ and strongly $\mathscr{C}$-Pareto dominates s^L.*

---

[20]Actually, Assumption 2 in Lee (2012) seems to exclude contests success functions for groups like the one in the example in Sect. 3.1 of Lee (2012). We are led to conjecture that the differentiability assumptions—which are not simply a detail—in Lee (2012) pertain only the topological interior of S_N.

Proof The joint strategy $o := (0, 0, 0, 0)$ where all players are inactive strongly $\mathscr{C}$-Pareto dominates s^L because

$$u_i(o) = \frac{1}{2} > \frac{1}{4} = u_i\left(s^L\right) \text{ for all } i \in N.$$

Thus we are done if we prove that o is a self-enforcing strategy for $\mathscr{C}$ (and a fortiori group-self-enforcing for $\mathscr{C}$). By way of contradiction, suppose o is not a self-enforcing strategy for $\mathscr{C}$. Then there is a deviating coalition D such that $\emptyset \subset D \subset N$ and that $(o_i)_{i\in D}$ is not a CP-Nash equilibrium for $\mathscr{C}|_{o,D}$: that is to say, there exists a self-enforcing strategy for $\mathscr{C}|_{o,D}$ that strongly $\mathscr{C}$-Pareto dominates $(o_i)_{i\in D}$ in $\mathscr{C}|_{o,D}$. We have three possible cases: (i) $G_1 = D$; (ii) $G_2 = D$; (iii) $G_1 \neq D$ and $G_2 \neq D$.

Case (i) and (ii). In both cases there should exist a self-enforcing joint strategy, say s^*, for $\mathscr{C}|_{o,D}$ that strongly $\mathscr{C}|_{o,D}$-Pareto dominates $(o_i)_{i\in D}$. Thus s^* should be a Nash equilibrium for $\mathscr{C}|_{o,D}$ such that $s^* \neq (0, 0)$. Let $j \in D$ be a player such that $s_j^* > 0$ and let $\{i\} = N\backslash\{j\}$. When j implements the strategy s_j^* the payoff to i is $\frac{1}{2}$ if $s_i = 0$ and $1 - s_i$ if $s_i > 0$. Consequently there is no best-reply in the game $\mathscr{C}|_{o,D}$ for i to s_j^*, and hence s^* cannot be a Nash equilibrium for $\mathscr{C}|_{o,D}$: a contradiction.

Case (iii). In this case either $\emptyset \neq G_1 \cap D \neq D$ or $\emptyset \neq G_2 \cap D \neq D$. Without loss of generality suppose $\emptyset \neq G_1 \cap D \neq D$. Thus there exists $j \in G_1\backslash D$ and $l \in G_1 \cap D$ and hence, whatever the joint strategy of the deviating coalition, the payoff to l can only be less than or equal to $1/2$ because $o_j = 0$. Thus there is no strategy for $\mathscr{C}|_{o,D}$ that strongly $\mathscr{C}|_{o,D}$-Pareto dominates $(o_i)_{i\in D}$, and hence case (iii) is impossible. □

Appendix

Lemma A1 *Let $\{f_i : \mathbb{R}_+ \to \mathbb{R}\}_{i=1}^k$ be a finite family of differentiable, increasing (resp. strictly increasing) and convex functions.*[21] *Then*

$$\gamma : \mathbb{R}_+ \to \mathbb{R}_+ : x \mapsto \int_0^x \max\left\{D^+ f_i(t) : i = 1, \ldots, k\right\} \mathrm{d}t$$

is differentiable, increasing (resp. strictly increasing) and convex.[22] *In particular,*

$$\gamma(0) = 0,$$

$$D\gamma(x) = \max\{Df_i(x) : i = 1, \ldots, k\} \textit{ for all } x > 0$$

[21] Thus $Df_i(x) = D^+ f_i(x)$ for all $x > 0$. Note that Df_i is not defined at 0.

[22] Thus $D\gamma(x) = D^+\gamma(x)$ for all $x > 0$. Note that $D\gamma$ is not defined at 0.

and

$$D^+\gamma(x) = \max\{D^+f_i(x) : i = 1, \ldots, k\} \text{ for all } x \geq 0.$$

Proof The continuous function

$$\phi : \mathbb{R}_+ \to \mathbb{R}_+ : t \mapsto \max\{D^+f_i(t) : i = 1, \ldots, k\}$$

is well-defined into $\mathbb{R}_+$ by the increasingness of each f_i,[23] and is increasing by the convexity of each f_i. By the Fundamental Theorem of Calculus the integral $\gamma(x)$ is well-defined as above for each $x \in \mathbb{R}_+$, γ is continuous and $D\gamma(x)$ is defined as above for all $x \in \mathbb{R}_{++}$. It is easily seen that γ is increasing (resp. strictly increasing) as ϕ is nonnegative (resp. positive on $\mathbb{R}_{++}$) and that $\gamma(0) = 0$. Since ϕ is increasing, γ must be convex. Thus, since $D^+\gamma$ is right-continuous and real-valued by the convexity of the increasing function γ, the right-hand derivative $D^+\gamma(0)$ is defined as above. □

Remark A1 Let f_1 and f_2 be the self-map on $\mathbb{R}_+$ respectively defined by $t \mapsto t$ and $t \mapsto t^2/2$. The function γ defined as in Lemma A1 is the self-map on $\mathbb{R}_+$ defined by

$$\gamma : t \mapsto \begin{cases} t & \text{if } t \leq 1, \\ \frac{1}{2} + \frac{t^2}{2} & \text{if } t \geq 1. \end{cases}$$

The function γ is (continuously) differentiable (with continuous derivative $t \mapsto \max(1, t)$) but is not twice differentiable.

Lemma A2 *Let $\Gamma = (N, (S_i)_{i\in N}, (u_i)_{i\in N})$ be a game with a set $N = \{1, \ldots, n\}$ of players where $S_i = \mathbb{R}_+$ for all $i \in N$. Let $k \in \mathbb{R}_+$ and $v_i \in \mathbb{R}_{++}$ for all $i \in N$. Suppose that, for all $i \in N$,*

$$u_i(s) = \begin{cases} \dfrac{\min\{s_l : l \in N\}}{k + \min\{s_l : l \in N\}} v_i - c_i(s_i) & \text{if } k + \min\{s_l : l \in N\} > 0, \\ 0 & \text{otherwise,} \end{cases}$$

where $c_i : \mathbb{R}_+ \to \mathbb{R}_+$ is a differentiable, convex and strictly increasing function that vanishes at 0.

(i) *Each Nash equilibrium for Γ, if any, is symmetric.*
(ii) *If $n = 1$ and $k = 0$ then u_i is a concave function with no maximizer.*
(iii) *If $n = 1$ and $k > 0$ then u_i is a concave function with exactly one maximizer.*
(iv) *If $n > 1$ and $k = 0$ then each $u_i(\cdot, s_{-i})$ is a concave function with exactly one maximizer.*

[23] Recall that each D^+f_i is continuous on $\mathbb{R}_+$ and that the pointwise maximum of finitely many continuous functions is continuous.

(v) *If $n > 1$ and $k > 0$ then each $u_i(\cdot, s_{-i})$ is a concave function with exactly one maximizer.*
(vi) *If $n = 1$ and $k = 0$ then there is no Nash equilibrium for Γ.*
(vii) *If $n = 1$ and $k > 0$ then there is exactly one Nash equilibrium for Γ.*
(viii) *If $n > 1$ then the origin is a Nash equilibrium for Γ.*
(ix) *If $n > 1$ then the set of Nash equilibria for Γ is convex.*
(x) *If $n > 1$ then the set of Nash equilibria for Γ is compact.*
(xi) *If $k + n > 1$ then there exists a greatest Nash equilibrium.*
(xii) *If e is a Nash equilibrium for Γ, $i \in N$ and $e_i > 0$ then $u_i(e) > 0$.*

Proof Proof of (i). Suppose that $s^* \in S_N$ is an asymmetric joint strategy for Γ. Then there exists a pair $(i, l) \in N \times N$ of players such that

$$s_i^* < s_l^*.$$

Thus $i \neq l$. But s^* cannot be a Nash equilibrium for Γ because

$$u_l(s^*) < u_l(s_l^{**}, s_{-l}^*) \text{ when } (s_l^* >) s_l^{**} := s_i^* \in S_l$$

in that c_l is strictly increasing and

$$\min\{s_j^* : j \in N\} = \min\{\{s_l^{**}\} \cup \{s_j^* : j \in N \setminus \{l\}\}\}.$$

Proof of (ii), (iii), (vi) and (vii). Suppose $n = 1$. If $k = 0$ then the payoff function of the unique player is discontinuous at the origin, lower semicontinuous on $\mathbb{R}_+$, strictly decreasing on $\mathbb{R}_{++}$ and concave on $\mathbb{R}_{++}$; thus in this case no maximizer of u_i (as well as no Nash equilibrium for Γ) exists and the payoff function of the unique player is concave. If $k > 0$ then the payoff function is strictly concave and continuous; thus in this case there exists exactly one maximizer for u_i (and hence exactly one Nash equilibrium for Γ) since the continuous and strictly concave payoff function vanishes at 0 and is eventually negative.

Proof of (iv) and (v). Suppose $n > 1$ and pick an arbitrary $i \in N$. If $\min\{s_j : j \in N \setminus \{i\}\} = 0$, the payoff function $u_i(\cdot, s_{-i})$ coincides with $-c_i$ and hence it is strictly decreasing and concave with exactly one maximizer. Henceforth suppose $\min\{s_j : j \in N \setminus \{i\}\} > 0$. The payoff function $u_i(\cdot, s_{-i})$ is continuous; since the payoff function $u_i(\cdot, s_{-i})$ vanishes at 0 and is eventually negative, there exists one maximizer; since the continuous payoff function $u_i(\cdot, s_{-i})$ is strictly concave on the (possibly nonproper) interval

$$\left[0, \min\{s_j : j \in N \setminus \{i\}\}\right]$$

and since it is both strictly decreasing and concave on

$$\left]\min\{s_j : j \in N \setminus \{i\}\}, +\infty\right[,$$

there exists exactly one maximizer for $u_i(\cdot, s_{-i})$; finally, as

$$D^+ u_i(\cdot, s_{-i})(s_i) \leq D^- u_i(\cdot, s_{-i})(s_i)$$

whenever $s_i > 0$, we are in a position to conclude that $u_i(\cdot, s_{-i})$ is concave.

Proof of (viii). Suppose $n > 1$. Since each payoff function u_i is concave in the i-th argument by parts (iv) and (v) of this Lemma A2, the fact that the origin is a Nash equilibrium is a consequence of the fact that, for all $i \in N$,

$$D^+ u_i(0, \ldots, 0, \cdot, 0, \ldots, 0)(0) = -D^+ c_i(0) \leq 0.$$

Proof of (ix). Suppose $n > 1$. By parts (i) and (viii) of this Lemma A2, it suffices to show that if $(e_1, \ldots, e_n)$ is a (necessarily symmetric) Nash equilibrium with

$$e_1 = \ldots = e_n > 0$$

then $(te_1, \ldots, te_n)$ is a Nash equilibrium for any $t \in]0, 1]$. Suppose $(e_1, \ldots, e_n)$ is a (necessarily symmetric) Nash equilibrium with

$$e_1 = \ldots = e_n > 0.$$

First note that, for all $i \in N$,

$$D^+ u_i(te_1, \ldots, te_{i-1}, \cdot, te_{i+1} \ldots, te_n)(te_i) = -D^+ c_i(te_i) < 0.$$

Then note that, for all $i \in N$,

$$D^- u_i(e_1, \ldots, e_{i-1}, \cdot, e_{i+1} \ldots, e_n)(e_i) = k\frac{v_i}{(k+e_i)^2} - D^- c_i(e_i) \geq 0$$

because $(e_1, \ldots, e_n)$ is a (necessarily symmetric) Nash equilibrium with $e_1 = \ldots = e_n > 0$ and because $u_i(e_1, \ldots, e_{i-1}, \cdot, e_{i+1} \ldots, e_n)$ is locally semidifferentiable at $e_i > 0$; therefore, for all $i \in N$,

$$D^- u_i(te_1, \ldots, te_{i-1}, \cdot, te_{i+1} \ldots, te_n)(te_i) = k\frac{v_i}{(k+te_i)^2} - D^- c_i(te_i) \geq 0$$

for any $t \in]0, 1]$, because c_i is convex on $\mathbb{R}_+$ (which implies that $]0, 1] \to \mathbb{R} : t \mapsto -D^- c_i(te_i)$ is decreasing in t) and because $k\frac{v_i}{(k+te_i)^2} \geq k\frac{v_i}{(k+e_i)^2}$. Thus, since each u_i is concave by parts (iv) and (v) of this Lemma A2, we can conclude that $(te_1, \ldots, te_n)$ is a Nash equilibrium for any $t \in]0, 1]$.

Proof of (x). Suppose $n > 1$. If there exists at most one Nash equilibrium, this part of the Lemma is trivially true. Suppose there are at least two Nash equilibria.

Then one Nash equilibrium must belong to $\mathbb{R}^n_{++}$ by part (i) of this Lemma A2. Pick an arbitrary i in N. Since

$$D^- u_i (t, \ldots, t, \cdot, t \ldots, t)(t) = k \frac{v_i}{(k+t)^2} - D^- c_i (t)$$

is negative for all sufficiently large $t > 0$, the set of Nash equilibria must be bounded. Thus we are done if we prove that the set of Nash equilibria is closed. Let

$$(\bar{t}, \ldots, \bar{t}, \ldots, \bar{t})$$

be any symmetric joint strategy other than the origin; note that such a strategy must belong to $\mathbb{R}^n_{++}$. It is easily seen that

$$D^+ u_i (\bar{t}, \ldots, \bar{t}, \cdot, \bar{t}, \ldots, \bar{t})(\bar{t}) = -D^+ c_i (\bar{t}) < 0.$$

As $u_i (\bar{t}, \ldots, \bar{t}, \cdot, \bar{t}, \ldots, \bar{t})$ is concave, parts (i), (viii) and (ix) of this Lemma A2 guarantee that we are done if the following implication is true:

$$D^- u_i (t, \ldots, t, \cdot, t, \ldots, t)(t) = k \frac{v_i}{(k+t)^2} - D^- c_i (t) \geq 0 \text{ for all } t \in]0, \bar{t}[$$

$$\Downarrow$$

$$D^- u_i (\bar{t}, \ldots, \bar{t}, \cdot, \bar{t}, \ldots, \bar{t})(\bar{t}) = k \frac{v_i}{(k+\bar{t})^2} - D^- c_i (\bar{t}) \geq 0.$$

Noting that $D^- c_i$ is left-continuous at $\bar{t}$, one easily verifies the validity of the previous implication.

Proof of (xi). When $n = 1$, part (xi) of Lemma A2 is a consequence of part (vii) of Lemma A2. When $n > 1$, part (xi) of Lemma A2 is a consequence of parts (i), (viii) and (x) of Lemma A2.

Proof of (xii). Suppose that e is a Nash equilibrium for Γ. Suppose $i \in N$ and $e_i > 0$. If i is the only player then, by parts (ii) and (iii) of this Lemma A2, the positive strategy e_i is the unique maximizer of u_i; since u_i vanishes at 0 we must have that $u_i (e) = u_i (e_i) > 0$. Henceforth suppose $n > 1$. By way of contradiction, suppose $u_i (e) \leq 0$; then

$$u_i (0, e_{-i}) = 0 \geq u_i (e)$$

in contradiction with parts (iv) and (v) of this Lemma A2 and with the fact that the strategy $e_i > 0$ maximizes $u_i (\cdot, e_{-i})$. □

The following Corollary A1 follows from parts (i), (viii)–(xi) of Lemma A2.

Corollary A1 *Let $\Gamma = (N, (S_i)_{i \in N}, (u_i)_{i \in N})$ be a game satisfying the conditions of Lemma A2. Suppose $n > 1$. Then there exists a greatest Nash equilibrium, say s^*,* *such that*

$$s^* = (\bar{t}, \ldots, \bar{t}, \ldots, \bar{t}) \text{ for some } \bar{t} \in \mathbb{R}_+$$

and the set of Nash equilibria for Γ is

$$\left\{(t, \ldots, t, \ldots, t) \in \mathbb{R}^n_+ : 0 \leq t \leq \bar{t}\right\}.$$

Theorem A1 *Let $\Gamma = (N, (S_i)_{i \in N}, (u_i)_{i \in N})$ be a game and $k \in \mathbb{R}_+$. Suppose $N = \{1, \ldots, n\}$ with $k + n > 1$ and that, for all $i \in N$:*

$$S_i = \mathbb{R}_+$$

and

$$u_i(s) = \frac{\min\{s_l : l \in N\}}{k + \min\{s_l : l \in N\}} v_i - c_i(s_i)$$

where $v_i \in \mathbb{R}_{++}$ and $c_i : \mathbb{R}_+ \to \mathbb{R}$ is a differentiable, convex and strictly increasing function that vanishes at 0. Then a greatest Nash equilibrium s^* for Γ exists.[24] *Suppose s^* is not the origin. Then $s^{**} \in S_N$ is the greatest Nash equilibrium for Γ if and only if*

$$s^{**} \in \mathbb{R}^n_{++} \text{ and } \min\left\{D^- u_i\left(\cdot, s^{**}_{-i}\right)\left(s^{**}_i\right) : i \in N\right\} = 0.$$

Proof When $n = 1$ (and hence $k > 0$ as $k + n > 1$), Theorem A1 is a consequence of part (iii) of Lemma A2. Henceforth suppose $n > 1$.

Proof of the if part. Pick a joint strategy $s^{**} \in \mathbb{R}^n_{++}$ such that

$$\min\left\{D^- u_i\left(\cdot, s^{**}_{-i}\right)\left(s^{**}_i\right) : i \in N\right\} = 0.$$

Note that

$$\max\left\{D^+ u_i\left(\cdot, s^{**}_{-i}\right)\left(s^{**}_i\right) : i \in N\right\} = \max\left\{-Dc_i\left(s^{**}_i\right) : i \in N\right\} < 0.$$

Since each $u_i\left(\cdot, s^{**}_{-i}\right)$ is concave by parts (iv) and (v) of Lemma A2, we can conclude that s^{**} is a Nash equilibrium for Γ. Since

$$\min\left\{D^- u_i\left(\cdot, s^{**}_{-i}\right)\left(s^{**}_i\right) : i \in N\right\} = 0$$

we must have that

$$D^- u_l\left(\cdot, s^{**}_{-l}\right)\left(s^{**}_l\right) = 0$$

[24]This is a consequence of part (xi) of Lemma A2.

for some $l \in N$. Since each Nash equilibrium must be symmetric by part (i) of Lemma A2 and since each u_i is concave in the i-th argument by parts (iv) and (v) of Lemma A2, in order to infer that no Nash equilibrium greater than s^{**} can exist it suffices to note that each function

$$\Delta_i^- : \left]-s_i^{**}, +\infty\right[\to \mathbb{R}$$

defined by[25]

$$\Delta_i^- : t \mapsto D^- u_i \left(s_1^{**} + t, \ldots, s_{i-1}^{**} + t, \cdot, s_{i+1}^{**} + t, \ldots, s_n^{**} + t\right) \left(s_i^{**} + t\right)$$

is strictly decreasing, and hence that

$$D^- u_l \left(s_1^{**} + t, \ldots, s_{l-1}^{**} + t, \cdot, s_{l+1}^{**} + t, \ldots, s_n^{**} + t\right) \left(s_l^{**} + t\right) < 0 \text{ for all } t > 0.$$

This proves only that s^{**} is maximal; by Corollary A1, s^{**} is indeed the greatest Nash equilibrium.

Proof of the only if part. Suppose $s^{**} \in S_N$ is the greatest Nash equilibrium. All Nash equilibria must be symmetric by part (i) of Lemma A1 and hence, as the greatest Nash equilibrium is not the origin by assumption, we must have that $s^{**} \in \mathbb{R}_{++}^n$. Since s^{**} is the greatest Nash equilibrium, the joint strategy

$$\left(s_i^{**} + t\right)_{i \in N}$$

is not a Nash equilibrium for all $t > 0$. Each function

$$u_i \left(s_1^{**} + t, \ldots, s_{i-1}^{**} + t, \cdot, s_{i+1}^{**} + t, \ldots, s_n^{**} + t\right)$$

is concave for all $t \geq 0$ by parts (iv) and (v) of Lemma A2 and

$$D^+ u_i \left(s_1^{**} + t, \ldots, s_{i-1}^{**} + t, \cdot, s_{i+1}^{**} + t, \ldots, s_n^{**} + t\right) \left(s_i^{**} + t\right) = -D^+ c_i \left(s_i^{**} + t\right) < 0$$

for all $i \in N$ and all $t \geq 0$, therefore the fact that $\left(s_i^{**} + t\right)_{i \in N}$ is not a Nash equilibrium for any $t > 0$ implies that for each $t > 0$ there exists $l_t \in N$ such that[26]

$$D^- u_{l_t} (s_1^{**} + t, \ldots, s_{l_t-1}^{**} + t, \cdot, s_{l_t+1}^{**} + t, \ldots, s_n^{**} + t)(s_{l_t}^{**} + t) < 0.$$

On the other hand, by Corollary A1 and by the semidifferentiability of each payoff functions in the i-th argument on $\mathbb{R}_{++}^n$, we must have that the Nash equilibrium

[25]Indeed, $\Delta_i^- (t) = \frac{k}{\left(k+s_i^{**}+t\right)^2} v_i - Dc_i \left(s_i^{**} + t\right)$.

[26]Note that l_t can depend on t.

conditions imply

$$D^{-}u_i\left(s_1^{**}+t,\ldots,s_{i-1}^{**}+t,\cdot,s_{i+1}^{**}+t,\ldots,s_n^{**}+t\right)\left(s_i^{**}+t\right)\geq 0$$

for all $i \in N$ and all $t \in \left]-s_i^{**},0\right]$. Thus, the continuity of each function Δ_i^- implies that

$$D^{-}u_i\left(s_1^{**},\ldots,s_{i-1}^{**},\cdot,s_{i+1}^{**},\ldots,s_n^{**}\right)\left(s_i^{**}\right)\geq 0$$

for all $i \in N$ and hence, in particular, the continuity of each function Δ_i^- implies $\min\left\{D^{-}u_i\left(\cdot,s_{-i}^{**}\right)\left(s_i^{**}\right): i \in N\right\}=0.$[27] □

Theorem A2 *Let* $\Gamma = (N, (S_i)_{i\in N}, (u_i)_{i\in N})$ *be a game and* $k \in \mathbb{R}_+$. *Suppose* $N = \{1,\ldots,n\}$ *and that, for all* $i \in N$*:*

$$S_i = \mathbb{R}_+$$

and

$$u_i(s) = \frac{\min\{s_l : l \in N\}}{k+\min\{s_l : l \in N\}} v_i - c_i(s_i)$$

where $v_i \in \mathbb{R}_{++}$ *and* $c_i : \mathbb{R}_+ \to \mathbb{R}$ *is a differentiable, convex and strictly increasing function that vanishes at* 0.

(i) *If* $k = 0$ *and* $n = 1$ *then no joint strategy can be a CP-Nash equilibrium for* Γ.
(ii) *If* $k = 0$ *and* $n > 1$ *then no joint strategy in* $\mathbb{R}^n_{++}$ *can be a CP-Nash equilibrium for* Γ.
(iii) *If* $k > 0$ *and* $n \geq 1$ *then a greatest Nash equilibrium for* Γ, *say* $\overline{e}$, *exists;* $\overline{e}$ *strongly* Γ*-Pareto dominates every other Nash equilibrium for* Γ *and is also the unique CP-Nash equilibrium for* Γ.

Proof Proof of (i). Suppose $k = 0$ and $n = 1$. Then part (i) of Theorem A2 is a consequence of part (vi) of Lemma A2.

Proof of (ii). Suppose $k = 0$ and $n > 1$. By way of contradiction, suppose s^* is a CP-Nash equilibrium for Γ that belongs to $\mathbb{R}^n_{++}$. As s^* must be also a Nash equilibrium for Γ, part (i) of Lemma A2 implies that

$$s^* = (z,\ldots,z) \text{ for some } z > 0;$$

[27] Otherwise $\min\left\{D^{-}u_i\left(\cdot,s_{-i}^{**}\right)\left(s_i^{**}\right): i \in N\right\} > 0$ and

$$D^{-}u_{l_t}\left(s_1^{**}+t,\ldots,s_{l_t-1}^{**}+t,\cdot,s_{l_t+1}^{**}+t,\ldots,s_n^{**}+t\right)\left(s_{l_t}^{**}+t\right) < 0 \text{ for all } t > 0$$

would contradict the continuity of each Δ_i^-.

but then, as $k = 0$ and c_i is strictly increasing, we would have

$$u_i\left(s_i^*, s_{-i}^*\right) = v_i - c_i(z) < v_i - c_i(z/2) = v_i - c_i\left(s_i^*/2\right) = u_i\left(s_i^*/2, s_{-i}^*\right)$$

in contradiction with the fact that s^* is a Nash equilibrium for Γ.

Proof of (iii). Suppose $k > 0$. When $n = 1$ part (iii) of Theorem A2 is a consequence of part (vii) of Lemma A2. Henceforth suppose $n > 1$. Every sufficiently large strategy is strictly dominated by the strategy 0: the strict increasingness and the convexity of each c_i imply the existence of $\overline{x}_i \in \mathbb{R}_{++}$ such that $c_i(s_i) \geq c_i(\overline{x}_i) > v_i$ for all $s_i \geq \overline{x}_i$ and that for all $\underline{s}_{-i} \in S_{N\setminus\{i\}}$:

$$u_i\left(0, \underline{s}_{-i}\right) = 0 > u_i\left(s_i, \underline{s}_{-i}\right) \text{ for all } s_i \geq \overline{x}_i, \text{ for all } i \in N.$$

Therefore the set of Nash equilibria for Γ and those of CP-Nash equilibria for Γ are unaltered when S_i is truncated to $\overline{x}_i$. Thus, if we had assumed since the beginning that each strategy set is the compact intervals $[0, \max\{\overline{x}_i : i \in N\}]$ none of the previous results would have changed. Therefore the best-reply correspondences of the "truncated" game (as well as those of the original game) are single-valued by part (v) of Lemma A2 (and hence all Nash equilibria for the "truncated" game are strict Nash equilibria); then note that each u_i of the "truncated" game (as well as each u_i of the original game) is increasing in all arguments other than the i-th one. Finally note that best-reply "functions" of the "truncated" game (as well as those of the original game) are increasing in all arguments since, for all $i \in N$:

$$\underline{s} \in S_N, \overline{s} \in S_N, \underline{s}_l \leq \overline{s}_l \text{ for all } l \in N\setminus\{i\} \text{ and } \{\beta\} = \arg\max u_i\left(\cdot, \underline{s}_{-i}\right)$$

$$\Downarrow$$

$$\overline{x}_i > \beta \leq \min\left\{\underline{s}_l : l \in N\setminus\{i\}\right\} \text{ and } u_i\left(\cdot, \underline{s}_{-i}\right)|_{[0,\beta]} = u_i\left(\cdot, \overline{s}_{-i}\right)|_{[0,\beta]}.$$

Conclude that the greatest Nash equilibrium is the only CP-Nash equilibrium of the "truncated" game by parts (vii) and (viii) of Theorem 1 in Quartieri (2013) and hence that $\overline{e}$ is the only CP-Nash equilibrium for Γ. □

Theorem A3 *Let $\mathscr{C}$ be a contest game and $\mathscr{F}$ be the associated fictitious game. Suppose there exists a group-self-enforcing strategy $s^* \in \mathbb{R}^n_{++}$ for $\mathscr{C}$ such that, for some $(x, y) \in \mathbb{R}^2_{++}$: $s_i^* = x$ for all $i \in G_1$ and $s_i^* = y$ for all $i \in G_2$. Then (x, y) is a Nash equilibrium for $\mathscr{F}$.*

Proof Since s^* is group-self-enforcing for the contest game $\mathscr{C}$, we have that $\left(s_i^*\right)_{i\in G_1}$ and $\left(s_i^*\right)_{i\in G_2}$ must be, respectively, CP-Nash equilibria for the game induced by coalition $G1$ and $G2$ at s^* from $\mathscr{C}$. Since x and y are positive, by part (iii) of Theorem A2, $\left(s_i^*\right)_{i\in G_1}$ and $\left(s_i^*\right)_{i\in G_2}$ are, respectively, the unique CP-Nash equilibria for the game induced by coalition $G1$ and $G2$ at s^* from $\mathscr{C}$. Consequently, again by part (iii) of Theorem A2, $\left(s_i^*\right)_{i\in G_1} \in \mathbb{R}^{|G_1|}_{++}$ and $\left(s_i^*\right)_{i\in G_2} \in \mathbb{R}^{|G_2|}_{++}$ are, respectively, the greatest Nash equilibrium for the game induced by coalition $G1$

and $G2$ at s^* from $\mathscr{C}$. Thus, by Theorem A1,

$$\min\left\{D^- u_i\left(\cdot, s^*_{-i}\right)\left(s^*_i\right) : i \in G_1\right\} = 0$$

and

$$\min\left\{D^- u_i\left(\cdot, s^*_{-i}\right)\left(s^*_i\right) : i \in G_2\right\} = 0.$$

Consequently,

$$D\upsilon_1\left(\cdot, y\right)\left(x\right) = \min\left\{v_i^{-1} \cdot D^- u_i\left(\cdot, s^*_{-i}\right)\left(s^*_i\right) : i \in G_1\right\} = 0$$

and

$$D\upsilon_2\left(x, \cdot\right)\left(y\right) = \min\left\{v_i^{-1} \cdot D^- u_i\left(\cdot, s^*_{-i}\right)\left(s^*_i\right) : i \in G_2\right\} = 0.$$

The payoff function of each player $i \in \{1, 2\}$ of $\mathscr{F}$ is concave in the i-th argument and hence (x, y) is a Nash equilibrium for $\mathscr{F}$. □

Theorem A4 *Let $\mathscr{C}$ be a contest game and $\mathscr{F}$ be the associated fictitious game. Suppose (x, y) is a Nash equilibrium for the associated fictitious game $\mathscr{F}$. Then there exists a group-self-enforcing strategy $s^* \in \mathbb{R}^n_{++}$ for $\mathscr{C}$ such that, for some $(x, y) \in \mathbb{R}^2_{++}$: $s^*_i = x$ for all $i \in G_1$ and $s^*_i = y$ for all $i \in G_2$.*

Proof Since each $\upsilon_i\left(\cdot, s_{-i}\right)$ has no maximizer when $s_{-i} = 0$, both fictitious players must be active in equilibrium. Thus

$$x > 0 < y,$$

and, as $\upsilon_i\left(\cdot, s_{-i}\right)$ is concave for both fictitious players, we must have that

$$D\upsilon_1\left(\cdot, y\right)\left(x\right) = 0 = D\upsilon_2\left(x, \cdot\right)\left(y\right).$$

for all $i \in \{1, 2\}$. Consequently, by (1),

$$\min\left\{v_i^{-1} \cdot D^- u_i\left(\cdot, s^*_{-i}\right)\left(s^*_i\right) : i \in G_1\right\} = D\upsilon_1\left(\cdot, y\right)\left(x\right) = 0$$

and

$$\min\left\{v_i^{-1} \cdot D^- u_i\left(\cdot, s^*_{-i}\right)\left(s^*_i\right) : i \in G_2\right\} = D\upsilon_2\left(x, \cdot\right)\left(y\right) = 0.$$

Therefore

$$\min\left\{D^- u_i\left(\cdot, s^*_{-i}\right)\left(s^*_i\right) : i \in G_1\right\} = 0$$

and

$$\min\left\{D^{-}u_i\left(\cdot,s_{-i}^{*}\right)\left(s_i^{*}\right):i\in G_2\right\}=0.$$

By Theorem A1, $\left(s_l^*\right)_{l\in G_1}$ and $\left(s_l^*\right)_{l\in G_2}$ are the greatest Nash equilibria for the games induced by, respectively, coalition G_1 and G_2 at s^{**} from $\mathscr{C}$. By part (iii) of Theorem A2, $\left(s_l^*\right)_{l\in G_1}$ and $\left(s_l^*\right)_{l\in G_2}$ are the only CP-Nash equilibria for the games induced by, respectively, coalition G_1 and G_2 at s^{**} from $\mathscr{C}$. Thus s^* is group-self-enforcing for $\mathscr{C}$. □

Acknowledgements The first author gratefully acknowledges financial support: part of his work was carried out in the frame of Programme STAR, financially supported by UniNA and Compagnia di San Paolo. The second author gratefully acknowledges JSPS KAKENHI Grant Number 15K03361.

References

Baik KH (1993) Effort levels in contests: the public-good prize case. Econ Lett 41:363–367
Baik KH (2008) Contests with group-specific public-good prizes. Soc Choice Welf 30:103–117
Bernheim BD, Peleg B, Whinston MD (1987) Coalition-proof Nash equilibria. I. Concepts. J Econ Theory 42:1–12
Choi JP, Chowdhury SM, Kim J (2015) Group contests with internal conflict and power asymmetry. Scand J Econ (Forthcoming). doi:10.1111/sjoe.12152
Chowdhury SM, Topolyan I (2015) The attack-and-defense group contests: best-shot versus weakest-link. Econ Inq (Forthcoming). doi:10.1111/ecin.12246
Chowdhury SM, Lee D, Sheremeta RM (2013) Top guns may not fire: best-shot group contests with group-specific public good prizes. J Econ Behav Org 92:94–103
Corchón LC (2007) The theory of contests: a survey. Rev Econ Des 11:69–100
Cornes R (1993) Dyke maintenance and other stories: some neglected types of public goods. Q J Econ 108:259–271
Cornes R, Hartley R (2005) Asymmetric contests with general technologies. Econ Theory 26:923–946
Hirshleifer J (1983) From weakest-link to best-shot: the voluntary provision of public goods. Public Choice 41:371–386
Katz E, Nitzan S, Rosenberg J (1990) Rent seeking for pure public goods. Public Choice 65:49–60
Kolmar M, Rommeswinkel H (2013) Contests with group-specific public goods and complementarities in efforts. J Econ Behav Org 89:9–22
Konishi H, Le Breton M, Weber S (1999) On coalition-proof Nash equilibria in common agency games. J Econ Theory 85:122–139
Lee D (2012) Weakest-link contests with group-specific public good prizes. Eur J Pol Econ 28:238–248
Milgrom P, Roberts J (1996) Coalition-proofness and correlation with arbitrary communication possibilities. Games Econ Behav 17:113–128
Münster J (2007) Simultaneous inter- and intra-group conflicts. Econ Theory 32:333–352
Quartieri F (2013) Coalition-proofness under weak and strong Pareto dominance. Soc Choice Welf 40:553–579
Quartieri F, Shinohara R (2015) Coalition-proofness in a class of games with strategic substitutes. Int J Game Theory 44:785–813

Shinohara R (2005) Coalition-proofness and dominance relations. Econ Lett 89:174–179

Szidarovszky F, Okuguchi K (1997) On the existence and uniqueness of pure Nash equilibrium in rent-seeking games. Games Econ Behav 18:135–140

Yamazaki T (2008) On the existence and uniqueness of pure-strategy Nash equilibrium in asymmetric rent-seeking contests. J Public Econ Theory 10:317–327

On the Nash Equilibrium of Asymmetric Public-Good Contests

Takeshi Yamazaki

Abstract In this paper we study a rent-seeking contest where several groups compete for a prize which is a public good among players in a group. In the contest players in a group may evaluate the prize of the contest differently. We prove that such an asymmetric public-good contest with a general contest success function possesses a unique pure-strategy Nash equilibrium, where the equilibrium is unique in the sense that the total expenditure of each group is uniquely determined. We also present some comparative statics in the model.

1 Introduction

In a traditional rent-seeking contest, it is assumed that at least two players, i.e. contestants, participate in the contest, that one player wins the contest with some probability which depends on the configuration of all players' expenditures, and that the winner gains the prize (rent) of the contest and the other players get nothing. In this setting, the prize is assumed to be a private good. However, the prize can be a public good. If a player wins a contest for a public good, some other players may be able to receive some benefits from the winner's victory of the contest. Katz et al. (1990), Nitzan (1991) and Gradstein (1993) among others analyze such a public-good contest with a specific contest success function. Yamazaki (2013b, chapter 6) analyzes an asymmetric public good contest with a general contest success function under usual assumptions and proves that there exists a unique pure-strategy Nash equilibrium in the model with two groups, assuming that production functions for lotteries are differentiable. In the equilibrium the total expenditure of each group is uniquely determined. However, each player's expenditure may not be determined in the equilibrium. He also reports some results of comparative statics in his model. In this paper, we will show that the same results can be established even in the model

T. Yamazaki
Faculty of Economics, Niigata University, 8050 Ikarashi 2-no-cho, Nishi-ku, Niigata-shi 950-2181, Japan
e-mail: tyamazak@econ.niigata-u.ac.jp

P. von Mouche, F. Quartieri (eds.), *Equilibrium Theory for Cournot Oligopolies and Related Games*, Springer Series in Game Theory,
DOI 10.1007/978-3-319-29254-0_16

of Yamazaki (2013b) with N groups, without assuming that the production functions for lotteries are differentiable.

This paper is organized as follows. Section 2 describes a general asymmetric public-good contest. Section 3 proves that there exists a unique pure-strategy Nash equilibrium in the general asymmetric public-good contest. Section 4 derives some comparative statics for the general asymmetric public-good contest. Section 5 concludes.

2 Model

Yamazaki (2013b, chapter 6) considers an asymmetric contest where only two groups seek for a prize or rent which can be interpreted as a public good among players in each group. In this paper, we assume that there are N groups competing for a predetermined prize $R > 0$, $N \geq 2$. Player i of group I expends x_i^I for group I to win the rent. Denote group I's total expenditure or effort as $X^I \equiv \sum_{i=1}^{n(I)} x_i^I$, where $n(I)$ is the number of players in group I. The contest regulator observes $\mathbf{X} \equiv \left(X^1, X^2, \cdots, X^N\right)$ and determines probability p^I that group I wins the contest, $I = 1, 2, \cdots, N$. Throughout this paper, we impose the following assumption.

Assumption 1 *Group I's winning probability is given by*

$$p^I = \frac{F_I\left(X^I\right)}{\sum_{J=1}^N F_J\left(X^J\right)},$$

where $F_I\left(X^I\right)$ is zero at $X^I = 0$, continuous, strictly increasing and weakly concave in X^I for any $I \in \{1, 2, \cdots, N\}$. If $\sum_{J=1}^N F_J\left(X^J\right) = 0$, p^I is defined to be $1/N$ or 0.

Following Szidarovszky and Okuguchi (1997), we call $F_I\left(X^I\right)$ in Assumption 1 group I's production function for lotteries. If group I's production function for lotteries $F_I\left(X^I\right)$ is differentiable for any $I \in \{1, 2, \cdots, N\}$, Assumption 1 can be rewritten as follows.

Assumption 2 *Group I's winning probability is given by*

$$p^I = \frac{F_I\left(X^I\right)}{\sum_{J=1}^N F_J\left(X^J\right)},$$

where $F_I(0) = 0$, $F_I'\left(X^I\right) > 0$ and $F_I''\left(X^I\right) \leq 0$ for any $I \in \{1, 2, \cdots, N\}$. If $\sum_{J=1}^N F_J\left(X^J\right) = 0$, p^I is defined to be $1/N$ or 0.

In the following analysis, after proving the unique existence of a Nash equilibrium under Assumption 2, we will briefly explain the reason why the unique existence can be established under Assumption 1. In the analysis of comparative statics, to simplify the analysis, we will impose Assumption 2. If one of the following assumptions is satisfied, it is clear that Assumption 2 is satisfied.

Assumption 3 *Group I's winning probability is given by*

$$p^I = \frac{\lambda_I F(X^I)}{\sum_{J=1}^{N} \lambda_J F(X^J)},$$

where $F(0) = 0$*,* $F'(X^I) > 0$*,* $F''(X^I) \leq 0$ *and* $\lambda_I > 0$ *for any* $I \in \{1, 2, \cdots, N\}$*. If* $\sum_{J=1}^{N} \lambda_J F(X^J) = 0$*,* p^I *is defined to be* $1/N$ *or* 0*.*

Assumption 4 *Group I's winning probability is given by*

$$p^I = \frac{\lambda_I (X^I)^r}{\sum_{J=1}^{N} \lambda_J (X^J)^r},$$

where $r \in (0, 1]$ *and* $\lambda_I > 0$ *for any* $I \in \{1, 2, \cdots, N\}$*. If* $\sum_{J=1}^{N} \lambda_J (X^J)^r = 0$*,* p^I *is defined to be* $1/N$ *or* 0*.*

In some comparative static analyses, Assumption 2 will be replaced by Assumption 3.

If $n(I) = 1$ for all I, that is, if the model reduces to a model of private-good contests, the contest success function in Assumption 2 is equivalent to the one in Szidarovszky and Okuguchi (1997). If $n(I) = 1$ for all I, the contest success function in Assumption 1 is equivalent to the one in Yamazaki (2008).[1] Assumption 4 is more general than an extended version of Tullock's (1980) contest success function $p^I = X^I \big/ \left(\sum_{J=1}^{N} X^J\right)$, which is assumed by Katz et al. (1990), Stein (2002) among others. Note that, under Assumptions 1, 2, 3 or 4, group I's production function for lotteries can vary across groups. Most researchers including Katz et al. (1990) assume that all players of group I assign the same value to the prize or rent R. In this paper we relax this traditional common-value assumption to assume that player i of group I puts player-specific value V_i^I on the rent R.

[1]For a private-good contest, Skaperdas (1996), Kooreman and Schoonbeek (1997) and Clark and Riis (1998) show that contest success function satisfies a set of reasonable axioms if and only if the contest success function has a functional form in Assumptions 1, 3 or 4, depending on the choice of the reasonable axioms.

3 Equilibrium Analysis

3.1 Non-cooperative Equilibrium

If players act non-cooperatively, player i of group I maximizes his or her expected utility:

$$u_i^I(\mathbf{x}) = p^I V_i^I - x_i^I \tag{1}$$

with respect to x_i^I, where p^I is given by the contest success function in Assumption 1, $\mathbf{x} = (\mathbf{x}^1, \mathbf{x}^2, \cdots, \mathbf{x}^N)$ and $\mathbf{x}^I = \left(x_1^I, x_2^I, \cdots, x_{n(I)}^I\right)$. If $\mathbf{x} = 0$, then u_i^I is defined to be zero.[2] Given X^J, $J \neq I$, the first term in the right hand side of (1) is strictly concave in X^I. The second term can be rewritten as $w_i^I - x_i^I$, because adding a fixed wealth w_i^I does not change the solution to the problem of maximizing (1). Hence, the right hand side of (1) can be interpreted as a quasi-linear utility function of player i of group I. The nonlinear part of this quasi-linear utility function measures the (expected) utility from the prize R, the public good for all players in group I, while the linear part measures the utility from a private good or a composite good other than the public good. Hence, given X^J, $J \neq I$, player i of group I privately contributes x_i^I for a pure public-good R in a familiar model with quasi-linear utility functions. Bergstrom et al. (1986) analyze such a model with quasi-linear utility functions. Their conclusions in our context are that only the total expenditure of group I, X^I, is determined, that X^I is contributed only by players with the highest valuation $V^I \equiv \max\left\{V_1^I, V_2^I, \cdots, V_{n(I)}^I\right\}$ and that given X^J, $J \neq I$, X^I is the unique solution to the problem of maximizing

$$u_I(\mathbf{X}) \equiv \frac{F_I(X^I)}{\sum_{J=1}^N F_J(X^J)} V^I - X^I \tag{2}$$

with respect to X^I, where $\mathbf{X} \equiv (X^1, X^2, \cdots, X^N)$. Hence, group I acts as if each group J consists of only one player J with V^J so that the model reduces to a private-good contest with player-specific I. For the private-good contest with $V^I = 1$ for all I, Szidarovszky and Okuguchi (1997) prove that there exists a unique pure-strategy Nash equilibrium in the model. Cornes and Hartley (2005) also prove the same result in a different way. The proof in the rest of this section uses the technique developed by Cornes and Hartley (2005) to prove that there exists a unique pure-strategy Nash equilibrium in our non-cooperative public-good contest under Assumption 2. After proving the unique existence of a Nash equilibrium under Assumption 2, we will explain that the unique existence can be established under Assumption 1.

[2]Assuming u_i^I to be V_i^I/N instead of zero does not affect the following analysis.

Define $Y_I \equiv F_I(X^I)$, $\mathbf{Y} \equiv (Y_1, Y_2, \cdots, Y_N)$ and $X^I = F_I^{-1}(Y_I) \equiv G_I(Y_I)$. By using these new notations, the expected payoff (2) can be rewritten as

$$U_I(\mathbf{Y}) = \frac{Y_I}{Y} V^I - G_I(Y_I) \tag{3}$$

where $Y \equiv \sum_{I=1}^{N} Y_I$. It is clear that maximizing (2) with respect to X^I is equivalent to maximizing (3) with respect to Y^I. Under Assumption 2, $G_I(\cdot)$ is twice continuously differentiable with respect to its argument, $G_I'(\cdot) > 0$ and $G_I''(\cdot) \geq 0$. The first-order condition of maximizing (3) is

$$\frac{\partial}{\partial Y_I} U_I(\mathbf{Y}) = \frac{Y_{-I}}{Y^2} V^I - G_I'(Y_I) = 0. \tag{4}$$

The first order condition (4) can be written as

$$\phi_I(Y, s_I, V^I) \equiv \left(\frac{1 - s_I}{Y}\right) V^I - G_I'(s_I Y) = 0 \tag{5}$$

where $s_I \equiv Y_I/Y$. For any given $Y > 0$ and $V^I > 0$, $\phi_I(Y, s_I, V^I)$ is strictly decreasing in s_I, $\lim_{s_I \to -\infty} \phi_I(Y, s_I, V^I) > 0$ and $\phi_I(Y, 1, V^I) \leq 0$. Hence, for any given $Y > 0$ and $V^I > 0$, there exists a unique solution $s_I \in (-\infty, 1]$ to (5). Denote the unique solution as a function of Y and V^I,

$$s_I = \sigma_I(Y, V^I). \tag{6}$$

Following Cornes and Hartley (2005), call this function group I's share function. By the definition of group I's share function,

$$\phi_I(Y, \sigma_I(Y, V^I), V^I) \equiv \left(\frac{1 - \sigma_I(Y, V^I)}{Y}\right) V^I - G_I'(\sigma_I(Y, V^I) Y) = 0 \tag{7}$$

must hold. Differentiating (7) with respect to Y proves

$$\frac{\partial}{\partial Y} \sigma_I(Y, V^I) = -\frac{(1 - s_I) V^I + s_I Y^2 G_I''}{Y(V^I + Y^2 G_I'')} < 0. \tag{8}$$

Hence, we have proved the following lemma.

Lemma 1 $\sigma_I(Y, V^I)$ *is strictly decreasing in* Y *for any* $I \in \{1, 2, \cdots, N\}$.

Since $\sigma_I(Y, V^I) \in (-\infty, 1]$ for any $Y > 0$ and $\sigma_I(Y, V^I)$ is decreasing in Y, $\lim_{Y\to 0} \sigma_I(Y, V^I)$ exists and $\lim_{Y\to 0} \sigma_I(Y, V^I) \le 1$. If $\lim_{Y\to 0} \sigma_I(Y, V^I) < 1$

$$\lim_{Y\to 0}\left[\left(\frac{1-\sigma_I(Y, V^I)}{Y}\right)V^I - G_I{}'(\sigma_I(Y, V^I)\,Y)\right] > 0,$$

which leads to a contradiction to (5). Hence, $\lim_{Y\to 0} \sigma_I(Y, V^I) = 1$ must hold. Define $\bar{Y}_I \equiv \max\{Y : \sigma_I(Y, V^I) \ge 0\}$. Since $\lim_{Y\to 0} \sigma_I(Y, V^I) = 1$, Lemma 1 implies $\bar{Y}_I \in (0, \infty]$. Summing up, we have proved the following lemma.

Lemma 2 *$\lim_{Y\to 0} \sigma_I(Y, V^I) = 1$ and $\sigma_I(\bar{Y}^I, V^I) = 0$ for some $\bar{Y}^I \in (0, \infty]$, $I = 1, 2, \cdots, N$.*

A Nash equilibrium Y corresponds to a solution to

$$1 = \sum\nolimits_{I=1}^{N} \sigma_I(Y, V^I). \tag{9}$$

Lemma 1 together with Lemma 2 ensures that there exists a unique positive solution to Eq. (9). Denote the unique solution as Y^*. This equilibrium Y^* uniquely determines all relevant equilibrium values: $p^{I*} = \sigma_I(Y^*, V^I)$, $Y_I^* = \sigma_I(Y^*, V^I)\,Y^*$ and $X^{I*} = G_I(Y_I^*)$ for all I. As already explained, only the equilibrium total expenditure of group I, X^I, is determined and provided only by players in a set $A(I) \equiv \{i : V_i^I = V^I\} \subseteq \{1, 2, \cdots, n(I)\}$, $I = 1, 2, \cdots, N$. Although the equilibrium value of x_i^I cannot be determined, $i \in A(I)$, we have proved the unique existence of a Nash equilibrium under Assumption 2.

As already explained, in the problem of maximizing (2), group I acts as if each group J consists of only one player J with V^J so that the model reduces to a private-good contest with player-specific V^I. If $n(I) = 1$ for all I, the contest success function in Assumption 2 is equivalent to the one in Szidarovszky and Okuguchi (1997). Szidarovszky and Okuguchi (1997) assume that all players assign the same value to the prize. Yamazaki (2013a) relaxes the common-value assumption to prove the unique existence of a Nash equilibrium in an asymmetric model of Szidarovszky and Okuguchi (1997). Hence, just applying the result of Yamazaki (2013a) to the problem of maximizing (2), we can claim that under Assumption 2, there exists a unique Nash equilibrium in our model.[3]

If $n(I) = 1$ for all I, the contest success function in Assumption 1 is equivalent to the one in Yamazaki (2008), who proves the unique existence of a Nash equilibrium in the model of Yamazaki (2013a) without assuming the differentiability of production functions for lotteries. Since the model reduces to a private-good

[3]Yamazaki (2013a) analyzes each player's share function to prove Szidarovszky and Okuguchi's result. The analysis of maximizing (2) is close to Yamazaki (2013a).

contest with player-specific V^I, just applying the result of Yamazaki (2008) to the problem of maximizing (2), we can prove the unique existence of a Nash equilibrium in our model under Assumption 1.

Proposition 1 *Under Assumption 1, there exists a unique pure-strategy Nash equilibrium where only the equilibrium total expenditure of group I, X^I, is determined and provided only by players in a set $A(I) \equiv \{i : V_i^I = V^I\} \subseteq \{1, 2, \cdots, n(I)\}$, $I = 1, 2, \cdots, N$.*

3.2 Cooperative Equilibrium and Other Equilibria

If players in group I lobby cooperatively, group I maximizes its expected payoff

$$u^I(\mathbf{X}) = \frac{F_I(X^I)}{\sum_{J=1}^N F_J(X^J)} V_I - X^I \tag{10}$$

with respect to X^I, where $V_I \equiv \sum_{i=1}^{n(I)} V_i^I$, and only X^I can be determined.[4] Note that it does not matter whether groups other than group I act cooperatively or not. Since group I's expected payoff (10) with V_I replaced by V^I is equivalent to the expected payoff (2) of player I with the highest valuation in group I, we can follow the same steps in the previous subsection to prove the unique existence of Nash equilibrium where players in some groups act cooperatively. As before, define $Y_I = F_I(X^I)$ and $s_I \equiv Y_I/Y$. If players in group I act cooperatively, group I's share function becomes

$$s_I = \sigma_I(Y, V_I). \tag{11}$$

Note that the above function σ_I itself is the same as the one in (6). Hence, by the same argument, the following two lemmas must hold under Assumption 2.

Lemma 3 *$\sigma_I(Y, V_I)$ is strictly decreasing in Y for any $I \in \{1, 2, \cdots, N\}$.*

Lemma 4 *$\lim_{Y \to 0} \sigma_I(Y, V_I) = 1$ and $\sigma_I(\bar{Y}_I, V_I) = 0$ for some $\bar{Y}_I \in (0, \infty]$, $I = 1, 2, \cdots, N$.*

Define C to be the set of cooperative groups and NC to be the set of non-cooperative groups. A Nash equilibrium Y corresponds to a solution to

$$1 = \sum_{I \in C} \sigma_I(Y, V_I) + \sum_{J \in NC} \sigma_J(Y, V^J). \tag{12}$$

If the set C is empty, Eq. (12) is equivalent to (9). Lemmas 1, 2, 3 and 4 prove the unique existence of a Nash equilibrium under Assumption 2, even if players in

[4] As before, if $\mathbf{X} = 0$, then u^I is defined to be zero or V_I/N.

some groups act cooperatively. As already mentioned, just applying the result of Yamazaki (2008) proves Proposition 1, which is a statement under Assumption 1. By the same reason, applying the result of Yamazaki (2008), we can claim that the following proposition must hold under Assumption 1.

Proposition 2 *Under Assumption 1, there exists a unique pure-strategy Nash equilibrium in an asymmetric public-good contest identified by the sets C and NC. In the unique equilibrium, only the equilibrium total expenditure of group I,* X^I*, is determined. If* $I \in NC$*,* X^I *is provided only by players in a set* $A(I) \equiv \{i : V_i^I = V^I\}$*,* $I = 1, 2, \cdots, N$*. If* $I \in C$*,* X^I *is cooperatively provided by all players in group I.*

4 Comparative Statics

In this section, we derive some results of comparative statics under Assumption 2. Differentiating (7) with respect to V^I proves that for any $I \in \{1, 2, \cdots, N\}$

$$\frac{\partial}{\partial V^I}\sigma_I\left(Y, V^I\right) = \frac{1 - s_I}{V^I + Y^2 G_I''} > 0. \tag{13}$$

Similarly, $\frac{\partial}{\partial V_I}\sigma_I(Y, V_I) > 0$ for any $I \in \{1, 2, \cdots, N\}$. Hence, we have proved the following lemma.

Lemma 5 *Under Assumption 2,* $\sigma_I(Y, V)$ *is strictly increasing in V, where V is* V^I *or* V_I*,* $I = 1, 2, \cdots, N$*.*

If V^I or V_I increases, Lemma 5 implies that the right hand side of (12) increases for any $Y > 0$. Since by Lemmas 1 and 2 $\sigma_I(Y, V)$ is strictly decreasing in Y and $\sigma_J(Y, V)$ does not depend on V^I or V_I for $J \neq I$, if the right hand side of (12) increases for any $Y > 0$, the equilibrium Y^* increases. Since $\sigma_J(Y, V)$ does not depend on V^I or V_I for $J \neq I$, Lemmas 1 or 3 implies that the equilibrium probability $p^{J*} = \sigma_J(Y^*, V)$ decreases for all $J \neq I$, whether V is equal to V^J or V_J. Since (12) holds at a new equilibrium, the equilibrium probability $p^{I*} = \sigma_I(Y^*, V)$ must increase. Hence, we have proved the following proposition.

Proposition 3 *Under Assumption 2, an increase in* V^I *or* V_I *increases* p^{I*} *and decreases* p^{J*} *for all* $J \neq I$*.*

Proposition 5 of Stein (2002) proves a similar result in a private-good ($n(I) = 1$ for all I) contest under Assumption 2 with $r = 1$. Proposition 3 extends Stein's Proposition 5 in two directions. Proposition 3 implies following propositions.

Proposition 4 *Under Assumption 2, cooperation among players in group I increase* p^{I*} *and decreases* p^{J*} *for all* $J \neq I$*.*

Proposition 5 *Suppose that Assumption 2 is satisfied. If players in group I act cooperatively, an increase in* $n(I)$ *increases* V_I*, increase* p^{I*} *and decreases* p^{J*} *for* *all $J \neq I$. If players in group I act non-cooperatively and if an increase in $n(I)$ increases V^I, an increase in $n(I)$ increases V^I increase p^{I*} and decreases p^{J*} for all $J \neq I$.*

Proposition 6 *Under Assumption 2, if players in group I act non-cooperatively and if mean-preserving spreads of $\{V_i^I\}_{i=1}^{n(I)}$ increase V^I, mean-preserving spreads of $\{V_i^I\}_{i=1}^{n(I)}$ increase p^{I*} and decreases p^{J*} for all $J \neq I$.*[5]

Proposition 5 means that if players in group I act cooperatively, the more the number of players in group I, the more likely group I wins. However, if players in group I act non-cooperatively, unless a player with valuation higher than V^I enters group I, an increase in $n(I)$ has no effect on equilibrium variables. By Proposition 6, if players in group I act non-cooperatively, mean-preserving spreads of $\{V_i^I\}_{i=1}^{n(I)}$ increases, does not change or even decreases p^{I*}, respectively, according as it increases, does not change or decrease V^I. However, if players in group I act cooperatively, we have the definite effect of mean-preserving spreads of $\{V_i^I\}_{i=1}^{n(I)}$ on the equilibrium variables. Since mean-preserving spreads of $\{V_i^I\}_{i=1}^{n(I)}$ do not change $V_I = n(I)\bar{V}^I$, where $\bar{V}^I \equiv (1/n(I)) \sum_{i=1}^{n(I)} V_i^I$, Proposition 3 implies the following proposition.

Proposition 7 *Under Assumption 2, if players in group I act cooperatively, mean-preserving spreads of $\{V_i^I\}_{i=1}^{n(I)}$ do not change p^{J*} for all $J = 1, 2, \cdots, N$.*

Under Assumption 3, we can examine in a concrete way how an increase in effectiveness of group I's expenditure changes equilibrium variables. Under Assumption 3, $Y_I = \lambda_I F(X^I)$, $X^I = F^{-1}(Y_I/\lambda_I) \equiv G(Y_I/\lambda_I)$ and group I's share function can be written as

$$s_I = \sigma_I(Y, V) = \sigma(Y, V, \lambda_I), \tag{14}$$

where V is V^I or V_I. Group I's share function satisfies

$$f(Y, \sigma(Y, V, \lambda_I), V) \equiv \left(\frac{1 - \sigma(Y, V, \lambda_I)}{Y}\right) V - G'\left(\frac{\sigma(Y, V, \lambda_I) Y}{\lambda_I}\right) \frac{1}{\lambda_I} = 0, \tag{15}$$

Differentiating (15) with respect to λ_I proves

$$\frac{\partial}{\partial \lambda_I} \sigma(Y, V, \lambda_I) = \frac{(\lambda_I G' + s_I Y G'') Y^{-1}}{(V\lambda_I + Y^2 G'') \lambda_I^2} > 0, \tag{16}$$

where V is V^I or V_I. Since the effect of increasing λ_I on group I's share function is the same as the one of increasing V, the following proposition holds.

---

[5]For the definition of mean preserving spreads, see Rothschild and Stiglitz (1970).

Proposition 8 *Under Assumption 3, an increase in* λ_I *increases* p^{I*} *and decreases* p^{J*} *for all* $J \neq I$.

Hence, the more effective group I's expenditure is, the more likely group I wins the contest. As Proposition 3 extends Proposition 5 of Stein (2002), Proposition 8 significantly extends Stein's Proposition 5'.

5 Conclusion

In this paper we have studied a rent-seeking contest where several groups compete for a prize which is a public good among players in a group. In the contest players in a group may evaluate the prize of the contest differently. We have proved that such an asymmetric public-good contest with a general contest success function possesses a unique pure-strategy Nash equilibrium, where the equilibrium is unique in the sense that the total expenditure of each group is uniquely determined. We have also derived some comparative statics in the asymmetric public-good contest.

Acknowledgements This research was in part supported by JSPS KAKENHI (Grants-in-Aid for Scientific Research) Numbers 19530151 and 24530194. A paper closely related with this paper was presented at PET15 (the 14th annual conference of the association for public economic theory) held in Luxembourg, 2015. I would like to thank an anonymous referee of this Festschrift and the participants in PET15 for many helpful comments and suggestions. Of course only I am responsible for any remaining errors and omissions.

References

Bergstrom T, Blume L, Varian H (1986) On The private provision of public goods. J Public Econ 29:25–49
Clark DJ, Riis C (1998) Contest success functions: an extension. Econ Theory 11:201–204
Cornes R, Hartley R (2005) Asymmetric contests with general technologies. Econ Theory 26:923–946
Gradstein M (1993) Rent seeking and the provision of public goods. Econ J 103:1236–1243
Katz E, Nitzan S, Rosenberg J (1990) Rent-seeking for pure public goods. Public Choice 65:49–60
Kooreman P, Schoonbeek L (1997) The specification of the probability functions in Tullock's rent-seeking contests. Econ Lett 56:59–61
Nitzan S (1991) Collective rent dissipation. Econ J 101:1522–1534
Rothschild M, Stiglitz JE (1970) Increasing risk 1: a definition. J Econ Theory 2:225–243
Skaperdas S (1996) Contest success functions. Econ Theory 7:283–290
Stein WE (2002) Asymmetric rent-seeking with more than two contestants. Public Choice 113:325–336
Szidarovszky F, Okuguchi K (1997) On the existence and uniqueness of pure Nash equilibrium in rent-seeking games. Games Econ Behav 18:135–140
Tullock G (1980) Efficient rent-seeking. In: Buchanan JM, Tollison RD, Tullock G (eds) Toward a theory of the rent-seeking society. Texas A&M Press, College Station

Yamazaki T (2008) On the existence and uniqueness of pure-strategy Nash equilibrium in asymmetric rent-seeking contests. J Public Econ Theory 10:317–327
Yamazaki T (2013a) On the unique existence of Nash equilibrium in smooth asymmetric contests. J Econ Niigata University 95, 99–115.
Yamazaki T (2013b) Aggregative games, lobbying models and endogenous tariffs. Niigata University Scholars Series 13, Graduate School of Modern Society and Culture, Niigata University

Heterogeneity and Number of Players in Rent-Seeking, Innovation, and Patent-Race Games

Noriaki Matsushima and Takeshi Yamazaki

Abstract Many economists have studied rent-seeking contests, innovation tournaments, and patent-race games independently. These three seemingly different games are known to be strategically equivalent under some reasonable assumptions. In these classic games, it is assumed that the value of a prize, i.e. the gain from rent-seeking, achieving an innovation, or obtaining a patent, is exogenously given and does not depend on the number of players, so that an increase in the number of players decreases the winning rate of each player. However, if players engage in R&D and then set quantities à la Cournot, the value of the prize in general depends on the number of players. In this paper, we set up a model with one efficient player and identical inefficient players to analyze how an increase in heterogeneity among players or the number of players changes the wining rate of the efficient player. One of the main results is that if the number of players is larger than some critical value, which can be less than two, an increase in the number of inefficient players always increases the winning rate of the efficient player.

1 Introduction

Many economists have studied rent-seeking contests, innovation tournaments, and patent-race games independently. Baye and Hoppe (2003) prove that these three seemingly different games are strategically equivalent under some reasonable assumptions. For example, they prove that in the limit as the interest rate approaches zero, the classic patent race game with a specific but reasonable hazard rate function is strategically equivalent to the Tullock (1980) rent-seeking game where firms in a market seek for a patent as a prize or rent. In the traditional models of rent-seeking contests, innovation tournaments, and patent-race games, it is assumed that the value

N. Matsushima (✉)
Institute of Social and Economic Research, Osaka University, 6-1 Mihogaoka, Ibaraki, Osaka 567-0047, Japan
e-mail: nmatsush@iser.osaka-u.ac.jp

P. von Mouche, F. Quartieri (eds.), *Equilibrium Theory for Cournot Oligopolies and Related Games*, Springer Series in Game Theory,
DOI 10.1007/978-3-319-29254-0_17

of a prize, i.e. the gain from rent-seeking, achieving an innovation, or obtaining a patent, is exogenously given and it does not depend on the number of players, so that an increase in the number of players decreases the winning rate of each player. However, if players engage in R&D and then set quantities à la Cournot, the value of the prize in general depends on the number of players. In this paper, we set up a simple Cournot model with one efficient firm and identical inefficient firms to analyze how an increase in heterogeneity among firms or the number of firms changes the wining rate of the efficient firm. One of the main results of this paper is that if the number of firms is larger than some critical value, which can be less than two, the entry of an inefficient firm always increases the winning rate of the efficient firm.

The remainder of the paper is organized as follows. Section 2 provides the basic model and shows the main result. Section 3 concludes the paper.

2 Model

2.1 *The Basic Setup of the Model*

Denote firm i's output as q_i $(i = 1, 2, \ldots, n)$. Define $Q \equiv \sum_{i=1}^{n} q_i$. The market inverse demand function is assumed to be linear:

$$P = P(Q) = a - bQ, \tag{1}$$

where a and b are positive constants. Firm i's cost C_i is assumed to be linear:

$$C_i = c_i q_i, \tag{2}$$

where

$$c_1 \leq c_2 = c_3 = \cdots = c_n \equiv c, \tag{3}$$

which means that firm 1 is an efficient firm (now, call it a dominant firm) and other $n - 1$ inefficient firms (now, call them minor firms) have the same cost condition.

In the second stage, firms set their output level à la Cournot. Under our assumptions, there exists a unique Cournot-Nash equilibrium in the second-stage game. It is easy to derive the equilibrium profits of firm 1 and each minor firm (firm $i, i = 2, \ldots, n$):

$$\pi_1 (c_1, \bar{c}, n) = \frac{\{(a - c_1) + (n - 1)(\bar{c} - c_1)\}^2}{b(n + 1)^2}, \tag{4}$$

$$\pi_i\left(c_1, c_i, \bar{c}, n\right) = \frac{\left\{(a - c_i) + (n-1)\left(\bar{c} - c_i\right) + \left(c_1 - c_i\right)\right\}^2}{b(n+1)^2}, \tag{5}$$

where $\bar{c} \equiv \sum_{i=2}^{n} c_i/(n-1)$.

In the first stage, firm i expends x_i to win a patent and reduce firm i's marginal cost by Δc_i. Assume

$$\Delta c_1 = \Delta c_2 = \cdots = \Delta c_n \equiv \Delta c > 0. \tag{6}$$

Furthermore, assume

$$\Delta c < (a - c) - (c - c_1)\,. \tag{7}$$

This assumption ensures that all minor firms attain non-negative profits no matter who wins the patent race. We also assume

$$\Delta c \leq 2\left(c - c_1\right). \tag{8}$$

This assumption is satisfied if $\Delta c \leq (c - c_1)$, that is, $c_1 \leq c - \Delta c$. Under assumption (8), after a minor firm wins the patent race, its marginal cost can be lower than firm 1's marginal cost c_1. However, assumption (8) requires that the patent does not make the winner of the patent race too strong. Define the cost difference between the dominant and each minor firm by

$$\delta \equiv (c - c_1)\,.$$

Since we can always find a positive number k such that

$$\Delta c = k\delta,$$

assumption (6) requires

$$k\delta > 0. \tag{6'}$$

Using these new notations, we can rewrite assumptions (7) and (8) as

$$\delta < \frac{a - c}{1 + k}, \tag{7'}$$

$$k \leq 2. \tag{8'}$$

Assume, given $\mathbf{x} \equiv (x_1, x_2, \ldots, x_n)$, probability p_i that firm i wins the patent and reduces its marginal cost is determined as follows.

$$p_i = \frac{(x_i)^r}{\sum_{j=1}^{n} (x_j)^r}, \quad r \in (0, 1]. \tag{9}$$

This assumption is standard in the literature of rent-seeking contests (See e.g., Tullock 1980; Perez-Castrillo and Verdier 1992; Skaperdas 1997; Kooreman and Schoonbeek 1997).

2.2 *The Expected Profit of the Firms*

We now derive the expected profits in the first stage to solve the rent-seeking game. Firm 1 maximizes its first-stage expected payoff

$$\begin{aligned} \Pi_1 &= p_1 \pi_1 (c_1 - \Delta c, c, n) + (1 - p_1) \pi_1 (c_1, c - \Delta c/(n-1), n) - x_1 \\ &= p_1 \Delta \pi_1 - x_1 + \pi_1 (c_1, c - \Delta c/(n-1), n) \end{aligned} \tag{10}$$

with respect to x_1, where

$$\begin{aligned} \Delta \pi_1 &= \pi_1 (c_1 - \Delta c, c, n) - \pi_1 (c_1, c - \Delta c/(n-1), n) \\ &= \frac{2\Delta c}{b(n+1)} \left\{ (a - c_1) + (n-1)(c - c_1) + (n-1) \frac{\Delta c}{2} \right\}. \end{aligned} \tag{11}$$

Similarly, for $i = 2, 3, \ldots, n$, firm i maximizes its first-stage expected payoff

$$\begin{aligned} \Pi_i &= p_i \pi_i (c_1, c - \Delta c, c - \Delta c/(n-1), n) + p_1 \pi_i (c_1 - \Delta c, c, c, n) \\ &\quad + (1 - p_i - p_1) \pi_i (c_1, c, c - \Delta c/(n-1), n) - x_i \\ &= p_i \Delta \pi_i + p_1 \Delta \pi_i' - x_i + \pi_i (c_1, c, c - \Delta c/(n-1), n) \end{aligned} \tag{12}$$

with respect to x_i, where

$$\begin{aligned} \Delta \pi_i &= \pi_i (c_1, c - \Delta c, c - \Delta c/(n-1), n) - \pi_i (c_1, c, c - \Delta c/(n-1), n) \\ &= \frac{2\Delta c}{b(n+1)} \left\{ (a - c) + (c_1 - c) + (n-1) \frac{\Delta c}{2} \right\}, \\ \Delta \pi_i' &= \pi_i (c_1 - \Delta c, c, c, n) - \pi_i (c_1, c, c - \Delta c/(n-1), n). \end{aligned} \tag{13}$$

We can show $\Delta\pi_i' = 0$ under assumptions (3) and (6), that is, which of other firms including firm 1 wins does not matter for firm i. Since $\Delta\pi_i' = 0$, we can rewrite (12) as

$$\Pi_i = p_i \Delta\pi_i - x_i + \pi_i\left(c_1, c, c - \Delta c/(n-1), n\right). \tag{12'}$$

Since the first-stage game can be interpreted as a rent-seeking contest with player specific valuations, Yamazaki (2008, Proposition 1) proves the following proposition.[1]

Proposition 1 *Under assumptions (3), (6), (7), (8), and (9), the two-stage game with linear demand and cost functions possesses a unique subgame perfect equilibrium.*

2.3 The Rent-Seeking Game

We now solve the rent-seeking game in the first stage. Define $y_i \equiv x_i^r$, $R \equiv 1/r$, and $\mathbf{y} \equiv (y_1, y_2, \cdots, y_n)$. Maximizing (10) or (12′) with respect to x_i is equivalent to maximizing

$$U_i(\mathbf{y}) = \frac{y_i}{\sum_{i=1}^{n} y_i} \Delta\pi_i - y_i^R, \tag{14}$$

with respect to y_i. The first-order condition of maximizing (14) is

$$\frac{\partial}{\partial y_i} U_i(\mathbf{y}) = \frac{Y_{-i}}{Y^2} \Delta\pi_i - R y_i^{R-1} = 0, \tag{15}$$

where $Y \equiv \sum_{i=1}^{n} y_i$ and $Y_{-i} \equiv \sum_{j \neq i} y_j$. Dividing (15) for firm i by (15) for firm 1 yields the following relation.

$$\frac{Y_{-i}}{Y_{-1}} \frac{\Delta\pi_i}{\Delta\pi_1} = \left(\frac{y_i}{y_1}\right)^{R-1}. \tag{16}$$

In equilibrium $y_2 = y_3 = \cdots = y_n \equiv y$. Using this relation, we can rewrite the above equation as

$$\left(\frac{y}{y_1}\right)^R - \left(\frac{n-2}{n-1}\right)\left(\frac{\Delta\pi}{\Delta\pi_1}\right)\left(\frac{y}{y_1}\right) - \left(\frac{1}{n-1}\right)\left(\frac{\Delta\pi}{\Delta\pi_1}\right) = 0. \tag{17}$$

where $\Delta\pi \equiv \Delta\pi_2 = \cdots = \Delta\pi_n$.

[1] Yamazaki (2008) proves the unique existence of the Nash equilibrium in a rent-seeking model with player specific valuations, which is a generalized version of the model analyzed by Szidarovszky and Okuguchi (1997).

Define $\theta \equiv \Delta\pi/\Delta\pi_1$. It is easy to show that

$$\begin{aligned}\theta &= \frac{(a-c)+(c_1-c)+(n-1)(\Delta c/2)}{(a-c_1)+(n-1)(c-c_1)+(n-1)(\Delta c/2)} \\ &= \frac{2(a-c)-2\delta+(n-1)k\delta}{2(a-c)+2n\delta+(n-1)k\delta} \equiv \theta(\delta,k,n).\end{aligned} \tag{18}$$

θ represents the relative significance of the investment incentives of firm 1 and each minor firm. A lower θ means that the investment incentive of firm 1 is relatively stronger than that of each minor firm. Note that if all firms are ex ante identical, that is, if $\delta = 0$, $\theta = 1$, that is, the investment incentives of firm 1 and each of the other firms are the same. If $\delta > 0$, $\theta < 1$. Hence, $\theta \leq 1$ under our assumptions.

From (17), y/y_i can be written as a function of θ, R and n:

$$y/y_1 = \phi(\theta, R, n). \tag{19}$$

ϕ represents the ratio of the investment levels of firm 1 and each minor firm. A low ϕ means that the investment level of firm 1 is relatively higher than that of each minor firm. Note that for any $R \geq 1$, $y/y_1 = 1$ satisfies (17) for $\theta = 1$, that is, $\delta = 0$. Hence, $\phi(1, R, n) = 1$. If $R = 1$, we can get the closed-form solution to (17):

$$\phi(\theta, 1, n) = \frac{\theta}{(n-1)-(n-2)\theta}. \tag{19$'$}$$

It is clear from (19′) that if $R = 1$, since $0 < \theta \leq 1$, $\phi(\theta, 1, n)$ is increasing in θ and decreasing in n. For the general cases $R \geq 1$, we derive the partial derivatives of θ:

$$\frac{\partial\phi}{\partial\theta} = \frac{\frac{n-2}{n-1}\left(\frac{y}{y_1}\right)+\frac{1}{n-1}}{R\left(\frac{y}{y_1}\right)^{R-1}-\frac{n-2}{n-1}\theta}, \tag{20}$$

$$\frac{\partial\phi}{\partial R} = \frac{-\left(\frac{y}{y_1}\right)^{R-1}\ln\left(\frac{y}{y_1}\right)}{R\left(\frac{y}{y_1}\right)^{R-1}-\frac{n-2}{n-1}\theta}, \tag{21}$$

and

$$\frac{\partial\phi}{\partial n} = \frac{-\frac{\theta}{(n-1)^2}\left(1-\frac{y}{y_1}\right)}{R\left(\frac{y}{y_1}\right)^{R-1}-\frac{n-2}{n-1}\theta}. \tag{22}$$

The following lemmas show how $\theta(\delta, k, n)$ and $\phi(\theta, R, n)$ defined in (18) and (19) respectively depend on their arguments.

Lemma 1 *Under assumptions (3), (6), (7), and (8),*

$$\left.\frac{\partial\theta\,(\delta,k,n)}{\partial\delta}\right|_{\Delta c=k\delta=0} < 0,\ \frac{\partial\theta\,(\delta,k,n)}{\partial k} \geq 0,\ \textit{and}\ \frac{\partial\theta\,(\delta,k,n)}{\partial n} \leq 0,$$

where the second and the third inequalities hold with equality if and only if $\delta = 0$.

Proof See Appendix.

Lemma 2 *Under assumptions (3), (6), (7), (8), and (9),* $y_1 \geq y$*, that is,* $\phi(\theta, R, n) \leq 1$ *in the unique subgame perfect equilibrium. Furthermore, for any* $R \geq 1$ *and* $n \geq 2$,

$$\frac{\partial\phi\,(\theta,R,n)}{\partial\theta} > 0,\ \frac{\partial\phi\,(\theta,R,n)}{\partial R} \geq 0,\ \textit{and}\ \frac{\partial\phi\,(\theta,R,n)}{\partial n} \leq 0.$$

Proof See Appendix.

The first inequality in Lemma 1 represents how an expansion of the ex ante cost difference between firm 1 and each minor firm influences the relative significance of their investment incentives. Intuitively, the expansion enhances (diminishes, respectively) the investment incentive of firm 1 (each minor firm, respectively), leading to a decrease in θ. The second inequality in Lemma 1, which is a flip-side of the first one, represents how an increase in the cost reduction through winning the patent race influences the relative significance of their investment incentives. This implies that the importance of the ex ante cost difference between firm 1 and each minor firm becomes relatively weaker than the cost reduction, leading to an increase in θ. The third inequality in Lemma 1, which is one of the key factors in Proposition 3 explained later, comes from the fact that an increase in the number of firms diminishes the residual demands for the existing firms, which decrease their gains from the cost reduction through winning the patent race. The market shrink through entry enhances the importance of the ex ante cost difference between firm 1 and each minor firm, which enlarges the ex ante cost superiority of firm 1. As a result, the normalized gain of firm 1 from winning the patent race, in terms of each minor firm's gain from winning the patent race, becomes higher, leading to a decrease in θ.

The first partial differential in Lemma 2 comes from the fact that an increase in the ratio of their profit increments weakens the investment incentive of firm 1, leading to an increase in the ratio of their investment levels. The second partial differential in Lemma 2 comes from the fact that a higher R, which means an increase in costs for investment, diminishes the investment incentive of firm 1 more because firm 1 sets a higher investment level. This leads to an increase in the ratio of their investment levels. The third partial differential in Lemma 2, which is one of the key factors in Proposition 3 explained later, comes from the fact that an increase in n diminishes the investment incentive of each minor firm more than firm 1 (this is related to the third inequality in Lemma 1). This leads to an decrease in the ratio of their investment levels.

2.4 The Winning Probability of the Dominant Firm

We now discuss the main concern in this paper, that is, we discuss the winning probability of the dominant firm (p_1). From the fact

$$p_1 = \frac{y_1}{Y} = \frac{y_1}{y_1 + (n-1)\,y} = \frac{1}{1 + (n-1)\,\phi}. \tag{23}$$

Lemmas 1 and 2, and the relation between p_1 and θ in (23), we find that an increase in $\delta = c - c_1$ or a decrease in Δc increases the probability the dominant firm wins the patent race, leading to the following proposition:

Proposition 2 *Under assumptions (3), (6), (7), (8), and (9), if R decreases, δ increases, or Δc decreases, then the winning probability of the dominant firm increases.*

We have already examined how the equilibrium depends on the heterogeneity of firms. To examine how entry of a minor firm changes the equilibrium variables, first prove that an increase in n always decreases the equilibrium ϕ.

Lemma 3 *Under assumptions (3), (6), (7), (8), and (9),*

$$\frac{d\phi}{dn} = \frac{\partial \phi}{\partial \theta}\frac{\partial \theta}{\partial n} + \frac{\partial \phi}{\partial n} < 0.$$

Proof See Appendix.

That is, the incentive of each minor firm to invest becomes relatively weaker than that of firm 1.

We now investigate the relation between p_1 and n. It is clear from (23) that the direct effect $\partial p_1/\partial n$ is negative. However, the indirect effect $(\partial p_1/\partial \phi)/(d\phi/dn)$ is positive (see (23) and Lemma 3). If the indirect positive effect dominates the direct negative one, an increase in n increases the equilibrium p_1. Formally speaking, $dp_1/dn > 0$ is positive (negative, respectively) if and only if

$$-\frac{\frac{d\phi}{dn}}{\phi} > (<\text{, respectively})\frac{1}{n-1}. \tag{24}$$

Totally differentiating (17) and using the relation (17) again, we can show that the relation (24) is equivalent to the following relation.[2]

$$G(\theta,\phi,R,n) \equiv -\frac{\frac{d\theta}{dn}}{\theta} - H(\phi,R,n) > (<\text{ , respectively})0, \tag{25}$$

where

$$H(\phi,R,n) \equiv \frac{1}{n-1}\left\{\frac{\phi}{(n-2)\phi+1} + (R-1)\right\}. \tag{26}$$

Substituting (18) into (25), we can show that (25) is equivalent to

$$\begin{aligned} L(\delta,k,\theta,\phi,R,n) \equiv\ & \delta\{(a-c)-(1+k)\delta\} \\ & -\frac{1}{4\theta}\{2(a-c)-2\delta+(n-1)k\delta\}^2 H(\phi,R,n) \\ & > (<\text{ , respectively})0. \end{aligned} \tag{27}$$

Now define

$$\begin{aligned} g(\delta,k,R,n) &\equiv G(\theta(\delta,k,n),\phi(\theta(\delta,k,n),R,n),R,n), \\ l(\delta,k,R,n) &\equiv (n-1)^2 L(\delta,k,\theta(\delta,k,n),\phi(\theta(\delta,k,n),R,n),R,n). \end{aligned}$$

---

[2] Totally differentiating (17), we can get

$$-\frac{\frac{d\phi}{dn}}{\phi} = \frac{\left\{\left(\frac{n-2}{n-1}\right)\phi + \left(\frac{1}{n-1}\right)\right\}\frac{d\theta}{dn} - \frac{\theta}{(n-1)^2}(1-\phi)}{R\phi^R - \left(\frac{n-2}{n-1}\right)\theta\phi}.$$

From (17),

$$\phi^R = \left(\frac{n-2}{n-1}\right)\theta\phi + \left(\frac{1}{n-1}\right)\theta.$$

Substituting it into the first equation, we have

$$-\frac{\frac{d\phi}{dn}}{\phi} = \frac{\{(n-2)\phi+1\}\frac{d\theta}{dn} - \frac{\theta}{n-1}(1-\phi)}{(R-1)(n-2)\theta\phi + R\theta}.$$

Substituting it into (24) leads to the equivalence between (24) and (25).

Substituting (19′) into (26), we can easily show the following:

$$
\begin{aligned}
H(\phi, 1, n) &= \frac{\theta}{(n-1)^2}, \\
l(\delta, k, 1, n) &= \alpha(n-1)^2 + \beta(n-1) + \gamma, \\
&\qquad \text{where } \alpha \equiv \delta\left\{(a-c) - (k+2)^2\delta/4\right\}, \\
&\qquad \beta \equiv -\{(a-c) - \delta\} k\delta, \\
&\qquad \gamma \equiv -\{(a-c) - \delta\}^2.
\end{aligned}
\tag{28}
$$

That is, when $R = 1$, $dp_1/dn > 0$ if and only if $l(\delta, k, 1, n) > 0$. Note that assumption (7) ensures that α is positive and also that β and γ are negative.

If $\delta = 0$, that is, if firms do not have any ex ante cost difference, l in (28) is equal to γ which is negative, implying that entry always decreases the probability the dominant firm (firm 1) wins the patent rance. However, if $\delta > 0$, there exists $\hat{n} \in (1, \infty)$ such that l in (28) is negative (positive, respectively) for any positive $n <$ (>, respectively)$\hat{n}$. Since H in (26) is continuous in R, if R is close enough to 1, the same property holds. Furthermore, because Lemma 2 implies

$$
\frac{d}{dR}H(\phi, R, n) = \frac{\partial H}{\partial \phi}\frac{\partial \phi}{\partial R} + \frac{\partial H}{\partial R} = \frac{1}{(n-1)\{(n-2)\phi + 1\}^2}\frac{\partial \phi}{\partial R} + \frac{1}{n-1} > 0,
$$

$$
\frac{d}{dR}g(\delta, k, 1, n) < 0,
$$

which in turn implies $\frac{d}{dR}l(\delta, k, 1, n) < 0$. Hence, an increase in R increases $\hat{n}$, as long as R is close enough to 1. Summing up, we have proved the following proposition.

Proposition 3 *Under assumptions (3), (6), (7), (8), and (9), if $\delta = 0$, entry of a minor firm always decreases the equilibrium winning probability of the dominant firm. Under assumptions (3), (6), (7), and (8), if $\delta > 0$ and R is sufficiently close to 1, there exists $\hat{n} \in (1, \infty)$ such that the winning probability of the dominant firm is decreasing (increasing, respectively) in the number of minor firms for any $n <$ (> , respectively)$\hat{n}$. Furthermore, if R increases from 1, $\hat{n}$ increases.*

The properties derived in the third inequalities in Lemmas 1 and 2 influence the nonmonotonicity between the investment incentive of the dominant firm and the number of firms in Proposition 3. An increase in the number of firms enhances the strategic effect of investments for cost reduction because a cost reduction of a firm diminishes the rivals' total quantities which increases with an increase in n. This effect works well for the dominant firm because the marginal gain from its investment is relatively larger due to the dominant firm's larger per unit profit, which comes from its ex ante cost advantage. As a result, for higher n, additional entry enhances the dominant firm's incentive to invest.

Note that $\hat{n}$ in Proposition 3 can be less than 2. If so, entry of a minor firm always increases the equilibrium winning probability of the dominant firm. In any case, if $\delta > 0$ and if the number of firms is large enough, entry of a minor firm increases the equilibrium winning probability of the dominant firm.

Note also that an increase in Δc increases $\hat{n}$ because

$$\frac{\partial}{\partial k} l(\delta, k, 1, n) = -\frac{k+2}{2}(n-1)^2\delta^2 - \{(a-c) - \delta\}(n-1)\delta < 0.$$

Because we can show

$$\left.\frac{\partial}{\partial \delta} l(\delta, k, 1, n)\right|_{\Delta c = k\delta = 0} = \left\{(a-c) - \frac{1}{4}(k+2)^2\delta\right\}(n-1)^2 + \delta k(n-1) + 2\{(a-c) - \delta\} > 0$$

for any $n \geq 0$, an increase in δ decreases $\hat{n}$. Hence, we have proved the following proposition.

Proposition 4 *Under assumptions (3), (6), (7), (8), and (9), a decrease in Δc or an increase in δ decreases $\hat{n}$.*

Those two changes enlarge the effect of the dominant firm's ex ante cost advantage, which enhances its incentive to invest.

3 Conclusion

This paper has analyzed how entry of a minor firm changes the winning probability of a dominant firm, i.e. the probability that the dominant firm obtains the patent, in the Tullock (1980) rent-seeking game where one dominant firm competes with several small homogeneous firms in a market. In the classic model of a patent race, which was formulated by Loury (1979) and Dasgupta and Stiglitz (1980), the firm that innovates first will obtain a patent, while all other firms gain nothing. Baye and Hoppe (2003) proved that in the limit as the interest rate approaches zero, the classic game of a patent race with a specific but reasonable hazard rate function is strategically equivalent to the Tullock (1980) rent-seeking game where firms in a market seek a patent as a prize or rent. We can interpret that the firm which gains the patent reduces its unit production cost. If firms patent-race and compete in a market, the value of gaining the patent in general depends on the number of firms. Ishida et al. (2011) prove that if the marginal cost of a dominant firm is sufficiently smaller than the identical marginal cost of other minor firms, entry of a minor firm always decreases the marginal cost of the dominant firm in a R&D model, where each firm's R&D expenditure reduces its constant marginal cost at the first stage and then firms play the Cournot game at the second stage. We have proved that if there exists an ex ante cost difference, which can be small, and if the number of firms is larger than some critical value, which can be less than two, entry of a minor firm always increases the probability that the dominant firm wins the patent and decreases its marginal cost. We have also proved that the critical value is decreasing in the ex ante cost difference. Hence, the result similar to Ishida et al. (2011) holds even in our patent-race game. Finally, we mention a remark on the property of strategic variables in the final stage. We have assumed that firms compete in quantity. We guess that if we consider a price competition model with differentiated products, the property of the main results is less likely to hold. This is because the benefit of cost reduction is weaker under price competition than under quantity competition because of the rival's aggressive reaction to such a cost reduction, which means that a firm's price cut through its cost reduction induces the rivals' price reductions which weaken the benefit of the cost reduction. This effect is contrast with that under quantity competition.

Acknowledgements This research was in part supported by JSPS KAKENHI (Grants-in-Aid for Scientific Research) Grant Number 19530151. An earlier version of this paper was presented at PET10 (the 9th annual conference of the association for public economic theory) held in Istanbul, 2010. We would like to thank a referee of this Festschrift and the participants in PET10 and the workshop held at Niigata University for many helpful comments and suggestions. Only the authors are responsible for any remaining errors and omissions.

Appendix

Proof of Lemma 1 Define $A \equiv 2(a-c) - 2\delta + (n-1)k\delta$ and $B \equiv 2(a-c) + 2n\delta + (n-1)k\delta$. Simple calculations lead to

$$\left.\frac{\partial\theta(\delta,k,n)}{\partial\delta}\right|_{\Delta c = k\delta = 0} = -\frac{2}{B^2}(nA+B) < 0,$$

$$\frac{\partial\theta(\delta,k,n)}{\partial k} = \frac{2\left(n^2-1\right)\delta^2}{B^2} \geq 0,$$

$$\frac{\partial\theta(\delta,k,n)}{\partial n} = -\frac{4\delta\left((a-c)-\delta(1+k)\right)}{B^2} \leq 0.$$

The last inequality holds from (7′). □

Proof of Lemma 2 The relation (16) implies

$$\left(\frac{y}{y_1}\right)^{R-1} = \frac{Y_{-i}}{Y_{-1}}\theta = \frac{y_1 + (n-2)\,y}{(n-1)\,y}\theta > \frac{n-2}{n-1}\theta,$$

note that $\Delta\pi/\Delta\pi_1 = \theta$. Because of this inequality and $R \geq 1$, the partial derivative in (20) is positive for any $n \geq 2$.

As already mentioned, $\phi(1, R, n) = 1$ for any $R \geq 1$ and

$$\frac{y}{y_1} = \phi(\theta, 1, n) = \frac{\theta}{(n-1) - (n-2)\theta}.$$

Since

$$\frac{\partial}{\partial \theta}\phi(\theta, 1, n) = \frac{n-1}{\{(n-1) - (n-2)\theta\}^2} > 0,$$

for $\theta < 1$,

$$\frac{y}{y_1} = \phi(\theta, 1, n) = \frac{\theta}{(n-1) - (n-2)\theta} < \frac{1}{(n-1) - (n-2)} = 1.$$

Since $\phi(\theta, R, n)$ is continuous in R for $R > 0$, $\phi(\theta, R, n) < 1$ for any R close enough to 1. By (21), if $\phi(\theta, R, n) < 1$ for some R_0 which is close to 1, $\dfrac{\partial \phi(\theta, R, n)}{\partial R} > 0$ at $R = R_0$ because $y/y_1 > 0$ and $\ln(y/y_1) < 0$. Hence, as long as $\phi(\theta, R, n) < 1$, an increase in R increases $\phi(\theta, R, n)$. If $\phi(\theta, R, n)$ reaches to 1 at some $R = R_1$, that is, $y/y_1 = 1$ at $R = R_1$. From (21), $\partial\phi/\partial R = 0$, thus $\phi(\theta, R, n) = 1$ for any $R \geq R_1$. Therefore, in any case, $\phi(\theta, R, n) \leq 1$ and $\dfrac{\partial \phi(\theta, R, n)}{\partial R} \geq 0$ for any $R \geq 1$.

From (22), $\dfrac{\partial \phi(\theta, R, n)}{\partial n} \leq 0$ for any $R \geq 1$ and $n \geq 2$. □

Proof of Lemma 3 Assumption (7′) implies

$$\frac{\partial}{\partial n}\theta(\delta, k, n) = \frac{-4\delta\{(a-c) - (1+k)\delta\}}{\{2(a-c) - 2\delta + (n-1)k\delta\}^2} < 0.$$

This fact together with Lemma 2 implies

$$\frac{d\phi}{dn} = \frac{\partial\phi}{\partial\theta}\frac{\partial\theta}{\partial n} + \frac{\partial\phi}{\partial n} < 0.$$

This proves Lemma 3. □

References

Baye MR, Hoppe HC (2003) The strategic equivalence of rent-seeking, innovation, and patent-race games. Games Econ Behav 44:217–226

Dasgupta P, Stiglitz J (1980) Uncertainty, industrial structure, and the speed of R&D. Bell J Econ 11:1–28

Ishida J, Matsumura T, Matsushima N (2011) Market competition, R&D and firm profits in asymmetric oligopoly. J Ind Econ 59:484–505

Kooreman P, Schoonbeek L (1997) The specification of the probability functions in Tullock's rent-seeking contests. Econ Lett 56:59–61

Loury GC (1979) Market structure and innovation. Q J Econ 93:395–410

Perez-Castrillo JD, Verdier T (1992) A general analysis of rent-seeking games. Public Choice 73:335–350

Skaperdas S (1997) Contest success functions. Econ Theory 7:283–290

Szidarovszky F, Okuguchi K (1997) On the existence and uniqueness of pure Nash equilibrium in rent-seeking games. Games Econ Behav 18:135–140

Tullock G (1980) Efficient rent-seeking. In: Buchanan JM, Tollison RD, Tullock G (eds) Toward a theory of the rent-seeking society. Texas A&M Press, College Station

Yamazaki T (2008) On the existence and uniqueness of pure-strategy nash equilibrium in asymmetric rent-seeking contests. J Public Econ Theory 10:317–327

MIX
Papier aus verantwortungsvollen Quellen
Paper from responsible sources
FSC® C105338

If you have any concerns about our products,
you can contact us on
ProductSafety@springernature.com

In case Publisher is established outside the EU,
the EU authorized representative is:
Springer Nature Customer Service Center GmbH
Europaplatz 3, 69115 Heidelberg, Germany

Printed by Libri Plureos GmbH
in Hamburg, Germany